DISCARD

Fundamentals of
   Linear Algebra

Fundamentals of Linear Algebra

Dennis B. Ames

Department of Mathematics
California State College, Fullerton

International Textbook Company
an Intext *publisher*
Scranton, Pennsylvania 18515

the Intext *series in*

Advanced Mathematics

under the consulting editorship of

Richard D. Anderson
Louisiana State University

Alex Rosenberg
Cornell University

Standard Book Number 7002 2263 4

Copyright ©, 1970, by International Textbook Company

All rights reserved. No part of the material protected by this copyright notice may be reproduced or utilized in any form or by any means, electronic or mechanical, including photocopying, recording, or by any informational storage and retrieval system, without written permission from the copyright owner. Printed in the United States of America by The Haddon Craftsmen, Inc., Scranton, Pennsylvania.

Library of Congress Catalog Card Number: 74-111939

To Lois and Judy

Preface

An early course in linear algebra seems to be rather generally accepted as a very useful, if not actually essential, part of the mathematical education of an undergraduate. The material itself is readily accessible without a high degree of prerequisite mathematical training and is fundamental in its importance for later work in algebra, analysis, and geometry. Moreover, the nature of the subject matter makes linear algebra peculiarly appropriate as a training ground for discovering the meaning of a mathematical proof. This favorable combination of being basically important for the calculus and subsequent advanced study and of its somewhat unique character (along with mathematical logic) in providing a rich mathematical experience in the comprehension of proofs gives to linear algebra a respected position in the traditional hierarchies of mathematical curricula.

In this book I have attempted to present the basic concepts and theories of linear algebra and to exhibit some, at least, of its many useful applications to mathematics as well as to other sciences.

The book closes with two chapters on tensor products and the exterior algebra. These chapters, along with the three sections on exact sequences, affine spaces, and modules, may very well be omitted—indeed, probably would have to be—from a one semester course, and no impairment of the rest of the presentation will result thereby. Nevertheless, they are important and interesting topics that naturally and logically find their place here, hopefully for purposes of additional study and reference. The author believes that a knowledge of tensor products and of the exterior algebra of a finite dimensional vector space has now been forced down into the undergraduate curriculum for mathematicians, physicists, and engineers alike.

A book should not be dogmatically required to restrict itself to only those topics that can safely be covered in the artificial time period of a one semester course. This is much too presumptuous and arrogates to course work a far too important role in education, since, of necessity, most of our learning occurs outside this narrow and rigid framework. Admittedly, there is more than enough material in the book for the time and pace of the usual linear algebra course; however, I am quite willing to leave to the instructor's discretion the selection of his own particular topics and the depth to which he wishes to pursue them.

I offer the following merely as suggestions:

An elementary introductory course at the freshman or sophomore level can very well include these portions of the book: Chapter 1; Chapter 2, Sections 2-1 – 2-3; Chapter 3, Sections 3-1 – 3-2; Chapter 4, Sections 4-1 – 4-3; Chapter 5; Chapter 6, Sections 6-1 – 6-4; Chapter 7; and Chapter 8, Sections 8-1 – 8-3.

A more advanced course could in addition include at least: Chapter 6, Section 6-5; Chapter 8, Sections 8-4 – 8-5; Chapter 9; and Chapter 10.

I am indebted to the publisher for many kindnesses.

Dennis B. Ames

December, 1969
Fullerton, California

Contents

Chapter 1   **Introduction ... 1**

    1-1   Sets 1
    1-2   Mappings 2
    1-3   Binary Operations 7
    1-4   The Integers 10
    1-5   Groups, Rings, Fields 12
    1-6   Polynomials 14

Chapter 2   **Vector Spaces ... 15**

    2-1   Definitions and Examples 16
    2-2   Subspaces 25
    2-3   Basis and Dimension of a Vector Space 30
    2-4   The Module 38
            Exercises 39

Chapter 3   **Linear Transformations ... 42**

    3-1   Definitions and Fundamental Properties 42
    3-2   Nonsingular Transformations 51
    3-3   Sets of Linear Transformations 54
    3-4   Affine Spaces 56
            Exercises 61

Chapter 4   **Quotient Spaces and Direct Sums ... 64**

    4-1   Quotient Spaces 64
    4-2   Equivalence Relations 68
    4-3   Direct Sums 71
    4-4   Exact Sequences 74
            Exercises 77

Chapter 5   **Linear Transformations and Matrices ... 80**

    5-1   Representation of a Linear Transformation by a Matrix 80
    5-2   Operations on Matrices 83

5-3	Change of Basis and Similar Matrices	90
5-4	The Vector Space Hom (V, W)	96
	Exercises	98

Chapter 6 Inner-Product Vector Spaces and Dual Spaces ... 100

6-1	The Inner Product	100
6-2	Orthonormal Basis	104
6-3	Isometries	110
6-4	Orthogonal Operators on Finite-Dimensional Inner-Product Vector Spaces	113
6-5	Dual Spaces	116
	Exercises	123

Chapter 7 Matrices and Determinants ... 125

7-1	Permutations	125
7-2	Rank of a Matrix	129
7-3	Elementary Matrices	131
7-4	The Determinant	136
7-5	Properties of the Determinant	140
7-6	Inverse of a Matrix	147
7-7	Systems of Linear Equations	149
	Exercises	153

Chapter 8 Eigenvalues and the Spectral Theorem ... 155

8-1	Definitions and Examples	155
8-2	The Minimal Polynomial	165
8-3	The Adjoint Operator	170
8-4	Hermitian Operators	174
8-5	The Spectral Theorem	178
	Exercises	184

Chapter 9 Bilinear and Quadratic Forms ... 185

9-1	Definitions and Notation	185
9-2	The Reduction of Quadratic Forms Under Groups	190
9-3	The Reduction of a Quadratic Form Under the Full Linear Group	194
9-4	Reduction of a Quadratic Form Under the Orthogonal Group	197
9-5	Positive Definiteness	201
9-6	The Simultaneous Reduction of Quadratic Forms	203
9-7	Hermitian Forms	205
	Exercises	206

Contents

Chapter 10   Canonical Forms for Linear Transformations ... 208

- 10-1   Invariant Subspaces ... 208
- 10-2   Decomposition Theorems ... 214
- 10-3   The Rational Canonical Form ... 220
- 10-4   The Jordan Canonical Form ... 223
-        Exercises ... 235

Chapter 11   The Tensor Product of Vector Spaces ... 237

- 11-1   Multilinear Mappings ... 237
- 11-2   The Tensor Product ... 239
- 11-3   Existence of the Tensor Product ... 243
- 11-4   Computation in $V \otimes W$ ... 245
- 11-5   Linear Transformations on $V \otimes W$ ... 246
- 11-6   Isomorphisms of Tensor Products ... 249
- 11-7   Direct Sums ... 251
-        Exercises ... 253

Chapter 12   The Exterior Algebra of a Vector Space ... 255

- 12-1   Algebras ... 255
- 12-2   The Algebra of Alternating Forms ... 259
- 12-3   The Exterior Algebra ... 262
- 12-4   The Algebra of Differential Forms ... 267
-        Exercises ... 272

Bibliography ... 275

Index ... 277

Fundamentals of Linear Algebra

Chapter **1**

Introduction

1-1 SETS

An extensive knowledge of set theory is not needed in this book, and in fact, even the most elementary set operations themselves are sparingly used. It is therefore out of place and imprudent to involve the reader in any maze of exercises about sets. However, the things we do stress in this section will be found useful in our later discussions.

A set or collection of objects must be defined in such a way that it is possible to state of an arbitrary object either that it is contained in the set or that it does not belong to the set.

In algebra it is customary to call the objects of a set its **elements.**

If x is an element of a set X, we write $x \in X$, and if x does not belong to X, we write $x \notin X$.

If X is a finite set whose elements are $x_1, x_2, \ldots, x_n$, we write
$$X = \{x_1, x_2, \ldots, x_n\}.$$

The **empty set** is defined as the set that contains no elements.

Examples of sets are at hand everywhere—for instance, the set of all double stars that are less than a billion light years from the earth. (Perhaps an astronomer could tell us whether this set is empty or not.)

We shall make very frequent use of the phrase "**if and only if**" applied to statements. If P and Q are statements, then the statement "P if and only if Q" means that P is true if Q is true and that P is true only if Q is true. In other words, "P if and only if Q" means that P implies Q and that Q implies P; that is, P and Q are both true or both false.

A set C is called a **subset** of a set A if every element of C is an element of A. We write $C \subset A$ to mean that C is a subset of A. The empty set is therefore a subset of every set.

If A and C are sets then we write $A = C$ if and only if $A \subset C$ and $C \subset A$.

The **intersection** $A \cap B$ of two sets A and B is defined to be the set consisting of all elements common to both sets A and B. Thus $x \in A \cap B$ if and only if $x \in A$ and $x \in B$. Clearly $B \cap A = A \cap B$, and $A \cap B$ is a subset of each of the sets A and B. Two sets are said to be **disjoint** if their intersection is empty.

The **union** $A \cup B$ of two sets A and B is defined to be the set of all elements that belong to A or to B or to both. Thus $A \cap B \subset A \subset A \cup B$ and $A \cap B \subset B \subset A \cup B$. Clearly $A \cup B = B \cup A$. The union $A \cup B$ can be described as the smallest set that contains both A and B, in the sense that if $A \subset C$, $B \subset C$, then $A \cup B \subset C$.

Example. If A is the set of all integers that are > -3 and if B is the set of all integers that are < 5, then $A \cap B = \{-2, -1, 0, 1, 2, 3, 4\}$, while $A \cup B$ is the set of all integers.

If A and B are nonempty sets, the **cartesian product** $A \times B$ of A and B is defined to be the set of all ordered pairs (a,b), where $a \in A$ and $b \in B$. The reader is no doubt familiar with this notation for it is used in analytic geometry and calculus, where, if R is the real line (the set of real numbers) then $R \times R$ is the set of all points (x,y) of the plane, and x and y are the coordinates of the point. Of course if $A \neq B$ then $B \times A \neq A \times B$. The notion of cartesian product extends readily to any finite number $n > 2$ of nonempty sets $A_1, A_2, \ldots, A_n$. We write $A_1 \times A_2 \times \cdots \times A_n$ for this cartesian product. It is the set of elements $(a_1, a_2, \ldots, a_n)$ where $a_i \in A_i$, $i = 1, 2, \ldots, n$.

1-2 MAPPINGS

A **mapping** f of a nonempty set A into a nonempty set B symbolized by $f:A \to B$, is a rule that assigns to each element of A a unique element of B. In this sense then a mapping is a **function.** The set A is called the **domain** of the mapping f and the set B is called the **codomain** or **range** of f.

If $a \in A$, we write af for the unique element of B into which the element a is mapped by f. We call af the **image** of a under f.

Example. If $A = \{a, b, c, d\}$, $B = \{1, 2, 3\}$ a mapping f of $A \to B$ is given by $af = 1$, $bf = 3$, $cf = 2$, $df = 3$. There are $3^4 = 81$ distinct mappings of $A \to B$.

Example. In analytic geometry we frequently meet mappings f of the set R of real numbers into itself, such as $f(x) = x^2 - x + 1$, $x \in R$.

Example. A mapping of the interior of a circle into the interior of a square can be pictured as shown in the sketch.

A more sophisticated definition of a mapping f of $A \to B$ is that f is a subset of the cartesian product $A \times B$ such that each $a \in A$ occurs in one and only one ordered pair (a,b) of this subset. On the other hand, there is no such restriction on the occurrence of the elements of B. In this notation (a,b), the element $b = af$.

Introduction

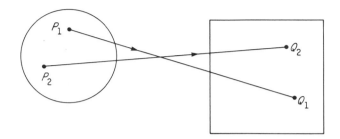

In all mappings it is assumed that the domains and codomains are nonempty sets.

We stress these facts about a mapping f of $A \to B$:

1. f must map every element of A into a unique element of B.
2. Distinct elements of A may map into the same element of B.
3. Not every element of B necessarily has to be the image of some element of A.

Of course we can have mappings of a set A into itself.

Two mappings f and g of $A \to B$ are **equal** if and only if $af = ag$ for every $a \in A$. We then write $f = g$.

If f maps $A \to B$ and g maps $B \to C$, we can define the **composite mapping** $f \circ g$ of $A \to C$ as follows:

$$a(f \circ g) = (af)g, \quad a \in A.$$

We can form the composite mapping $f \circ g$ if and only if the codomain of f is a subset of the domain of g.

We shall frequently abbreviate $f \circ g$ to simply fg.

Thus if f maps $a \in A$ into $b \in B$ and g maps $b \in B$ into $c \in C$, then, by definition, $a(fg) = (af)g = bg = c$, and so fg maps $a \in A$ into $c \in C$.

Two very natural questions that arise in connection with a mapping f of $A \to B$ are: (i) if $b = af$, is a the only element of A that is mapped into the element b of B by f? (ii) is every element of B the image of some element of A?

These questions lead to the following two properties of a mapping that will be of particular importance to us in our later work.

Definition. A mapping f of $A \to B$ is called **injective** (one-to-one) if distinct elements of A are always mapped into distinct elements of B. This means that if a and a' are in A and $a \neq a'$, then $af \neq a'f$.

We isolate the following simple criterion for an injective mapping since it is a very common way of proving injectivity.

Lemma 1. A mapping f of $A \to B$ is injective if and only if $af = a'f$ implies $a = a'$.

Proof: Assume f is injective and let $af = a'f$. Then $a = a'$. For if $a \ne a'$ then, since f is injective, $af \ne a'f$ which is a contradiction. Conversely assume that whenever $af = a'f$ we have $a = a'$. To show f is injective, suppose $a \ne a'$. Then if $af = a'f$ we would have $a = a'$. Hence if $a \ne a'$ we must have $af \ne a'f$ which proves f is injective.

Definition. A mapping f of $A \to B$ is called **surjective** if *every* element $b \in B$ is the image under f of at least one element $a \in A$, that is $b = af$ for at least one $a \in A$.

The symbols $f(A)$ or im f are used to denote the *image of* A *under* f.

Since $f(A) \subset B$, another way of saying f is surjective is to write $f(A) = B$. This means f maps A onto all of B; no element of B is left out.

Definition. A mapping f of $A \to B$ that is both injective and surjective is called **bijective**. Another name for a bijection is **one-to-one correspondence** of the sets A and B.

We shall use the notation 1_A to denote the **identity mapping** on a set A. It is the mapping of $A \to A$ for which $a1_A = a$, for all $a \in A$.

If f is a bijective mapping of $A \to B$, then to each $b \in B$ corresponds a unique element $a \in A$ determined by $b = af$. Thus the mapping f induces a mapping f^{-1} of $B \to A$ defined by $bf^{-1} = a$, where $af = b$. Since $a(ff^{-1}) = (af)f^{-1} = bf^{-1} = a$ we see that $ff^{-1} = 1_A$. Similarly it follows that $f^{-1}f = 1_B$. Thus a bijective mapping f of $A \to B$ determines a mapping f^{-1} of $B \to A$ such that $ff^{-1} = 1_A$ and $f^{-1}f = 1_B$. The mapping f^{-1} is called the **inverse** of the mapping f (likewise, of course, f is the inverse of f^{-1}) and f is said to be an **invertible mapping.**

We now prove that f^{-1} is a bijective mapping of $B \to A$. If a is any element of A, then $a = a1_A = a(ff^{-1}) = (af)f^{-1}$ and therefore f^{-1} is surjective. Next if $bf^{-1} = b'f^{-1}$, then $(bf^{-1})f = (b'f^{-1})f$, that is, $b(f^{-1}f) = b'(f^{-1}f)$ and hence $b = b'$, so that f^{-1} is injective.

THEOREM 1

If f is a mapping of $A \to B$, then the inverse mapping f^{-1} of $B \to A$ exists if and only if the mapping f is bijective.

Proof: All that is left for us to prove of this theorem is that if f^{-1} exists then f is bijective. This follows at once, as above, from the two equations $ff^{-1} = 1A$ and $f^{-1}f = 1_B$.

Exercise. If a mapping has an inverse, prove the inverse is unique.

If f is an injective mapping of $A \to B$ then f is clearly a bijective mapping of $A \to$ im f, where im f is the subset of B into which f maps A.

Hence there exists an inverse mapping f^{-1} of im $f \to A$ such that ff^{-1} is the identity mapping on A and $f^{-1}f$ is the identity mapping on im f. For this reason an injective mapping is also called **invertible,** but it must be understood that by the inverse is here meant a mapping of im $f \to A$.

In some illustrative examples we assume a knowledge of addition and multiplication of integers and fractions (rational numbers), although their formal introduction does not come until later. Since this is done only for the purposes of illustration it should not be objectionable.

Example 1. If A and B are two sets, then the mapping of $A \times B \to A$ defined by $(a, b) \to a$, $a \in A$, $b \in B$, is surjective, but is not injective unless $B = \{b\}$. For if $b \neq b'$, (a, b) and (a, b') both map into a. This map is called the **projection** of $A \times B$ on A.

Example 2. Let Z be the set of integers $0, \pm 1, \pm 2, \ldots$.
(i) The mapping f of $Z \to Z$ defined by $n \to 2n - 1$ that is $(n)f = 2n - 1$, $n \in Z$, is injective but not surjective.
(ii) The mapping g of $Z \to Z$ defined by $(n)g = n + 1$ is bijective. It therefore has the inverse g^{-1} defined by $(n)g^{-1} = n - 1$.
(iii) The mapping h of $Z \to Z$ defined by $(2n)h = n$, $(2n - 1)h = n$ is surjective but not injective. For while h maps Z onto all of Z, we do have, for instance, $(4)h = 2 = (3)h$ and so h is not injective.

We showed that if a mapping f has an inverse f^{-1} then f is bijective. We can now generalize this to prove the following lemma, in which the method of proof is the same.

Lemma 2. Let f be a mapping from $A \to B$.
(i) If there exists a mapping g from $B \to A$ such that $f \circ g = 1_A$, then f is injective.
(ii) If there exists a mapping h from $B \to A$ such that $h \circ f = 1_B$, then f is surjective. (g is called a **right inverse** of f and h is called a **left inverse** of f. If $g = h$ then g is called an **inverse** of f and f is now bijective.)

Proof: (i) Let $af = a'f$ for two elements $a, a' \in A$. Then $a = a1_A = a(f \circ g) = (af)g = (a'f)g = a'(f \circ g) = a'1_A = a'$. Hence f is injective.
(ii) Let b be any element of B. Then $b = b1_B = b(h \circ f) = ((b)h)f$ and $(b)h \in A$. Hence f is surjective.

This last lemma is often very useful.

Let f be a mapping of $A \to B$ and let C be a nonempty subset of A. We can define a mapping g of $C \to B$ by $xg = xf$, for all $x \in C$. This mapping g is called the **restriction** of the mapping f to C. That is g is a **restriction** of f to C if $g = f$ on C, where $C \subset A$.

Likewise if f is a mapping of $A \to B$ and if A is a subset of a set D, then a mapping h of $D \to B$ for which $yh = yf$ whenever $y \in A$, is called an **extension** of f to D. That is h is an extension of f to D if $h = f$ on A, where $A \subset D$.

If C is a subset of the set A, we can define a mapping of $C \to A$ by $x \to x$, for all $x \in C$. This mapping is called the **inclusion mapping** of $C \to A$.

Let f be a mapping of $A \to B$ and suppose that C is a nonempty subset of B. We define the set $f^{-1}(C)$ to be that subset of A for which $x \in f^{-1}(C)$ if and only if $xf \in C$. This means that $f^{-1}(C)$ is the subset of all elements of A that are mapped under f into the subset C of B.

WARNING. The symbol f^{-1} alone is used for the inverse of a mapping f, if f has an inverse. Thus f^{-1} is a mapping. However, $f^{-1}(C)$ is not a mapping, it is a set.

Example 3. Let Z be the set of integers and define a mapping f of $Z \to Z$ by

$$xf = \frac{x}{2} \text{ for } x = 0, \pm 2, \pm 4, \pm 6, \ldots$$

$$xf = x + 1 \text{ for } x = \pm 1, \pm 3, \pm 5, \pm 7, \ldots.$$

Then

$$f^{-1}\{0\} = \{0, -1\}$$
$$f^{-1}\{1, 2, 3\} = \{2, 4, 1, 6\}.$$

Thus f is not injective, but it is surjective.

EXERCISES

1. Z is the set of integers. Which of the following mappings of $Z \to Z$ are injective or surjective or bijective?

(a) $n \to |n|$
(b) $n \to n + 2$
(c) $n \to n^2$
(d) $n \to n^3$
(e) $n \to 2n + 1$
(f) $n \to n^3 + 1$
(g) $n \to n$, for n even
  $n \to -n$, for n odd

(h) $n \to n!$ for $n \geq 0$
  $n \to n$ for $n < 0$
(i) $2n \to n$
  $2n - 1 \to 0$
(j) $n \to n \quad n < -2$
  $n \to -n \quad -2 \leq n \leq 3$
  $n \to n \quad n > 3$

2. Find both composite mappings in Exercise 1 for the pairs (a) 1(a), 1(e); (b) 1(g), 1(h); (c) 1(i); 1(f); (d) 1(c), 1(j).

3. Let f be a mapping (function) of the sets $A \to B$. If C and D are subsets of A prove

$$f(C \cup D) = f(C) \cup f(D)$$
$$f(C \cap D) \subset f(C) \cap f(D).$$

If H and K are subsets of B, prove
$$f^{-1}(H \cup K) = f^{-1}H \cup F^{-1}K$$
$$f^{-1}(H \cap K) = f^{-1}H \cap f^{-1}K.$$

4. The diagram of sets and mappings given here is commutative, which means $f = gh$. If f is injective, prove that g is injective and if f is surjective prove that h is surjective.

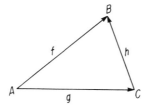

5. For $A \xrightarrow{f} B \xrightarrow{g} A$, prove that if fg is injective and gf is surjective, then f is bijective.

1-3. BINARY OPERATIONS

Let A be a nonempty set. A mapping f of $A \times A \to A$ defines what is called a **binary operation** in the set A. This means that if x and y are any two elements of A, then $(x, y)f$ is an element of A. Thus starting with any two elements of A the mapping f sends them into a third element of A. Hence we can write

$$(x, y)f = x * y$$

and think of $x * y$ as being the unique element of A manufactured out of the two elements x and y of A. The asterisk * is used to symbolize the "combining" of these two elements (that is the binary operation) to form an element of A.

For example, if A is a set of real numbers and x and y are real numbers that are in A, then $x * y$ could possibly mean the sum $x + y$ or the product xy or the difference $x - y$. If A is a set of positive integers, then $x * y$ might be x^y. We prefer to attach no specific meaning to $x * y$, since it is precisely this generality that we want to preserve. In this way we derive results that are valid for many different interpretations of the binary operation symbolized by the asterisk *. Frequently, for simplicity, we write xy for $x * y$ and still intend that it is to be regarded as a general binary operation; this is true sometimes even when we call it a product of x and y. It will be the author's responsibility to stipulate unequivocally when a specific concrete binary operation is being used.

It will be assumed in a set A with a binary operation that equality "=" has the following three properties:
1. it is *reflexive*, that is, $x = x$ for all $x \in A$
2. it is *symmetric*, that is, if $x = y$ then $y = x$
3. it is *transitive*, that is, if $x = y$ and $y = z$, then $x = z$.

Example 4. Let Z be the set of all integers $0, \pm 1, \pm 2, \ldots$. We give examples of binary operations in Z.
   (a) the mapping of $Z \times Z \to Z$ defined by $(m, n) \to m + n$
   (b) the mapping of $Z \times Z \to Z$ defined by $(m, n) \to mn$
   (c) the mapping of $Z \times Z \to Z$ defined by $(m, n) \to m - n$
   (d) the mapping of $Z \times Z \to Z$ defined by $(m, n) \to m + mn + n$.

Example 5. Let P be the set of positive integers $1, 2, 3, \ldots$. Examples of binary operations in P are:
   (a) the three binary operations 4(a), 4(b), 4(d) (but not 4(c), since it is not a mapping of $P \times P \to P$; for example; $(2, 3) \to 2 - 3 = -1$ does not belong to P).
   (b) the mapping of $P \times P \to P$ defined by $(m, n) \to m^n$
   (c) the mapping of $P \times P \to P$ defined by $(m, n) \to m$.

Definition. Let A be a set. A binary operation in the set A is called **commutative** if
$$x * y = y * x, \quad \text{for all } x, y \in A.$$

Examples 4(a), 4(d) are commutative binary operations and so is 5(a); however, 4(c), 5(b), 5(c) are not.

Definition. Let A be a set. A binary operation on the set A is called **associative** if
$$(x * y) * z = x * (y * z)$$
for all $x, y, z \in A$.

Examples 4(a), 4(b), 5(c) are examples of associative binary operations, while the binary operations in 4(c), 4(d), and 5(b) are not associative.

If a binary operation is not known to be associative then we must distinguish between $(x * y) * z$ and $x * (y * z)$, since they may differ if the binary operation is not associative. On the other hand, if a binary operation is known to be associative we can write the triple product as $x * y * z$ without any danger of ambiguity.

Observe that in Examples 4(a) and 4(b) the binary operations are both commutative and associative, while in 4(c) and 5(b) they are neither commutative nor associative. In Example 4(d) the binary operation is commutative but not associative, while in Example 5(c) it is associative but not commutative.

Introduction

A very important example of a binary operation that is associative but not commutative is **map composition.** Let E be the set of all mappings of a set A into itself. Then if $f, g \in E$, clearly the composite mappings $f \circ g$ and $g \circ f$ exist and belong to E. Moreover in general we cannot expect these two mappings to be equal, that is $f \circ g \neq g \circ f$. However, map composition is associative. This is easy to prove. Let $f, g, h \in E$. Form $(f \circ g) \circ h$. It also belongs to E. Let $a \in A$, then

$$a[(f \circ g) \circ h] = [a(f \circ g)]h = [(af)g]h = (af)g \circ h = a[f \circ (g \circ h)].$$

This is true of every $a \in A$. Hence

$$(f \circ g) \circ h = f \circ (g \circ h).$$

Let A be a set with a binary operation $*$ defined on A. An element $e \in A$ is called a **neutral** or **identity element** of the binary operation if

$$e * a = a = a * e \quad \text{for each } a \in A.$$

If a neutral element e exists then it is unique. For if e' is a second neutral element then we have $e' = e * e' = e$.

Let A be a set with a binary operation that has a neutral element e. Let $a \in A$. If there exists $b \in A$ such that $a * b = e = b * a$, we call the element b the **inverse** of the element a. It is customary to denote the inverse of an element a by a^{-1}.

If the binary operation is associative we can prove that if an element a has an inverse a^{-1}, then a^{-1} is unique for this element. For suppose $c \in A$ is a second inverse of the element a. Then $(a^{-1} * a) * c = e * c = c$. Since $(a^{-1} * a) * c = a^{-1} * (a * c) = a^{-1} * e = a^{-1}$, we have $c = a^{-1}$.

EXERCISES

1. A is a set and a binary operation is defined in A by a mapping f of $A \times A \to A$. Let B be a set that is bijective to A and let g be the bijection of $A \to B$. Denote by $g \times g$ the mapping of $A \times A \to B \times B$ defined by $(a, a') \to (ag, a'g)$ where $a, a' \in A$. Let h be the mapping of $B \times B \to B$ defined by $(g \times g)h = fg$.

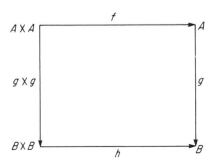

Because $fg = (g \times g)h$, the diagram is called **commutative**.

(a) Prove that h defines a binary operation in B that is commutative and/or associative if the binary operation in A is commutative and/or associative.

(b) If the binary operation in A has a neutral element e, prove that eg is the neutral element for the binary operation in B.

(c) If $a \in A$ has an inverse $a^{-1} \in A$, show that the element $b = ag$ of B has the inverse $b^{-1} = a^{-1}g$.

2. Let Z be the set of integers with ordinary multiplication as the binary operation. Find the binary operations defined by the following bijections g of Z.

(a) $Z \to Z$ where $(n)g = n - n_o, n \in Z$ and n_o is a fixed integer.
(b) $Z \to Z$ where $(n)g = n - 1, n \in Z$.
(c) $Z \to E$ where E is the set of even integers and $(n)g = 2n, n \in Z$.
(d) $Z \to E$ where $(n)g = 2n - 2, n \in Z$.

1-4 THE INTEGERS

The integers $0, \pm 1, \pm 2, \ldots$ form a set with two binary operations, addition and multiplication, and these binary operations are both commutative and associative. The integers are totally ordered; that is, for each pair of integers m and n, either $m = n$ or $m > n$ or $n > m$.

A further property of the set P of positive integers is the

Finite Induction Axiom

If T is a subset of P such that
(i) $1 \in T$
(ii) $n \in T$ implies $n + 1 \in T$
then $T = P$.

The finite induction axiom can be proved to be equivalent to the following

Well-Ordering Axiom

If T is a nonempty subset of the set P of positive integers, then T contains a least positive integer.

This axiom means that there exists an $n_o \in T$ such that for every $n \in T, n \geq n_o$; and the set of positive integers is said to be **well-ordered**.

Finite Induction Principle

Suppose there is associated with each positive integer n a statement $S(n)$ that is either true or false. For some fixed positive integer n_0, let $S(n_0)$ be true. If for each positive integer $k \geq n_0$ the truth of $S(k)$ implies the truth of $S(k + 1)$, then $S(n)$ is true for all positive integers $n \geq n_0$. (The assumption that $S(k)$ is true is called the **induction hypothesis**.)

For $n_0 = 1$, this follows at once from the finite induction axiom by taking T to be the set of positive integers for which $S(n)$ is true. The generalization to an arbitrary positive integer n_0 is easily done.

Example. Let $S(n)$ be the statement that

$$(1) \qquad 1 + 2 + 3 + \cdots + n = \frac{n(n+1)}{2}.$$

For $n = 1$, this yields $1 = 1$ and hence $S(1)$ is true. We take $n_0 = 1$. Our induction hypothesis is that $S(k)$ is true and so

$$1 + 2 + \cdots + k = \frac{k(k+1)}{2}.$$

Add $k + 1$ to both sides of this equation and we get

$$1 + 2 + \cdots + k + k + 1 = \frac{k(k+1)}{2} + k + 1$$

$$= \frac{(k+1)(k+2)}{2}.$$

We see by putting $n = k + 1$ in (1) that this is $S(k + 1)$. Hence if $S(k)$ is true then $S(k + 1)$ is true. Therefore $S(n)$ is true for all positive integers n.

Second Form of the Finite-Induction Principle

Suppose there is associated with each positive integer n a statement $S(n)$ that is either true or false. For some fixed positive integer n_0, let $S(n_0)$ be true. If for each positive integer $k > n_0$, the truth of $S(i)$ for all $n_0 \leq i < k$ implies the truth of $S(k)$, then the statement $S(n)$ is true for all positive integers $n \geq n_0$. (The assumption that $S(i)$ is true for all $n_0 \leq i < k$ is called the **induction hypothesis**.)

Example. A positive integer p is called a **prime** if $p > 1$ and if the only positive integral divisors of p are 1 and p. For example, 2, 17, 11, 5, 101 are primes.

Let $S(n)$ be the statement that the positive integer n is either a prime or is equal to the product of primes. $S(1)$ is false but $S(2)$ is true. Take $n_0 = 2$. Let k be a positive integer > 2. Our induction hypothesis is that every positive integer i, for which $2 \leq i < k$, is either a prime or a product of primes.

If k is a prime, then $S(k)$ is true. If k is not a prime then k can be factored into

$$k = bc$$

where b and c are positive integers > 1. Since $b < k$ the induction hypothesis states that b is either a prime or a product of primes, and since

$c < k$ the same is true of c. Since $k = bc$ it follows that k is a product of primes. Hence $S(k)$ is true in this case.

Thus every positive integer ≥ 2 is either a prime or a product of primes.

1-5 GROUPS, RINGS, FIELDS

Groups, rings and fields are such fundamental algebraic systems that the reader should become casually acquainted with them as early as possible. Moreover, they are easy to define and many examples of them are already familiar to the reader. We shall not be concerned with their theory and so their introduction here is entirely for the purpose of giving our discussion conciseness and coherence. There is no need for the reader to memorize these definitions. They are here for reference.

Definition. A **group** G is a set with a binary operation such that
 (i) the binary operation is associative,
 (ii) the binary operation has a neutral element,
 (iii) every element of G has an inverse in G.

If the binary operation of a group is commutative the group is called a **commutative group.**

Examples of groups are:
1. The integers $0, \pm 1, \pm 2, \ldots$ under addition form a commutative group.
2. The set of fractions m/n, where m and n are integers and $n \neq 0$, form a commutative group under addition.
3. The set of nonzero fractions m/n form a commutative group under multiplication.
4. The set of all bijective mappings of $S \to S$, where S is any nonempty set, forms a noncommutative group under the binary operation of map composition. For map composition is associative but not commutative. The neutral element is the identity map on S and bijective mappings always have inverse mappings that likewise are bijective.

A ring is an algebraic system with two binary operations.

Definition. A **ring** A with a unit element 1 is a set with two binary operations, addition and multiplication, such that
 (i) Addition is both commutative and associative and A is a commutative group under addition. We write 0 for the neutral element of the group and $-x$ for the inverse of $x \in A$.
 (ii) Multiplication is associative but not necessarily commutative and for any $x, y, z \in A$,

$$x(y + z) = xy + xz$$
$$(x + y)z = xz + yz$$
(distributive laws)

(iii) for any $x \in A$, $\quad x \cdot 1 = x = 1 \cdot x$.

A ring is called **commutative** if its multiplication is commutative.

We assume $1 \neq 0$ in order to prevent the ring degenerating into a ring containing only 0.

Examples of rings are:
1. The set of integers is a commutative ring with a unit element.
2. The set of even integers is a commutative ring without a unit element.
3. The set of real numbers is a commutative ring with a unit element.

We shall see later that the set of 2×2 matrices over a field is a non-commutative ring with a unit element.

Such important results for a ring A as

(1) $\qquad 0 \cdot a = a \cdot 0 = 0$, for all $a \in A$,
(2) $\qquad (-a)(-b) = ab, a, b \in A$,

are easy derivations from the axioms for a ring.

We come finally to a field.

Definition. A **field** F is a commutative ring with unit element such that every nonzero element of F has a multiplication inverse.

This means that if $x \in F$ and $x \neq 0$ then there exists an element $x^{-1} \in F$ such that $x \cdot x^{-1} = 1 = x^{-1} \cdot x$. Loosely speaking, this says that a field is a commutative ring with division.

The field is especially important to us in this study, since it can be combined with a commutative additive group to produce a "hybrid" algebraic system called a **vector space**. The elements of the group will be the vectors and the elements of the field will be the scalars.

In a field then we have a commutative and associative addition and multiplication, obeying the two distributive laws; a zero element 0; a unit element 1; as well as division, if $b \neq 0$, $\frac{a}{b} = ab^{-1} = b^{-1}a$.

By the **field of rational numbers** we mean the ordered field of all fractions of the form $\frac{m}{n}$, where m and n are integers and $n > 0$. We define $\frac{m}{n} > \frac{m'}{n'}$ if and only if $mn' > m'n$.

By the **field of real numbers** we mean an ordered field containing the rational field as a subfield, and having the property that every nonempty subset, that is bounded above, has a least upper bound (supremum).

A field is said to be of **characteristic** p, where p is a positive integer if
(i) $px = 0$ for all $x \in E$,
(ii) p is the least positive integer for which (i) is true.

If such a positive integer p exists for a field, then it can be proved that p must be a prime. If no such positive integer exists for a field, the field is said to be of **characteristic zero**. The rational field and real field are both fields of characteristic zero.

It is possible for a ring A to have elements b and c such that $bc = 0$, without either b or c being 0. Such elements are called **divisors of zero**.

It is easy to prove that a field has no divisors of zero.

1-6 POLYNOMIALS

For later use we summarize here some information about polynomials.

Let F be a field and let A be a commutative ring containing F. Let x be an element of A. By a **polynomial in x over F** is meant an expression of the form

$$a_0 + a_1 x + a_2 x^2 + \cdots + a_k x^k$$

where the coefficients $a_0, a_1, \ldots, a_k$ are elements of the field F.

If $a_k \neq 0$ it is called a **polynomial of degree k** and a_k is called its **leading coefficient**, and if $a_k = 1$ it is said to be a **monic polynomial**.

We shall assume that a polynomial in x over F is equal to 0 if and only if each of its coefficients is 0. In this case x is said to be **a variable over F**.

We shall also assume that the reader is familiar with the addition and multiplication of polynomials over a field F. Both these binary operations are commutative and associative.

Let $F[x]$ denote the set of all polynomials in x over F. Then $F[x]$ is a commutative ring. Its unit element is the polynomial 1 and its zero element is the polynomial 0. $F[x]$ is called a **polynomial ring in the variable x**.

A polynomial in x over F of degree > 1 is said to be **reducible over F** if it can be factored into a product of two polynomials in x over F, each of degree ≥ 1. A polynomial that is not reducible over F is said to be **irreducible over F**.

We shall make use of the following property of polynomials in $F[x]$. If $f(x) \neq 0, g(x) \neq 0$ are polynomials in $F[x]$ then there exist polynomials $q(x)$ and $r(x)$ in $F[x]$ such that

$$f(x) = q(x)g(x) + r(x),$$

where either $r(x) = 0$ or degree $r(x) <$ degree $g(x)$. This is called the **division algorithm**. If $r(x) = 0$, $g(x)$ is said to be a **divisor** of $f(x)$.

Chapter 2

Vector Spaces

Often one's first acquaintance with vectors is in the guise of "quantities" having both magnitude and direction. They are represented as "arrows" or "directed line segments" and, when referred to a rectangular coordinate system, a vector **v** is an arrow drawn from the origin in a given direction and having a specified length.

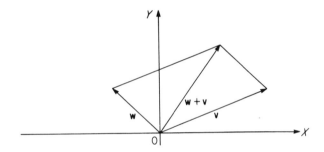

The sum of two vectors **v** and **w** is defined by the parallelogram law which, in two dimensions, is illustrated by the sketch.

This sum can be proved geometrically to be both commutative and associative. The zero vector is one of zero length. If **v** is multiplied by a real number (scalar) c then **cv** is a vector whose magnitude is c times that of **v** and if $c > 0$, **cv** has the same direction as **v**, while if $c < 0$, **cv** has the opposite direction to **v**.

In algebra vectors are defined as elements of an algebraic system called a **vector space** and their properties are enunciated by the set of axioms of this algebraic system. In this way great generality and power are realized. The dependency of vectors on a coordinate system is removed and their defining properties are abstracted from the intuitive but crude concept of a vector as an arrow. A vector space then emerges as a mathematical system occurring in a great many varieties. Needless to say, this greatly increases both out comprehension of the significance of vectors and the scope of their usefulness and their applications.

2-1 DEFINITIONS AND EXAMPLES

A nonempty set V forms a vector space if the set possesses certain properties. A statement of a property that a vector space must have is what is called an **axiom** for the vector space. The axioms are the conditions that must be satisfied in order for V to be a vector space.

Definition. A **vector space** (or **linear space**) V over a field F is a set of elements V that satisfies the following axioms:
1. V is a commutative group under addition. This means
   (a) if $\alpha, \beta \in V$, then $\alpha + \beta \in V$ and
   $$\alpha + \beta = \beta + \alpha;$$
   (b) if $\alpha, \beta, \gamma \in V$ then $(\alpha + \beta) + \gamma = \alpha + (\beta + \gamma);$
   (c) there exists an element of V, denoted by $\bar{0}_V$, such that for any $\alpha \in V$, $\alpha + \bar{0}_V = \alpha = \bar{0}_V + \alpha;$
   (d) for each $\alpha \in V$, there exists an inverse element, denoted by $-\alpha$, such that $\alpha + (-\alpha) = \bar{0}_V = (-\alpha) + \alpha.$
2. There is given a mapping f of $F \times V \to V$ which defines what is called **scalar multiplication** in V; that is, for $x \in F$, $\alpha \in V$, $(x, \alpha)f \in V$ and $(x, \alpha)f$ is called the **scalar product** of x and α. We denote this scalar product by $x\alpha$, that is we put $(x, \alpha)f = x\alpha$. (This of course is merely a notational convenience.)
3. The scalar multiplication obeys the following rules or axioms.
   (a) $(x_1 + x_2)\alpha = x_1\alpha + x_2\alpha$, for all $x_1, x_2 \in F$, and $\alpha \in V$.
   (b) $x(\alpha + \beta) = x\alpha + x\beta$, for all $x \in F$, and $\alpha, \beta \in V$.
   (c) $x(y\alpha) = (xy)\alpha$, for all $x, y \in F$, and $\alpha \in V$.
   (d) $1 \cdot \alpha = \alpha$, for all $\alpha \in V$, $1 \in F$.

The elements of a vector space V are called **vectors** and the elements of the field F, associated with the vector space V, are called **scalars.**

We remind the reader of all the things we can do in a field. First of all, addition and subtraction of scalars are possible, and indeed a field forms a commutative group under addition with the neutral or identity element 0. Also multiplication and division of scalars are possible (of course we have no division by the scalar 0), and indeed the nonzero scalars form a commutative group under multiplication with the identity element 1.

With the exception of the **zero vector,** denoted by $\bar{0}_V$, we shall use Greek letters $\alpha, \beta, \gamma, \ldots$ to denote vectors and English letters to denote scalars. The symbol 0 is the zero scalar, that is the zero element of the field. The symbol 1 is the identity scalar, that is the identity element of the field.

Another very important notational convenience that we shall adopt is to write $\alpha - \beta$ for $\alpha + (-\beta)$, that is,
$$\alpha - \beta = \alpha + (-\beta).$$

In Axiom 2, $F \times V$, as we know, stands for the cartesian product of F and V and is therefore the set of all ordered pairs (x, α), $x \in F$, $\alpha \in V$. It is the mapping f of $F \times V \to V$ that defines the scalar multiplication and f maps (x, α) into the vector denoted by $x\alpha$, the scalar product of x and α. The conspicuous feature to be noted here is that multiplying a vector on the left by a scalar yields a vector. We express this by saying that V is **closed** under scalar multiplication.

Note that if we put $\beta = \bar{0}_V$ in Axiom 3(b), we get $x(\alpha + \bar{0}_V) = x\alpha$, that is $x(\alpha) = x\alpha$. Similarly from Axiom 3(a), we see that $(x)\alpha = x\alpha$.

In order to determine whether a given set V and field F constitute a vector space over F, we must first ascertain if V forms a commutative group under a definition of addition. Then next we look for a definition of scalar multiplication and check to see if the four axioms under (3) are fulfilled. Our axioms tell us precisely what tests must be conducted.

The paramount advantages of the system of axioms are that (1) it specifies precisely the properties that a given set V must possess in order to be a vector space over some given field F; (2) at the same time it does not limit the concept of a vector to any special kind of concrete object, but treats vectors as abstract elements that have certain properties, thus permitting greater generality and usefulness, and hence a freedom of interpretation. This is the power of the axiomatic method. Theorems proved about such an axiom system remain true in any concrete situation or application that satisfies the system.

Strictly speaking, we have defined what is called a **left vector space** V over F, in that the scalar x appears to the left of the vector α in the scalar product $x\alpha$. This is caused by the mapping f, which determines the scalar multiplication, being taken from $F \times V \to V$. If this mapping is assumed to be a mapping of $V \times F \to V$, the scalar product would appear as αx, and V would be called a **right vector space** over F. The simple modifications in the axioms that are necessary to define V as a right vector space over F are very easily made. For instance, Axiom 3(c) would become $(\alpha x)y = \alpha(xy)$. Later we shall see, as we might quite reasonably expect, that algebraically it is immaterial whether we choose to regard V as a left or right vector space over the same field F.

Example 1. Complex numbers are those of the form $x + yi$, where x and y are real numbers and $i = \sqrt{-1}$. The sum of two complex numbers

$$(x_1 + y_1 i) + (x_2 + y_2 i) = (x_1 + x_2) + (y_1 + y_2)i$$

is a complex number.

The complex numbers form a commutative group under addition with the complex number 0 as the neutral element and with the complex number $(-x) + (-y)i$ as the additive inverse of $x + yi$.

If u is any real number, then we have

$$u(x + yi) = (ux) + (uy)i$$

which is also a complex number. This can be easily verified to be an acceptable definition of scalar multiplication, with the real numbers as scalars and the complex numbers as vectors. The reader can check through Axioms 3 for a vector space.

Hence the complex numbers form a vector space over the real field.

Since a real number is also a complex number $x + 0i$, this vector space has the unusual feature that the real numbers play a double role. They are both scalars and vectors.

Example 2. If R is the real field, the set $R[x]$ of polynomials in the variable x with real coefficients has elements of the form (See 1, Sec. 1-6)

$$f(x) = a_0 + a_1 x + \cdots + a_n x^n.$$

The sum and difference of two such polynomials is a polynomial in $R[x]$ and $R[x]$ forms a commutative group under addition.

We can regard these polynomials as vectors over the real field R, with the scalar multiplication taken to be

$$bf(x) = ba_0 + ba_1 x + \cdots + ba_n x^n,$$

where b is any real number. It is easy to verify that Axioms 3 for scalar multiplication are satisfied. For $b, c \in R$ and $f(x), g(x) \in R[x]$, these are

$$(b + c)f(x) = bf(x) + cf(x),$$
$$b[f(x) + g(x)] = bf(x) + bg(x),$$
$$b[cf(x)] = bcf(x),$$
$$1 f(x) = f(x).$$

Hence $R[x]$ is a vector space over R.

Example 3. Let F be any field and form the cartesian product $F \times F$, that is the set of all pairs (x, y), where x and y are any elements of F. Two pairs shall be defined as equal, $(x_1, y_1) = (x_2, y_2)$ if and only if $x_1 = x_2$ and $y_1 = y_2$. Let us next define an addition of pairs by

$$(x_1, y_1) + (x_2, y_2) = (x_1 + x_2, y_1 + y_2),$$

where x_1, x_2, y_1, y_2, are any elements of F. Since F is a field $x_1 + x_2 = x_2 + x_1$ and $y_1 + y_2 = y_2 + y_1$. Hence

$$(x_2, y_2) + (x_1, y_1) = (x_2 + x_1, y_2 + y_1)$$
$$= (x_1 + x_2, y_1 + y_2) = (x_1, y_1) + (x_2, y_2).$$

This proves the addition of pairs is commutative. Again, since F is a field, the addition of elements of F is associative, and so for any $x_1, x_2, x_3, y_1, y_2, y_3 \in F$,

$$[(x_1, y_1) + (x_2, y_2)] + (x_3, y_3) = (x_1 + x_2, y_1 + y_2) + (x_3, y_3)$$
$$= (x_1 + x_2 + x_3, y_1 + y_2 + y_3),$$

and similarly

$$(x_1, y_1) + [(x_2, y_2) + (x_3, y_3)] = (x_1 + x_2 + x_3, y_1 + y_2 + y_3).$$

Thus the addition of pairs is associative. Since $(x,y) + (0,0) = (x,y) = (0,0) + (x,y)$, it follows that $(0,0)$ is the neutral element for addition of pairs. Moreover $(x, y) + (-x, -y) = (x - x, y - y) = (0,0)$ and hence the pair $(-x, -y)$ is the additive inverse of the pair (x,y), that is,

$$(-x, -y) = -(x,y).$$

We have now proved that the set of pairs forms a commutative group under our definition of the addition of pairs. Let us now designate this group of pairs by $V_2(F)$. For any $z \in F$, let us consider the following as a possible definition of scalar multiplication,

$$z(x,y) = (zx, zy), \qquad x, y, z \in F.$$

Certainly the right-hand side is meaningful, since we are merely multiplying elements of the field. Now using this definition we find

$$(z_1 + z_2)(x,y) = ((z_1 + z_2)x, (z_1 + z_2)y)$$
$$= (z_1 x + z_2 x, z_1 y + z_2 y)$$
$$= (z_1 x, z_1 y) + (z_2 x, z_2 y)$$
$$= z_1(x,y) + z_2(x,y).$$

Thus Axiom 3(a) is satisfied. Also

$$z[(x_1, y_1) + (x_2, y_2)] = z(x_1 + x_2, y_1 + y_2)$$
$$= (zx_1 + zx_2, zy_1 + zy_2)$$
$$= (zx_1, zy_1) + (zx_2, zy_2)$$
$$= z(x_1, y_1) + z(x_2, y_2)$$

and so Axiom 3(b) is seen to be satisfied. Moreover,

$$z_1(z_2(x,y)) = z_1(z_2 x, z_2 y)$$
$$= (z_1 z_2 x, z_1 z_2 y) = z_1 z_2 (x,y),$$

and hence Axiom 3(c) is satisfied. Finally, since $1(x,y) = (1x, 1y) = (x,y)$, Axiom 3(d) is satisfied. Thus $V_2(F)$ is a vector space over the field F.

If we specialize the field F in our example to be the field R of real numbers, then the vectors of $V_2(R)$ are the pairs (x,y) of real numbers.

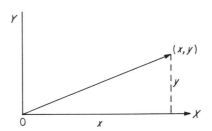

We can therefore introduce a rectangular coordinate system and let the vector (x,y) be geometrically represented by the "arrow" drawn from the origin to the point (x,y).

Elementary plane geometry will enable us to prove that if (x_1, y_1) and (x_2, y_2) are two vectors of $V_2(R)$, then forming the parallelogram as illustrated, the arrow drawn from 0 along the diagonal of the parallelogram is the vector that is the sum of the two given vectors. This means

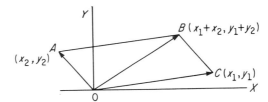

we can prove that the point B has the coordinates $(x_1 + x_2, y_1 + y_2)$. (Draw perpendiculars from the points A, B, C to the two axes and use geometry to prove that certain triangles are congruent.) We shall not continue any further with vectors as "arrows" but these arrows can be shown to have properties corresponding to the axioms for a vector space. This geometrical picture is also successful in three dimensions, with simple modifications.

Returning now to our general field F, it is readily seen that the vector space $V_2(F)$ can be very easily generalized to a vector space $V_n(F)$, whose vectors are now n-tuples $(x_1, x_2, \ldots, x_n)$, where the $x_i \in F$. Addition is defined, as before, by

$$(x_1, x_2, \ldots, x_n) + (y_1, y_2, \ldots, y_n)$$
$$= (x_1 + x_2, y_1 + y_2, \ldots, x_n + y_n)$$

and scalar multiplication by

$$y(x_1, x_2, \ldots, x_n) = (yx_1, yx_2, \ldots, yx_n), y \in F.$$

It is a simple matter to verify that $V_n(F)$ is actually a vector space over F.

As before, two pairs are defined as equal,

$$(x_1, x_2, \ldots, x_n) = (y_1, y_2, \ldots, y_n)$$

if and only if $x_i = y_i$ for each $i = 1, 2, \ldots, n$.

The scalars $x_1, \ldots, x_n$ are called the **coordinates** or **components** of the vector $(x_1, x_2, \ldots, x_n)$.

Example 4. Consider the set M of all functions defined on some closed interval of real numbers, say $[0, 1]$, whose values are real numbers (that is, the set of all mappings of $[0, 1]$ into the field of real numbers). We call these real-valued functions.

Let us define the sum $f + g$ of two such functions f and g, by

$$(f + g)(x) = f(x) + g(x), \qquad x \in [0, 1].$$

Clearly this makes the sum $f + g$ a function belonging to M. Moreover, since addition of real numbers is commutative and associative, we see from our definition that addition of two functions is commutative and associative. Let θ denote the zero function of M, that is the function defined on $[0, 1]$ whose values are all zero. Hence

$$\theta(x) = 0, \qquad \text{for all } x \in [0, 1].$$

Clearly $f + \theta = \theta + f = f$, for all $f \in M$. Hence θ is the neutral element for the addition of functions.

If $f \in M$, define $-f$ by

$$(-f)(x) = -f(x), \qquad \text{for all } x \in [0, 1].$$

Clearly then $-f \in M$ and furthermore we easily see that $f + (-f) = (-f) + f = \theta$, the zero function. This justifies our definition of the inverse $-f$ of the function f.

The upshot of all this is that M forms a commutative group under our addition.

We next obtain a scalar multiplication in M by defining yf, $y \in R$, $f \in M$, by

$$(yf)(x) = yf(x), \qquad x \in [0, 1].$$

Clearly this makes yf a function in M. It is now easy to verify that all of Axiom 3 is satisfied and hence that M is a vector space over the real field R. We leave the rest of the details to the reader.

Consider the subset C of M consisting of all functions that are continuous on the interval $[0, 1]$. Since the sum of two continuous functions is also a continuous function, and since multiplying a continuous function by a real number again yields a continuous function, it follows that the set C forms likewise a vector space over R. In virtue of the fact that C is a subset of M, it is called a **subspace** of M.

In M the vectors are functions and in C the vectors are continuous functions.

Following the same procedure, it can be shown that the set of function from any nonempty set into a field forms a vector space over the field.

Example 5. Consider the set P_4 of all polynomials in a variable x over a field F whose degrees are ≤ 4. All such polynomials are expressions of the form

$$a_0 + a_1 x + a_2 x^2 + a_3 x^3 + a_4 x^4$$

where the coefficients a_0, a_1, a_2, a_3, a_4 are elements of F.

If $b_0 + b_1 x + b_2 x^2 + b_3 x^3 + b_4 x^4$ is also a polynomial in P_4, then we can define the sum of the two polynomials, in the usual way, by

$$(a_0 + b_0) + (a_1 + b_1)x + \cdots + (a_4 + b_4)x^4.$$

Note that the sum of two polynomials may be a polynomial of degree less than the degree of either summand. Clearly the sum belongs to P_4 and the addition is commutative and associative. The zero element of F is the neutral element, that is, the zero polynomial. Also the inverse of a polynomial is merely the polynomial with the signs of all its coefficients changed. For $c \in F$ we define

$$c(a_0 + a_1 x + \cdots + a_4 x^4) = c a_0 + c a_1 x + \cdots + c a_4 x^4.$$

This can be verified as an acceptable definition of scalar multiplication. Hence P_4 is a vector space over the field F of coefficients. In this case the vectors are polynomials of degrees ≤ 4.

The reader will easily observe that we can generalize this vector space to P_n, the set of all polynomials over F of degrees $\leq n$, where n is any positive integer, and prove that P_n is a vector space over F.

Example 6. The "general" solution of a second order homogeneous linear differential equation

$$\frac{d^2 y}{dx^2} + f(x)\frac{dy}{dx} + g(x)y = 0$$

has the form $y = c_1 y_1(x) + c_2 y_2(x)$, where c_1 and c_2 are arbitrary complex numbers and $y_1(x), y_2(x)$ are independent solutions. The set of all solutions $c_1 y_1(x) + c_2 y_2(x)$ is clearly a vector space over the field of complex numbers. The vectors in this space are these linear combinations

$$c_1 y_1(x) + c_2 y_2(x) \text{ of } y_1(x) \text{ and } y_2(x).$$

We come now to five very fundamental and elementary lemmas dealing with the arithmetic of vectors and scalars. They will be used constantly throughout the proofs of theorems. V is taken to be a vector space over some field F.

Lemma 1. For every $\alpha \in V, 0\alpha = \bar{0}_V$.

Proof: $\bar{0}_V = 0\alpha + (-0\alpha) = (0 + 0)\alpha + (-0\alpha)$
$= (0\alpha + 0\alpha) + (-0\alpha), \quad$ by Axiom 3(a).

Since addition in the group is associative, we have
$$\bar{0}_V = 0\alpha + (0\alpha + (-0\alpha))$$
$$= 0\alpha + \bar{0}_V = 0\alpha.$$

Lemma 2. For every scalar $x \in F, x\bar{0}_V = \bar{0}_V$.

Proof: Since $\bar{0}_V = \bar{0}_V + \bar{0}_V$, we have
$$x\bar{0}_V = x(\bar{0}_V + \bar{0}_V) = x\bar{0}_V + x\bar{0}_V, \quad \text{by Axiom 3(b).}$$

Now adding the inverse $(-x\bar{0}_V)$ of $x\bar{0}_V$ to both sides, we get
$$(-x\bar{0}_V) + x\bar{0}_V = (-x\bar{0}_V) + x\bar{0}_V + x\bar{0}_V,$$

that is,
$$\bar{0}_V = \bar{0}_V + x\bar{0}_V = x\bar{0}_V.$$

Lemma 3. For every vector $\alpha \in V$ and every scalar $x \in F$,
$$(-x)\alpha = -x\alpha = x(-\alpha).$$

Proof: Since $\bar{0}_V = \alpha + (-\alpha)$, we have by Lemma 1
$$\bar{0}_V = x\bar{0}_V = x(\alpha + (-\alpha))$$
$$= x\alpha + x(-\alpha), \quad \text{by Axiom 3(b).}$$

Add the inverse $-x\alpha$ of $x\alpha$ to both sides and we have
$$-x\alpha + \bar{0}_V = -x\alpha + x\alpha + x(-\alpha),$$

that is,
$$-x\alpha = \bar{0}_V + x(-\alpha) = x(-\alpha).$$

Also $x + (-x) = 0$ and therefore by Lemma 1,
$$\bar{0}_V = 0a = (x + (-x))\alpha$$
$$= x\alpha + (-x)\alpha, \quad \text{by Axiom 3(a).}$$

Again adding the inverse $-x\alpha$ of $x\alpha$ to both sides we obtain
$$-x\alpha = (-x)\alpha.$$

Lemma 4. For every scalar x and for every finite set of vectors $\alpha_1, \alpha_2, \ldots, \alpha_n$,

$$x(\alpha_1 + \alpha_2 + \cdots + \alpha_n) = x\alpha_1 + x\alpha_2 + \cdots + x\alpha_n.$$

Proof: The method of proof here is finite induction on the number of vectors n. First we observe that if $n = 1$, the lemma is true, since $x(\alpha_1) = x\alpha_1$. [In fact, by Axiom 3(b) the lemma is true when $n = 2$.] Our induction hypothesis is that the lemma is true for $n - 1$ vectors. We shall then prove it to be true for n vectors.

Consider $x(\alpha_1 + \alpha_2 + \cdots + \alpha_{n-1} + \alpha_n)$. The sum $\alpha_1 + \alpha_2 + \cdots + \alpha_{n-1}$ is a vector. Hence regarding this sum as a single vector, we have, by Axiom 3(b), that

$$x(\alpha_1 + \cdots + \alpha_{n-1} + \alpha_n) = x(\alpha_1 + \cdots + \alpha_{n-1}) + x\alpha_n.$$

The induction hypothesis states that

$$x(\alpha_1 + \cdots + \alpha_{n-1}) = x\alpha_1 + \cdots + x\alpha_{n-1}.$$

Using this in the right-hand side of the previous equation, we get

$$x(\alpha_1 + \cdots + \alpha_{n-1} + \alpha_n) = x\alpha_1 + \cdots + x\alpha_{n-1} + x\alpha_n.$$

Thus the induction hypothesis implies the lemma is true for n vectors. This means then that the lemma is true for any finite number of vectors.

Lemma 5. For any vector α and any finite number of scalars $x_1, x_2, \ldots, x_n$,

$$(x_1 + x_2 + \cdots + x_n)\alpha = x_1\alpha + x_2\alpha + \cdots + x_n\alpha.$$

Proof: Again we use finite induction. For $n = 1$, we know $(x_1)\alpha = x_1\alpha$. (By Axiom 3(a) the lemma is true for $n = 2$.) Our induction hypothesis is that the lemma is true for $n - 1$ scalars, that is

$$(x_1 + \cdots + x_{n-1})\alpha = x_1\alpha + \cdots + x_{n-1}\alpha.$$

We then consider $(x_1 + x_2 + \cdots + x_{n-1} + x_n)\alpha$. Regarding the sum $x_1 + x_2 + \cdots + x_{n-1}$ as being a single scalar, we have, by Axiom 3(a),

$$(x_1 + \cdots + x_{n-1} + x_n)\alpha = (x_1 + \cdots + x_{n-1})\alpha + x_n\alpha.$$

Substituting the induction hypothesis in the right-hand side of this last equation, we get

$$(x_1 + \cdots + x_n)\alpha = x_1\alpha + \cdots + x_n\alpha,$$

that is, the lemma is true for n scalars. Hence the lemma is true for any finite number of scalars.

Observe that if $\alpha \neq \bar{0}_V$, then $x\alpha = \bar{0}_V$ if and only if $x = 0$. If $x = 0$ then $x\alpha = \bar{0}_V$, by Lemma 1. If $x\alpha = \bar{0}_V$, then if $x \neq 0$, we have $x^{-1}x\alpha = x^{-1}\bar{0}_V = \bar{0}_V$, where x^{-1} is the multiplication inverse of x. Therefore $1 \cdot \alpha = \bar{0}_V$ and therefore $\alpha = \bar{0}_V$ by Axiom 3(d). This contradicts the hypothesis $\alpha \neq \bar{0}_V$. Hence $x \neq 0$ leads to a contradiction. Therefore $x = 0$.

EXERCISES

1. Define a vector.
2. α and β are vectors. Prove (a) $\alpha + \beta = \alpha$ implies $\beta = \bar{0}_V$. (b) $\alpha + \beta = \bar{0}_V$ implies $\beta = -\alpha$.
3. α is a vector and $\alpha = -\alpha$. Prove $\alpha = \bar{0}_V$.
4. If x is a scalar and if α and β are vectors, prove that $x(\alpha - \beta) = x\alpha - x\beta$.
5. S is a set of two elements $\{a,b\}$. F is the field of residue classes modulo 3. Define a vector space over F on the set V of all functions of $S \to F$.

    How many distinct vectors are there in V?
6. The rational field Q is a subfield of the real field R. If V is a vector space over R, show that V is a vector space over Q.
7. Does the set of all polynomials of degree 5 over the real field R form a vector space over R?
8. Find an algebraic system that satisfies all the axioms for a vector space except the Axiom 3(d).

2-2 SUBSPACES

As we have seen in Example 4, Sec 2-1, some subsets of a vector space V over a field form vector spaces themselves over the same field and they are called **subspaces**. Valuable and useful information about the structure of a vector space is given by a knowledge of its subspaces. As we shall see later, the subspaces constitute component parts of the space itself.

Definition. Let U be a nonempty subset of a vector space V over a field F. The subset U is called a **subspace** of V if U is a vector space over F under the same vector addition and scalar multiplication as for V.

This definition means that the subset U of V is a subspace if the set U satisfies the axioms for a vector space over F. This entails U being a subgroup of the commutative group V which has the property that if $\alpha \in U$, $x \in F$, then $x\alpha \in U$. We state this as a theorem.

THEOREM 1

A subgroup U of a vector space V over F is a subspace of V if U is a subgroup that is closed under the scalar multiplication defined in V.

Proof: Axiom 1 is satisfied, since U is a commutative group under addition. Axiom 2 is satisfied since U is closed under scalar multiplication. This means that the mapping f of $F \times V \to V$ maps $F \times U \to U$. Axiom 3 is satisfied for U, since the scalar multiplication for U is the same as that for V.

THEOREM 2

A nonempty subset U of a vector space V over F is a subspace of V if (1) when $\alpha, \beta \in U$, then $\alpha + \beta \in U$ (that is U is closed under vector addition), (2) when $\alpha \in U$, $x \in F$, then $x\alpha \in U$ (that is, U is closed under scalar multiplication).

Proof: If $\alpha \in U$ then, by (2), $(-1)\alpha = -\alpha \in U$. Hence $\alpha + (-\alpha) = \alpha - \alpha = \bar{0}_V \in U$, by (1). This proves U is a subgroup of V. Hence by Theorem 1, U is a subspace of V.

A vector space V is trivially a subspace of itself. Also the subset of V consisting of the single element $\bar{0}_V$, the zero vector, is obviously a subspace of V. A subspace $U \neq V$ and containing a nonzero vector is frequently called a **proper subspace** of V.

The set of all points (x,y,z) on a plane $Ax + By + Cz = 0$ through the origin forms a subspace of $V_3(R)$. For the sum of two such points on the plane is a point on the plane and if (x,y,z) is on the plane then we see that $r(x,y,z) = (rx,ry,rz)$ for all $\mathbf{r} \in R$ also satisfies the equation and hence is on the plane.

The following are some examples of subspaces of $V_2(R)$, where R is the real field. The vectors here are ordered pairs (x,y) of real numbers.

Example 7. The set of all pairs $(0,y)$, where y is an arbitrary real number, forms a subspace of $V_2(R)$. According to Theorem 2, to prove this, all we have to show is that the sum of two such pairs and the scalar product of such a pair form again pairs whose first members are 0. This is easy, since

$$(0, y_1) + (0, y_2) = (0, y_1 + y_2)$$
$$z(0,y) = (0, zy)$$

where $y_1, y_2, y, z \in R$.

Example 8. In the same way we can prove that the set of all pairs of real numbers of the form $(x, -x)$ forms a subspace of $V_2(R)$.

The same is true of the set of all pairs of the form $(x, 2x)$, where x is an arbitrary real number.

Example 9. On the other hand, the set of all pairs of $V_2(R)$ of the form (x,y), where both x and y are integers, is not a subspace of $V_2(R)$. While the sum of two such vectors is a vector of this type, the scalar product, in general, is not.

Definition. A **linear combination** of vectors of a nonempty subset S of a vector space V over F is a finite sum of vectors of the form

$$x_1\alpha_1 + x_2\alpha_2 + \cdots + x_n\alpha_n,$$

where the $x_i \in F$, the $\alpha_i \in S$, and n is any positive integer.

NOTATION. We shall denote the set of all linear combinations of vectors of a nonempty subset S of V by $\mathbf{L(S)}$.

THEOREM 3

$L(S)$ is a subspace of V.

Proof: We use Theorem 2 to prove this theorem. Let $\alpha = x_1\alpha_1 + \cdots + x_n\alpha_n$ and $\beta = y_1\beta_1 + \cdots + y_m\beta_m$ be vectors in $L(S)$. Then the vectors α_i and β_i are vectors in S, while the x_i and y_i are scalars. Since $\alpha + \beta = x_1\alpha_1 + \cdots + x_n\alpha_n + y_1\beta_1 + \cdots + y_m\beta_m$, it follows that $\alpha + \beta \in L(S)$. Moreover if $\alpha = x_1\alpha_1 + \cdots + x_n\alpha_n \in L(S)$, and if $y \in F$, then $y\alpha = yx_1\alpha_1 + \cdots + yx_n\alpha_n$ belongs to $L(S)$. Hence by Theorem 2, $L(S)$ is a subspace of V.

Definition. If S is a nonempty subset of a vector space V over F, then the subspace $L(S)$ **is called the subspace of** V **generated or spanned by the subset** S.

Clearly, $S \subset L(S)$.

For convenience, if $\square$ is the empty set, we define $L(\square) = \bar{0}_V$.

If U is a subspace of V, then the sum of any finite number of vectors of U is a vector of U; also the scalar product $x\alpha$, $x \in F$, $\alpha \in U$, is a vector of U. Hence any linear combination of vectors of a subspace U is a vector of U.

Lemma 6. If U is a subspace of a vector space V and if S is a subset of U, then $L(S) \subset U$.

Proof: Any linear combination of vectors of S would have to be a vector of U.

Example 10. The subspace of $V_3(R)$ generated by the vectors $(1, 0, 1)$ and $(-1, 1, 0)$ is the set of all linear combinations $x(1, 0, 1) + y(-1, 1, 0) = (x - y, y, x)$, where x and y are real numbers. Hence all vectors of the form $(x - y, y, x)$ form a subspace of $V_3(R)$.

Let I be an arbitrary nonempty set, finite or infinite, and consider a family of subspaces U_i, $i \in I$, of a vector space V; that is, I is the index set of the family. The intersection $\bigcap_{j \in I} U_i$ of the family is the subset of all vectors of V that belong to each member U_i of the family of subspaces. This intersection is nonempty, since it will always contain at least the zero vector $\bar{0}_V$.

THEOREM 4

The intersection $\bigcap_{i \in I} U_i$ of an arbitrary nonempty family of subspaces of V is a subspace of V.

Proof: This theorem is proved by use of Theorem 2. Let $\alpha, \beta \in \bigcap U_i$. Then $\alpha, \beta \in U_i$ for each $i \in I$. Since each U_i is a subspace, $\alpha + \beta \in U_i$ for each $i \in I$. Hence $\alpha + \beta \in \bigcap U_i$. Next let $\alpha \in \bigcap U_i$ and let x be any scalar. Then $\alpha \in U_i$ for each $i \in I$ and hence, since each U_i is a subspace, $x\alpha_i \in U_i$ for each $i \in I$. Hence $x\alpha_i \in \bigcap U_i$. Therefore, by Theorem 2, the intersection $\bigcap U_i$ is a subspace of V.

In particular, then, the intersection of any finite number of subspaces of V is a subspace of V.

THEOREM 5

If S is any nonempty subset of a vector space V, then the subspace $L(S)$ spanned by S is the intersection of all subspaces of V that contain S.

Proof: Let U be the intersection of all subspaces of V that contain S. Since $L(S)$ contains S, it follows that $U \subset L(S)$. On the other hand, any subspace that contains S would have to contain all finite linear combinations of the vectors of S, and hence would have to contain $L(S)$. Since U contains S, it follows that U contains $L(S)$. Thus $L(S) \subset U$. Since $L(S) \subset U$, it follows that $U = L(S)$, and this proves the theorem.

The proof of Theorem 5 tells us that we can now describe $L(S)$ as the least or smallest subspace of V that contains the subset S. In particular if S is itself a subspace of V, then $L(S) = S$.

If U and W are subspaces of a vector space V, denote by $U + W$ the set of all vectors of the form $\alpha + \beta$, where $\alpha \in U$ and $\beta \in W$. By Theorem 2 it follows that $U + W$ is a subspace of V. It is called the **linear sum** of the subspaces U and W. Clearly the linear sum $U + W$ contains both U and W. Moreover, any subspace of V containing U and W would obviously have to contain all the vectors of $U + W$. Hence the linear sum $U + W$ is the smallest subspace of V that contains both U and W. Hence the linear sum $U + W$ can be described as the intersection of all subspaces of V that contain both U and W. It can also be described as the subspace generated by the set-union of U and W.

Example 11. The following are five different subspaces of the vector space $V_n(F)$, where F is an arbitrary field.

All vectors of $V_n(F)$ of the forms
(a) $(0, x_2, x_3, \ldots, x_n)$ where the x_i are arbitrary elements of F.
(b) $(x, x, \ldots, x)$, where $x \in F$ is arbitrary.
(c) $(x_1, -x_1, x_2, \ldots, x_n)$, the x_i are arbitrary.

(d) $(x_1, 2x_1, x_3, \ldots, x_n)$ the x_i are arbitrary.
(e) $(x_1, 0, 0, \ldots, 0, 0)$, x_1 is arbitrary.

Example 12. The following are three different subspaces of the vector space of all real functions, defined on the interval $[0, 1]$. (See Example 4, Sec. 2-1.)
   (a) the continuous real functions on $[0, 1]$;
   (b) the real functions on $[0, 1]$ that have continuous first derivatives;
   (c) the real functions on $[0, 1]$ that are constants; (if $f(x)$ is a constant for all x of $[0, 1]$, call f a **constant function**).

Example 13. If m is a fixed positive integer, then the set of all polynomials of degree $\leq m$ is a subspace of the vector space of all polynomials. (See Sec. 2-1, Example 2.) Note however that the set of all polynomials of degree m is not a subspace.

EXERCISES

1. Which of the following sets of vectors form a subspace of $V_n(R)$,
   (a) all vectors $(x_1, \ldots, x_n)$ for which $x_1 + x_2 = 0$?
   (b) all vectors for which $2x_1 + 3x_2 = 1$?
   (c) all vectors for which either $x_1 = 0$ or $x_2 = 0$?
   (d) all vectors for which $x_1 x_2 = 1$?

2. Let V be an additive commutative group and let F be a field. Let scalar multiplication be defined in V by $x\alpha = \bar{0}_V$, for all $x \in F$ and all $\alpha \in V$, where $\bar{0}_V$ is the neutral element of the group. Check the axioms to determine if V is a vector space.

3. Prove the uniqueness of the zero vector $\bar{0}_v$ and the uniqueness of the inverse $-\alpha$ of a vector $\alpha \in V$.

4. Define a vector and a scalar.

5. Prove $-(-\alpha) = \alpha$ for all $\alpha \in V$.

6. Prove
$$x(\alpha_1 + \alpha_2 + \cdots \alpha_n) = x\alpha_1 + x\alpha_2 + \cdots + x\alpha_n$$
$$(x_1 + x_2 + \cdots + x_n)\alpha = x_1\alpha + x_2\alpha + \cdots + x_n\alpha$$
where the α_i are vectors and the x_i are scalars.

7. Show there are five ways of associating the four vectors $\alpha, \beta, \gamma, \delta$ to find their sum. Assuming the associative law for three vectors prove $\alpha + ((\beta + \gamma) + \delta) = \alpha + \beta + (\gamma + \delta)$.

8. Which of the following are subspaces of $V_3(F)$, where F is the real field? Describe them.
   (a) The set of all vectors of the form $(x, x + y, y)$.
   (b) The set of all vectors of the form $(x + 1, x, -x)$.
   (c) The set of all vectors of the form $(x, x^2, 0)$.

(d) The set of all vectors of the form
$$(x + y + z, x - z, y + z).$$
(e) The set of all vectors (x, y, z) for which x, y, z are rational.

9. If S and T are subsets of a vector space V and if $S \subset T$ and $T \subset L(S)$, the subspace spanned by S, prove $L(S) = L(T)$.

10. Give the proofs for Examples 1, 2, 5, 6 of Sec. 2-1.

11. Show that in the axioms defining a vector space the field F can be replaced by a division ring, that is a "noncommutative" field.

2-3 BASIS AND DIMENSION OF A VECTOR SPACE

The concept of an independent set of vectors is fundamental in linear algebra, for it leads to the definition of a basis of a vector space. The significance of this will be readily appreciated after a few preliminary theorems have been proved.

Definition. A subset S of a vector space V over F is called an **independent set of vectors** (or a set of **independent vectors**) if
$$x_1\alpha_1 + x_2\alpha_2 + \cdots + x_n\alpha_n = \bar{0}_V,$$
where the $\alpha_i \in S$ and the $x_i \in F$, implies that each scalar coefficient $x_i = 0$.

This means that a finite linear combination of vectors of an independent set of vectors is equal to the zero vector if and only if every scalar coefficient in the linear combination is zero.

A subset S of vectors is called a **dependent set of vectors** if it is not an independent set. S is a dependent set if and only if there are distinct vectors $\alpha_1, \alpha_2, \ldots, \alpha_n$ of S, and scalars $x_1, x_2, \ldots, x_n$ not all zero, such that
$$x_1\alpha_1 + x_2\alpha_2 + \cdots + x_n\alpha_n = \bar{0}_V.$$

A set that consists of a single nonzero vector α is independent, since $x\alpha = \bar{0}_V$ if and only if $x = 0$. Any set of vectors that contains the zero vector $\bar{0}_V$ is dependent, since $x\bar{0}_V = \bar{0}_V$ for every scalar x. The empty set is trivially independent.

Example 14. If $\alpha = x\beta$, where x is a scalar, then the set consisting of the two vectors α and β is dependent.

Example 15. The vectors $(1, -1)$ and $(0, 2)$ form an independent set of two vectors in $V_2(R)$. For if $x(1, -1) + y(0, 2) = \bar{0}_V = (0, 0)$, then $(x, -x + 2y) = (0, 0)$. Hence $x = 0$ and $-x + 2y = 0$. The only solution of this pair of equations is $x = y = 0$. Hence the set is independent.

Vector Spaces

On the other hand the pair of vectors $(-2,3)$ and $(4,-6)$ are dependent, since
$$2(-2,3) + 1 \cdot (4,-6) = (0,0) = \bar{0}_V,$$
that is, not all the scalar coefficients of this linear combination are zero.

The set of the three vectors $(-2,3)$, $(4,-6)$, and $(1,0)$ is a dependent set, since $2(-2,3) + 1 \cdot (4,-6) = \bar{0}_V$.

Example 16. In the vector space $V_3(R)$, the vectors $(1, 2, 3)$, $(-2, 1, 4)$ and $(-1, -\frac{1}{2}, 0)$ form a dependent set, since
$$4(1, 2, 3) - 3(-2, 1, 4) + 10(-1, -\frac{1}{2}, 0) = (0, 0, 0) = \bar{0}_V.$$

On the other hand, the vectors $(1, -1, 0)$, $(2, 0, 1)$ and $(1, 1, -1)$ form an independent set. For if x, y, z are real scalars, assume
$$x(1, -1, 0) + y(2, 0, 1) + z(1, 1, -1) = (0, 0, 0) = \bar{0}_V.$$
That is, $(x + 2y + z, -x + z, y - z) = (0, 0, 0)$ if and only if $x + 2y + z = 0$, $-x + z = 0$, and $y - z = 0$. The only solution of this system of equations is $x = y = z = 0$.

Definition. A vector α is said to be **dependent on** a set S of vectors if $\alpha \in L(S)$; that is, if α is a finite linear combination of vectors in the set S. Thus the zero vector is dependent on any nonempty set of vectors.

THEOREM 6

A set S of vectors is dependent if and only if there is a vector $\alpha \in S$ such that α is dependent on the other vectors of S.

Proof: Let $\alpha \in S$ and assume $\alpha \in L(S')$ where S' is the subset of S resulting from the removal of α from S. If $\alpha = \bar{0}_V$ then S is a dependent set. Assume $\alpha \neq \bar{0}_V$, then $\alpha = \sum_1^k x_i \beta_i$, where the x_i are scalars and $\beta_1, \ldots, \beta_k$ are vectors in S'. Thus $\alpha - \sum_1^k x_i \beta_i = \bar{0}_V$ and this is a linear combination of vectors of S whose scalar coefficients are not all zero. Hence, by definition, S is a dependent set of vectors.

Conversely, assume S is a dependent set. Then there exists a linear combination $\sum_1^k x_i \alpha_i$ of vectors of S such that $\sum_1^k x_i \alpha_i = \bar{0}_V$, $\alpha_i \in S$, and not all $x_i = 0$. Suppose $x_j \neq 0$; then, solving for α_j,
$$\alpha_j = -x_j^{-1}(x_1 \alpha_1 + \cdots + x_{j-1}\alpha_{j-1} + x_{j+1}\alpha_{j+1} + \cdots + x_k \alpha_k)$$
and hence α_j is dependent on the other vectors of S.

Lemma 7. Let S and T, where $S \subset T$, be subsets of a vector space V. If $S \subset T \subset L(S)$, then $L(T) = L(S)$.

Proof: Since $L(T)$ is the smallest subspace containing T, we have $L(T) \subset L(S)$. Let $\alpha \in L(S)$. Then α is a linear combination of vectors from S and hence α is a linear combination of vectors from T. Therefore $\alpha \in L(T)$, and so $L(S) \subset L(T)$. Thus $L(S) = L(T)$.

THEOREM 7

Every finite set S of vectors contains an independent subset S' such that $L(S) = L(S')$.

Proof: Use induction on the number of vectors in S. If S contains only one vector α, then if $\alpha = \bar{0}_V$, take S' to be the empty set (a subset of any set) and $L(S') = \bar{0}_V$. If $\alpha \neq \bar{0}_V$, take $S' = S$. Thus the theorem is true for a set S with only one element. Now assume it true for all sets containing less than k vectors. Let S be a set of k distinct vectors $\alpha_1, \alpha_2, \ldots, \alpha_k$. If S is independent, take $S' = S$ and we have $L(S) = L(S')$. If S is a dependent set, then for some j, α_j is dependent on the remaining vectors. By reindexing the α_i, if necessary, we can take $j = k$. Then α_k is dependent on the set S'' of vectors $\alpha_1, \alpha_2, \ldots, \alpha_{k-1}$. By the induction hypothesis, S'' contains an independent set S' such that $L(S') = L(S'')$. Since $\alpha_k \in L(S'')$, we have $S'' \subset S \subset L(S'')$. Therefore by Lemma 7, $L(S) = L(S'')$. Hence $L(S) = L(S')$. Thus the theorem is true for a set of k vectors. Hence we have shown by induction that the theorem is true for any finite set of vectors.

Theorem 7 states that a nonempty finite set S of vectors always contains an independent subset that spans the same subspace that S does.

Lemma 8. If $\alpha_1, \ldots, \alpha_n$ is a sequence of vectors forming a dependent set, then for some j, $1 \leq j \leq n$, there exists a vector α_j such that α_j is dependent on (that is, is a linear combination of) the vectors $\alpha_1, \alpha_2, \ldots, \alpha_{j-1}$.

Proof: Being a dependent set of vectors, there exists a linear combination $\Sigma x_i \alpha_i$ of the vectors $\alpha_1, \ldots, \alpha_n$ such that $\Sigma x_i \alpha_i = \bar{0}_V$ and some of the scalar coefficients x_i are not 0. Let x_j be that coefficient with the largest subscript, which is not 0. Then solving for α_j,

$$\alpha_j = -x_j^{-1}(x_1 \alpha_1 + \cdots + x_{j-1} \alpha_{j-1}).$$

THEOREM 8

In a vector space *spanned (generated)* by k vectors $\alpha_1, \ldots, \alpha_k$ there are at most k independent vectors. (That is, any independent set of vectors in the vector space is finite and contains at most k vectors.)

Proof: Note that this theorem differs from Theorem 7 which deals only in finite sets. If the field of scalars is infinite, this theorem asserts that the infinite set of vectors $L(\alpha_1, \alpha_2, \ldots, \alpha_k)$ contains at most k independent vectors. We shall prove the theorem by showing that any set of $k + 1$ vectors $\beta_1, \beta_2, \ldots, \beta_k$ of the vector space $L = L(\alpha_1, \ldots, \alpha_k)$ are dependent.

If any $\beta_i = \bar{0}_V$ then of course the $k + 1$ vectors are dependent. Suppose then that no $\beta_i = \bar{0}_V$. Since $\beta_1 \in L$, it follows that β_1 is a linear combination of the vectors $\alpha_1, \alpha_2, \ldots, \alpha_k$. Hence the sequence of vectors $\beta_1, \alpha_1, \ldots, \alpha_k$ is a dependent set. Applying Lemma 8, we obtain a vector α_j that is linearly dependent on $\beta_1, \alpha_1, \ldots, \alpha_{j-1}$. (Note that this vector cannot be β_1, since $\beta_1 \neq \bar{0}_V$.) If necessary let us reindex the α_i so that the new α_k is this α_j. Then $\alpha_k \in L(\beta_1, \alpha_1, \ldots, \alpha_{k-1})$ and therefore $L = L(\beta_1, \alpha_1, \ldots, \alpha_{k-1})$. Now $\beta_2 \in L$ and hence $L = L(\beta_2, \beta_1, \alpha_1, \ldots, \alpha_{k-1})$. Thus the sequence of vectors $\beta_2, \beta_1, \alpha_1, \ldots, \alpha_{k-1}$ is a dependent set. Again by Lemma 8, there is a vector of this sequence that is linearly dependent on its predecessors in the sequence. Again reindexing the sequence, if necessary, we can suppose that this vector is α_{k-1}. Thus $\alpha_{k-1} \in L(\beta_2, \beta_1, \alpha_1, \ldots, \alpha_{k-2})$ and hence $L = L(\beta_2, \beta_1, \alpha_1, \ldots, \alpha_{k-2})$.

Continuing in this way we add a β_i and simultaneously discard an α_j each time. Eventually then we reach $L = L(\beta_k, \beta_{k-1}, \ldots, \beta_1)$. Since $\beta_{k+1} \in L$ we have a contradiction of the assumption that the $k + 1$ vectors $\beta_1, \beta_2, \ldots, \beta_k$ constitute an independent set. Thus $L(\alpha_1, \ldots, \alpha_k)$ cannot contain an independent set of $k + 1$ vectors, which implies that any independent set of vectors in $L(\alpha_1, \ldots, \alpha_k)$ can contain, at most, k vectors.

Definition. A subset B of a vector space V over F is called a **basis** of V if B is an independent set and if $V = L(B)$.

Definition. A vector space V over F is said to be **finite-dimensional** if there is a finite subset S of vectors of V such that $V = L(S)$. A vector space that is not finite-dimensional is called **infinite-dimensional.**

A subspace U of a finite-dimensional vector space V is finite-dimensional. For if V is spanned by n vectors, then U can contain at most n independent vectors.

It is not hard to see that the n vectors $(1, 0, 0, \ldots, 0), \ldots, (0, 0, \ldots, 0, 1)$ span the vector space $V_n(R)$. Hence $V_n(R)$ is a finite-dimensional vector space. In Example 6, Section 1, we see that the independent solutions $y_1(x)$ and $y_2(x)$ form a basis of two vectors for that vector space. However, in Examples 2 and 4 of Sec. 2-1 it is not hard to show that the vector spaces are infinite-dimensional.

It has been proved that every vector space, whether finite or infinite-dimensional, has a basis. However, we shall prove here that only a finite-dimensional vector space has a basis.

THEOREM 9

A finite-dimensional vector space V has a basis. Any basis of such a vector space is finite and all bases contain the same number of elements.

Proof: Since V is finite-dimensional, there is a finite set S' of vectors such that $V = L(S')$. By Theorem 7, S' contains an independent subset of vectors S such that $L(S) = L(S')$, and so S is a basis for V.

Let S'' be a second basis for V. By Theorem 8, S'' is a finite set. If the first basis S contains k vectors, then S'' contains $h \leq k$ vectors. But S is also an independent set in $L(S'')$ and therefore $k \leq h$. Thus $h = k$.

A direct and natural consequence of Theorem 9 is the following:

Definition. The number of vectors in a basis of a finite dimensional vector space is called the **dimension** of the space.

Example 17. If $A^2 + B^2 + C^2 \neq 0$, then the set of points (x,y,z) on the plane $Ax + By + Cz = 0$ through the origin is a two-dimensional subspace of $V_3(R)$.

The next theorem demonstrates clearly the importance of a basis in a vector space.

THEOREM 10

If $\alpha_1, \ldots, \alpha_k$ is a basis for a vector space V over F, then every vector $\alpha \in V$ is a unique linear combination of the basis vectors.

Proof: Since $\alpha \in L(\alpha_1, \alpha_2, \ldots, \alpha_k)$, α is a linear combination of the α_i, $\alpha = \sum_{i=1}^{k} x_i \alpha_i$, $x_i \in F$. Suppose α has another such representation, $\alpha = \sum_{i=1}^{k} y_i \alpha_i$, $y_i \in F$. Then

$$\sum_{1}^{k} (x_i - y_i) \alpha_i = \bar{0}_V.$$

Since the α_i are independent, each scalar coefficient must be 0. Hence $y_i = x_i$, $i = 1, 2, \ldots, k$. This proves the uniqueness.

Definition. Let $\alpha_1, \alpha_2, \ldots, \alpha_n$ be a basis of the finite-dimensional vector space V. Then any vector $\alpha \in V$ is a unique linear combination

$$\alpha = \sum_{i=1}^{n} x_i \alpha_i, \quad x_i \in F$$

of these basis vectors. The scalars $x_1, x_2, \ldots, x_n$ are called the **components** of the vector α with respect to this basis.

Example 18. The vectors $\epsilon_1 = (1, 0, \ldots, 0)$, $\epsilon_2 = (0, 1, 0, \ldots, 0)$, $\ldots, \epsilon_n = (0, 0, \ldots, 0, 1)$ form a basis, called the **standard basis**, for the vector space $V_n(F)$, where F is the real field. Thus the dimension of $V_n(F)$ is n.

Example 19. The vector space of all polynomials in x over a field F of degrees $\leq m$ is a vector space of dimension $m + 1$. A basis of this vector space is the set of vectors $1, x, x^2, \ldots, x^m$.

Example 20. Find bases and dimensions of the following subspaces of $V_4(R)$:

(a) U is the subset of all vectors of the form (x_1, x_2, x_3, x_4), where $x_1 + x_2 + x_3 + x_4 = 0$. First we prove U is a subspace. Let $(y_1, y_2, y_3, y_4) \in U$. Then $y_1 + y_2 + y_3 + y_4 = 0$. $(x_1, x_2, x_3, x_4) + (y_1, y_2, y_3, y_4) = (x_1 + y_1, x_2 + y_2, x_3 + y_3, x_4 + y_4)$, $x_1 + y_1 + x_2 + y_2 + x_3 + y_3 + x_4 + y_4 = (x_1 + x_2 + x_3 + x_4) + (y_1 + y_2 + y_3 + y_4) = 0 + 0 = 0$. Hence the sum of two vectors of U is a vector of U. Also if $z \in R$, then $z(x_1, x_2, x_3, x_4) = (zx_1, zx_2, zx_3, zx_4)$; $zx_1 + zx_2 + ax_3 + ax_4 = z(x_1 + x_2 + x_3 + x_4)$. Hence $x_1 + x_2 + x_3 + x_4 = 0$ implies $z(x_1 + x_2 + x_3 + x_4) = 0$; that is, if $(x_1, x_2, x_3, x_4) \in U$ then $z(x_1, x_2, x_3, x_4) \in U$. Hence by Theorem 2, U is a subspace of $V_4(R)$.

If $(x_1, x_2, x_3, x_4) \in U$, then $(x_1, x_2, x_3, x_4) = (x_1, x_2, x_3, -x_1 - x_2 - x_3)$ where x_1, x_2, x_3 are arbitrary real numbers. Taking successively $x_1 = 1, x_2 = x_3 = 0$; $x_1 = x_3 = 0$, $x_2 = 1$; $x_1 = x_2 = 0$, $x_3 = 1$, we get the vectors $(1,0,0,-1)$, $(0,1,0,-1)$, $(0,0,1,-1)$ and we claim these vectors form a basis of U. For if (x_1, x_2, x_3, x_4) is any vector of U, then

$$(x_1, x_2, x_3, x_4) = (x_1, x_2, x_3, -x_1 - x_2 - x_3)$$
$$= x_1(1,0,0,-1) + x_2(0,1,0,-1) + x_3(0,0,1,-1).$$

Moreover, the vectors $(1,0,0,-1)$, $(0,1,0,-1)$ and $(0,0,1,-1)$ are independent, since if

$$a(1,0,0,-1) + b(0,1,0,-1) + c(0,0,1,-1) = (0,0,0,0),$$

where $a, b, c \in R$, then

$$(a,b,c,-a,-b,-c) = (0,0,0,0).$$

Hence $a = b = c = 0$. Thus the dimension of U is 3.

(b) U is the subset of all vectors of $V_4(R)$ of the form $(x, 2x, 3x, 4x)$, where x is an arbitrary real number. In exactly the same way as in part (a) we can show U is a subspace. Clearly, taking $x = 1$ (or taking x equal to any nonzero real number) we get the vector $(1,2,3,4)$ as a basis of U. Hence the dimension of U is one.

(c) U is the subset of all vectors of the form (x_1, x_2, x_3, x_4), where $x_1 + x_2 = 0$ and $x_3 - x_4 = 0$. Again we easily show U is a subspace and we can write

$$(x_1, x_2, x_3, x_4) = (x_1, -x_1, x_3, x_3),$$

where x_1 and x_3 are arbitrary real numbers. For $x_1 = 1$ (or any nonzero real number), $x_3 = 0$ and $x_1 = 0, x_3 = 1$ we obtain a basis of U consisting of the vectors $(1, -1, 0, 0)$ and $(0, 0, 1, -1)$. This can and should be verified by the reader. Hence the dimension of U is 2.

(d) The subset U of all vectors of the form (x_1, x_2, x_3, x_4) where $x_1 + x_2 + x_3 + x_4 = 1$ (or any nonzero real number) is however not a subspace of $V_4(R)$. For it is easy to see that neither the sum of two such vectors nor the scalar product of such a vector belongs to U.

Lemma 9. If U is a subspace of a finite-dimensional vector space V then any basis $\alpha_1, \alpha_2, \ldots, \alpha_k$ of U can be enlarged to a basis for V.

Proof: Let $\beta_1, \beta_2, \ldots, \beta_n$ be a basis for V. Write down the sequence of vectors $\alpha_1, \alpha_2, \ldots, \alpha_k, \beta_1, \beta_2, \ldots, \beta_n$. If β_1 depends on $\alpha_1, \ldots, \alpha_k$, cast it out and if not, retain it. If β_2 depends on $\alpha_1, \alpha_2, \ldots, \alpha_k$ (or on $\alpha_1, \alpha_2, \ldots, \alpha_k, \beta_1$, if β_1 has been retained) then cast out β_2. If β_2 is not dependent on its predecessors, then retain it. Continuing in this way, since every basis has the same number of vectors, we eventually add $n - k$ of the vectors from $\beta_1, \beta_2, \ldots, \beta_n$ to the vectors $\alpha_1, \ldots, \alpha_k$ to obtain a basis for V.

The method explained in this lemma for enlarging the basis of a subspace is both simple and practical, and hardly requires any illustration

Note that a basis for a vector space V need not contain a basis for a subspace U of V.

For instance, the vectors $(1, 0, 0), (0, 1, 0)$ and $(0, 0, 1)$ form a basis for $V_3(R)$, but no two of these vectors will serve as a basis for the two-dimensional subspace of $V_3(R)$ spanned by the independent vectors $(1, 0, 1)$ and $(0, 1, 1)$.

Example 21. Let S be a set of n elements and R the real field. Let V be the set of all mappings of $S \to R$. Take the elements of S to be $a_1, a_2, \ldots, a_n$. For $f, g \in V$, define $f + g$ by $a_i(f + g) = a_i f + a_i g$. For $f \in V$ and $x \in R$, define xf by $a_i(xf) = x(a_i f)$. As we have seen before, it is easy to show, with these definitions of sum and scalar product, that V is a vector space over R.

For $i = 1, 2, \ldots, n$, define $f_i \in V$ by $a_i f_i = 1, a_j f_i = 0, j \neq i$. Let f be any element of V. Then we see that we can express f as follows:

$$f = f_1(a_1 f) + f_2(a_2 f) + \cdots + f_n(a_n f);$$

that is, we claim for each $i = 1, 2, \ldots, n$,
$$a_i f = (a_i f_1)(a_1 f) + (a_i f_2)(a_2 f) + \cdots + (a_i f_n)(a_n f).$$
Now $a_i f_j = 0, i \neq j$, and $a_i f_i = 1$, hence
$$(a_i f_1)(a_1 f) + (a_i f_2)(a_2 f) + \cdots + (a_i f_n)(a_n f) = (a_i f_i)(a_i f) = a_i f.$$

Moreover, $f_1, f_2, \ldots, f_n$ are independent. For suppose $c_1 f_1 + c_2 f_2 + \cdots + c_n f_n = 0$, where 0 is the zero function of V (it maps every a_i into zero) and the c_i are real numbers. Then
$$a_i[c_1 f_1 + \cdots + c_n f_n] = 0, \quad i = 1, 2, \ldots, n$$
$$a_i(c_1 f_1) + \cdots + a_i(c_n f_n) = 0$$
$$c_1(a_i f_1) + \cdots + c_n(a_i f_n) = 0.$$
Since $a_i f_j = 0, j \neq i$ and $a_i f_i = 1$, we see that $c_i = 0, i = 1, 2, \ldots, n$.
Thus $f_1, f_2, \ldots, f_n$ is a basis of V and hence the dimension of V is n.
We give two examples of infinite-dimensional vector spaces.

Example 22. The vector space of polynomials in Example 2. From the definition of the equality of polynomials it follows that $a_0 + a_1 x + \cdots + a_n x^n = 0$ if and only if each scalar coefficient $a_i = 0$. Thus the vectors $1, x, \ldots, x^n$ are linearly independent and this is true for every positive integer n. Hence there can exist no finite basis for this space.

Example 23. The vector space M of functions in Example 4. For each $a \in [0, 1]$ define a function $f_a \in M$ by
$$f_a(x) = 0, \quad x \neq a$$
$$f_a(a) = 1.$$
We show that for any $n > 0$, the functions $f_{a_1}, \ldots, f_{a_n}$, where $a_1, \ldots, a_n$ are distinct points of $[0, 1]$, are linearly independent. Suppose
$$c_1 f_{a_1} + \cdots + c_n f_{a_n} = 0,$$
where 0 is the zero function of M and the c_i are real numbers. Then
$$(a_i(c_1 f_{a_1} + \cdots + c_n f_{a_n}) = c_1 f_{a_1}(a_i) + \cdots + c_i f_{a_i}(a_i)$$
$$+ \cdots + c_n f_{a_n}(a_i) = 0.$$
Now $f_{a_j}(a_i) = 0$ for $j \neq i$ and $f_{a_i}(a_i) = 1$. Hence $c_i = 0$. Similarly, we see that all the $c_i = 0$, and therefore the $f_{a_i}, i = 1, 2, \ldots, n$, are linearly independent. This is true for any $n > 0$, and hence there can exist no finite basis for the vector space M.

In most physics applications the mathematics is involved in the context of infinite-dimensional vector spaces. Examples of this are Fourier series and Hilbert space, of which the latter is important in quantum theory.

As a very useful application of Lemma 9 we prove the following:

Lemma 10: If U and W are finite-dimensional subspaces of a vector space V then

$$\dim U + \dim W = \dim (U \cap W) + \dim (U + W).$$

Proof: $U + W$ is the join or linear sum of U and W. Let $\alpha_1, \ldots, \alpha_r$ be a basis of $U \cap W$. Complete this basis to a basis $\alpha_1, \ldots, \alpha_r, \beta_1, \ldots, \beta_s$ of U and to a basis $\alpha_1, \ldots, \alpha_r, \gamma_1, \ldots, \gamma_k$ of W. (Lemma 9.) Now the vectors $\alpha_1, \ldots, \alpha_r, \beta_1, \ldots, \beta_s, \gamma_1, \ldots, \gamma_k$ clearly span $U + W$. We claim they form a basis of $U + W$. Suppose

$$(1) \quad x_1\alpha_1 + \cdots + x_r\alpha_r + y_1\beta_1 + \cdots + y_s\beta_s + z_1\gamma_1 + \cdots + z_k\gamma_k = \overline{0}_V,$$

where the x_i, y_i, z_i are scalars.
Then

$$\sum_{i=1}^{s} y_i\beta_i = -\sum_{i=1}^{r} x_i\alpha_i - \sum_{i=1}^{k} z_i\gamma_i \in W.$$

Since

$$\sum_{i=1}^{s} y_i\beta_i \in U, \text{ we have } \sum_{i=1}^{s} y_i\beta_i \in U \cap W. \text{ Hence}$$

$$\sum_{i=1}^{n} y_i\beta_i = \sum_{i=1}^{r} w_i\alpha_i,$$

where the w_i are scalars. But the vectors α_i and β_i are independent, since they form a basis of U. Hence all the y_i (and the w_i) are 0. In the same way we show $\sum_{i=1}^{k} z_i\gamma_i \in U \cap W$, and also in the same way we prove all the z_i are 0. If all the y_i and z_i are 0 in (1), then all the x_i must be 0. Hence the $\alpha_i, \beta_i, \gamma_i$ are linearly independent and therefore form a basis of $U + W$.

Now $\dim U = r + s$, $\dim W = r + k$, $\dim (U \cap W) = r$, $\dim (U + W) = r + s + k$. Since $r + s + r + k = r + (r + s + k)$, we have proved $\dim U + \dim W = \dim (U \cap W) + \dim (U + W)$.

2-4 THE MODULE

A module is a simple but far-reaching generalization of a vector space over a field. If in the axioms for a vector space we replace the scalar field

Vector Spaces

by a (not necessarily commutative) ring with unit element we get an algebraic system called a **module over a ring.** All the axioms remain the same. The actual change from a field to a ring is simple enough, for after all a field is a ring. In fact, one might call a module a *vector space over a ring*. Nevertheless some very significant differences result in the theory of modules and the module turns out to be one of the most important algebraic systems.

There is no division assumed in a ring; that is, all nonzero elements do not have multiplication inverses. Hence some theorems, true for vector spaces, are false for modules; while others that remain true may require modifications in their proofs.

It was proved by use of division that a set of vectors D is dependent if and only if one of the vectors of D is a linear combination of the remaining vectors. This led to the result that any finite set D of vectors contains a subset of independent vectors generating the same subspace as is generated by D. In turn this enabled us to prove that a finitely generated vector space has a basis. However, this result is not true for modules and one of the most important differences is that even a finitely generated module may not have a basis. In fact a submodule of a finitely generated module need not be itself a finitely generated module. On the contrary any vector space, whether finite-dimensional or not, can be proved to have a basis. Thus any theorem about a vector space whose proof invokes the basis property is suspect as a theorem about modules. On the other hand, many theorems in Chapter 3, with obvious modifications, do hold for modules.

We shall find the module useful in the discussion in Chapter 12 of the exterior algebra of differential forms.

We briefly mention another generalization of a vector space over a field, that of replacing the field by a division ring. A division ring is a noncommutative field—that is, multiplication is not assumed commutative. Since division still holds in a division ring, this change in the nature of the scalars of the vector space does not give rise to the differences that characterize a module.

EXERCISES

1. If α,β form a basis of a two-dimensional vector space V, prove that $\bar{\alpha} = \alpha + 2\beta$, $\bar{\beta} = 3\alpha - \beta$, also form a basis of V. If x_1, x_2 are the components of the vector $\zeta \in V$ relative to the basis α,β (i.e., $\zeta = x_1\alpha + x_2\beta$) find the components of ζ relative to the basis $\bar{\alpha},\bar{\beta}$.

2. Prove that the vectors $(1,0,2,3)$ and $(0,-1,1,0)$ of $V_4(R)$ are independent. Extend this pair of vectors to a basis for $V_4(R)$.

3. Which of the following sets of functions are linearly independent

vectors of the vector space M of real-valued functions on the closed interval $[0,1]$? (See Example 4, Sec 2-1.)
- (a) $f_1(x) = x^2$, $f_2(x) = x$, $f_3(x) = 1$
- (b) $f_1(x) = x^2 + 1$, $f_2(x) = x(x-1)$, $f_3(x) = 2$
- (c) $f_1(x) = x$, $f_2(x) = \sin x$, $f_3(x) = \cos x$
- (d) $f_1(x) = x^3$, $f_2(x) = x^4 + 3$, $f_3(x) = x^2(1-2x)$, $f_4(x) = 3x^2$

4. Is the following set of vectors linearly independent in $V_3(R)$, where R is: (a) the rational field? (b) the real field?

$$(2,2,-2), \quad (10,-2,5), \quad (\sqrt{2}, 1, -\sqrt{2})$$

5. Let N be the subset of the vector space M of real-valued functions on $[0,1]$ defined by: $f \in N$ if and only if all except a finite number of values $f(x), x \in [0,1]$, of f are zero.
- (a) Prove N is a subspace of M.
- (b) For $x \in [0,1]$ define f_x of $[0,1] \longrightarrow R$ by $f_x(x) = 1, f_x(y) = 0, y \in [0,1]$ and $y \neq x$. Prove these functions form a basis for N.
- (c) Is N finite-dimensional?

6. U is a subspace of a finite-dimensional vector space V. If $\dim U = \dim V$, prove $U = V$.

7. Prove that every n-dimensional vector space contains subspaces of dimension m, $m \leq n$.

8. Find the dimension of the vector space of polynomials of degree $\leq n$ over a field F, where n is a positive integer.

9. If U is a subspace of a finite-dimensional vector space, prove that U is finite-dimensional.

10. Find the dimension of the subspace of $V_4(F)$, spanned by the four vectors $(1,2,1,0)$, $(-1,1,-4,3)$, $(2,3,3,-1)$ and $(0,1,-1,1)$ when (a) F is the rational field (b) F is the real field (c) F is the field Z_3 of integers modulo 3.

11. If $\dim V = n$ and if S is a set of k vectors such that $L(S) = V$ prove $k \geq n$. If $k = n$ prove that the set S is independent.

12. Find the dimensions of the following vector spaces:

(a) The set of vectors $(x_1, x_2, \ldots, x_n)$ of $V_n(F)$, $x_i \in F$, such that $\sum_1^n x_i = 0$

(b) The set of vectors of $V_n(F)$ such that $\sum_1^k x_i = 0$, where $k < n$.

(c) The set of vectors of $V_n(F)$ such that $x_i = 0$ for $i < k$.

13. Find the subspaces of $V_3(R)$ generated by each of the following sets of vectors.
- (a) $(2,-1,0)$, $(1,0,1)$, $(1,-2,-3)$.
- (b) $(0,1,1)$, $(-1,2,1)$, $(0,0,3)$.

14. Find the subspace of the vector space M of illustrative example 4 that is generated by the subset of polynomial functions on $[0,1]$. (A polynomial func-

tion f is a function whose value at any point $x \in [0, 1]$ has the form
$$f(x) = a_0 + a_1 x + \cdots + a_n x^n$$
where the a_i are real numbers and $n \geq 0$.)

15. Are the vectors $(1, 4, 7, 20)$ and $(-3, -2, -1, 6)$ of $V_4(R)$ in the subspace spanned by the vectors $(2, 1, 0, -3)$ and $(1, 0, -1, 0)$?

16. Prove that the dimension of the subspace of a vector space, that is spanned by a finite set S of vectors of V, is equal to the maximal number of linearly independent vectors in S.

Chapter **3**

Linear Transformations

3-1 DEFINITIONS AND FUNDAMENTAL PROPERTIES

Among the great variety of possible mappings of one vector space into another, by far the most important are linear transformations. They have so many desirable and useful properties that we shall study them here to the exclusion of almost all other types of mappings.

Definition. Let V and W be vector spaces over the same field F. A mapping T of $V \longrightarrow W$ is called a **linear transformation** if T has the following two properties.
  (i) $(\alpha + \beta)T = \alpha T + \beta T$, for all $\alpha, \beta \in V$.
  (ii) $(x\alpha)T = x(\alpha T), x \in F, \alpha \in V$.

A linear transformation is also often called a **vector space homomorphism**. Like all mappings a linear transformation may be injective or surjective or bijective or, of course, may have none of these properties.

If $\alpha \in V$, the vector $\alpha T \in W$ is called the **image** of α under T.

Under a linear transformation, therefore, the image of a sum is the sum of the images and the scalar multiplication is preserved.

A bijective linear transformation T of $V \longrightarrow W$ is called an **isomorphism**, and in this case the vector spaces V and W are said to be **isomorphic**.

We write $V \approx W$ if V and W are isomorphic vector spaces.

Exercise. Use induction on n to prove that if $x_1, x_2, \ldots, x_n$ are scalars and $\alpha_1, \alpha_2, \ldots, \alpha_n$ are vectors of V, then

$$\left(\sum_{i=1}^{n} x_i \alpha_i \right) T = \sum_{i=1}^{n} x_i (\alpha_i T),$$

where T is any linear transformation of $V \longrightarrow W$.

Some examples of linear transformations of $V_2(R) \longrightarrow V_2(R)$ are:
  (i) the reflections of the plane given by

  $$(x,y)T = (-x,y) \quad \text{and} \quad (x,y)T = (x,-y);$$

  (ii) the reflection about the origin, $(x,y)T = (-x,-y)$;

Linear Transformations

(iii) the plane dilations $(x,y)T = (kx,y)$ and $(x,y)T = (x,ky)$, where k is a fixed real number;

(iv) the shears parallel to the x and y axes given by

$$(x,y)T = (x + ky, \ y)$$

and $$(x,y)T = (x, \ y + kx),$$

where again k is a fixed real number. Other examples of linear transformations are:

(v) the projection P of $V_3(R)$ onto $V_2(R)$ given by

$$(x,y,z)P = (x,y);$$

(vi) the set of real-valued "functions" $f(x)$ with continuous first derivatives $f'(x)$ on the closed interval $[0,1]$ form a real vector space V. The transformation D defined by $D(f(x)) = f'(x)$, $x \in [0,1]$ is linear;

(vii) the transformation T on the vector space V of continuous "functions" $f(t)$, defined on $[-1,2]$, given by $T(f(t)) = \int_0^1 f(x)\,dx$ is linear on V to V.

Example 1. The reader is no doubt familiar with the formulas

$$X = x \cos \theta + y \sin \theta$$
$$Y = -x \sin \theta + y \cos \theta$$

for the rotation through an angle θ of the axes in the plane.

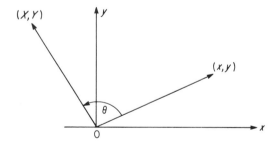

If $\alpha = (x,y) \in V_2(R)$, these equations can be regarded as defining a transformation T of $V_2(R) \longrightarrow V_2(R)$ for which

$$(X,Y) = \alpha T = (x \cos \theta + y \sin \theta, \ -x \sin \theta + y \cos \theta)$$

is a rotation of the vector α through the angle θ into the vector $\beta = (X,Y)$.

There is no difficulty in verifying that T satisfies the two conditions (i) and (ii) above. We need simply apply the rules for vector addition and scalar multiplication as defined for the vector space $V_2(R)$. (See Sec. 2-1.)

This linear transformation T is of a type known as an orthogonal transformation, which we shall study later.

Note that if we solve the above equations for x and y we obtain

$$x = X \cos \theta - Y \sin \theta$$
$$y = X \sin \theta + Y \cos \theta.$$

We therefore obtain a second linear transformation S of $V_2(R) \rightarrow V_2(R)$ defined by

$$(X, Y)S = (X \cos \theta - Y \sin \theta, \quad X \sin \theta + Y \cos \theta).$$

We therefore have $\alpha T = \beta$ and $\beta S = \alpha$. S is called the "**inverse**" of T and is usually denoted by T^{-1}. (Of course T is therefore the inverse of S.)

Definition. The **kernel**, ker T, of a linear transformation T of $V \rightarrow W$ is the subset of all vectors $\alpha \in V$ such that $\alpha T = \bar{0}_W$.

The kernel of a linear transformation T is therefore the set of all vectors of V that are mapped by T into the zero vector of W.

If T is a linear transformation of $V \rightarrow W$, we denote by im T the set of all vectors $\alpha' \in W$ for which $\alpha' = \alpha T$ for at least one vector $\alpha \in V$. Thus im T is the subset of all vectors of W that are images under T of vectors of V. If U is any subset of V, denote by $T(U)$ the subset of W into which U is mapped by T. Clearly if T is surjective then im $T = T(V) = W$.

A surjective linear transformation is called an **epimorphism**, and an injective linear transformation is called a **monomorphism**.

Naturally we can also have linear transformations of a vector space into itself. A linear transformation of $V \rightarrow V$ is called an **endomorphism of V**, and, if it is bijective, it is called an **automorphism of V**.

Definition. An **endomorphism** of a vector space V is called a **linear operator on V**.

We begin next the derivation of the basic properties of linear transformations.

THEOREM 1

If T is a linear transformation of $V \rightarrow W$ then (i) $\bar{0}_V T = \bar{0}_W$, (ii) if U is a subspace of V then $T(U)$ is a subspace of W.

Proof: (i) For $\alpha \in V$, $(\alpha + \bar{0}_V)T = \alpha T$. Also $(\alpha + \bar{0}_V)T = \alpha T + \bar{0}_V T$. Hence $\alpha T = \alpha T + \bar{0}_V T$. Adding the inverse $-(\alpha T)$ to both sides, we get $\bar{0}_W = \bar{0}_W + \bar{0}_V T = \bar{0}_V T$.

(ii) Let $\alpha', \beta' \in T(U)$. Then there exist $\alpha, \beta \in U$ such that $\alpha' = \alpha T$ and $\beta' = \beta T$. Hence $\alpha' + \beta' = (\alpha + \beta)T$. U is a subspace and therefore $\alpha + \beta \in U$. Hence $\alpha' + \beta' \in T(U)$. Next let $x \in F$

$\alpha' \in T(U)$. Then $\alpha' = \alpha T$, $\alpha \in U$. Hence $x\alpha' = x(\alpha T) = (x\alpha)T$. Since U is a subspace, $x\alpha \in U$, and therefore $x\alpha' \in T(U)$. By Theorem 2, Chapter 2, this proves $T(U)$ is a subspace of W.

Corollary. Im T is a subspace of W.

THEOREM 2

If T is a linear transformation of $V \longrightarrow W$, then ker T is a subspace of V.

Proof: Again we use Theorem 2, Chapter 2. Let $\alpha, \beta \in$ ker T. Then $(\alpha + \beta)T = \alpha T + \beta T = \bar{0}_W + \bar{0}_W = \bar{0}_W$. Hence $\alpha + \beta \in$ ker T. Moreover for $x \in F$, $\alpha \in$ ker T, $(x\alpha)T = x(\alpha T) = x\bar{0}_W = \bar{0}_W$. Hence $x\alpha \in$ ker T. This proves ker T is a subspace of V.

The next theorem states a very useful criterion for a linear transformation to be injective.

THEOREM 3

A linear transformation T of $V \longrightarrow W$ is injective if and only if ker $T = \{\bar{0}_V\}$.

Proof: Let T be injective and let $\alpha \in$ ker T. Then $\alpha T = \bar{0}_W$. By Theorem 1, $\bar{0}_V T = \bar{0}_W$. Hence $\alpha = \bar{0}_V$; that is, the only vector in ker T is the zero vector.

Conversely assume ker $T = \{\bar{0}_V\}$. Suppose $\alpha T = \beta T$. Then $(\alpha - \beta)T = \bar{0}_W$. Hence $\alpha - \beta \in$ ker T, and therefore $\alpha - \beta = \bar{0}_V$. Thus $\alpha = \beta$. This proves T is injective.

Example 2. A field F can be regarded as a vector space over itself. In fact it is a vector space over F of dimension one, since any nonzero element of F constitutes a basis of F.

Consider the mapping T of the two vector spaces $V_n(F) \longrightarrow F$, defined by

$$(x_1, x_2, \ldots, x_n)T = x_1$$

(or for that matter $(x_1, x_2, \ldots, x_n)T = x_i$, for any fixed value of the subscript i from 1 to n). T is a linear transformation. This is easily proved by applying the two criteria for such a transformation. This transformation is surjective, im $T = F$. For if y is any element of F, then y is the image under T of any vector of $V_n(F)$ of the form $(y, x_2, \ldots, x_n)$. This remark also proves T is not injective. In fact $(0, x_2, \ldots, x_n)T = 0$ for arbitrary values of $x_2, x_3, \ldots, x_n$.

We have shown that T is an epimorphism.

Example 3. The mapping T of $F \longrightarrow V_n(F)$ defined by $xT = (x, 0, 0, \ldots, 0)$, $x \in F$, is an injective linear transformation (a **monomorphism**, as it is called). This can be easily verified. In fact $xT = (0, 0, \ldots, 0)$ if and only if $x = 0$. The kernel of T consists only of the zero vector 0 of F. T is not surjective, since, for example, the vector $(1, 1, 0, 0, \ldots, 0)$ of $V_n(F)$ is not the image of any element (vector) of F. Putting it another way, im $T \subset V_n(F)$ but im $T \neq V_n(F)$.

Still another way of proving T is injective is this: $(x, 0, 0, \ldots, 0) = (y, 0, 0, \ldots, 0)$ if and only if $x = y$.

We have proved T is a monomorphism.

Example 4. The mapping T of $V_n(F) \longrightarrow V_n(F)$ defined by

$$(x_1, x_2, \ldots, x_n)T = (x_n, x_{n-1}, \ldots, x_1)$$

is linear. Again we appeal to the two criteria for a linear transformation and we find we can quickly verify the claim.

Moreover, $(x_n, x_{n-1}, \ldots, x_1) = (0, 0, \ldots, 0)$ implies $x_n = x_{n-1} = \cdots = x_1 = 0$ and this in turn implies $(x_1, x_2, \ldots, x_n) = (0, 0, \ldots, 0)$. Thus the kernel of T consists only of the zero vector of $V_n(F)$. Hence T is injective.

Also T is surjective; that is, im $T = V_n(F)$. For if $(y_1, y_2, \ldots, y_n)$ is any vector of $V_n(F)$, then by the definition of T we have

$$(y_1, y_2, \ldots, y_n) = (y_n, y_{n-1}, \ldots, y_1)T.$$

This means the vector $(y_1, y_2, \ldots, y_n)$ is the image under T of a vector of $V_n(F)$. Hence T is surjective. Thus T is bijective.

We have therefore proved that T is an automorphism of $V_n(F)$.

Example 5. Let R be the real field and let S be an arbitrary nonempty set. The symbol R^S denotes the set of all mappings of $S \longrightarrow R$. We have seen earlier how, with the usual definitions of addition of mappings and of the scalar product of a mapping, R^S is a vector space over R. We repeat these definitions. For $f, g \in R^S$, define $f + g$ by $(f + g)(s) = f(s) + g(s), s \in S$. For $f \in R^S, x \in R$, define xf by $(xf)(s) = xf(s)$ $s \in S$. Observe that the zero vector of R^S is the mapping θ of $S \longrightarrow R$ defined by $\theta(s) = 0$, for all $s \in S$.

The following mappings T of $R^S \longrightarrow R^S$ defined by
(a) $T(f) = -f, \quad f \in R^S$,
(b) $T(f) = 2f, \quad f \in R^S$,
are easily shown to be automorphisms of R^S.

Let us consider still another mapping T of $R^S \longrightarrow R^S$. Choose any element of S and let us denote it by s_0. Now define T as follows:

$$T(f) = f', \quad f \in R^S$$

where f' is defined by

$$f'(s_0) = 0,$$
$$f'(s) = f(s), \quad s \in S, \quad s \neq s_0.$$

Clearly $f' \in R^S$. We claim T is a linear transformation.

To see this, let $f, g \in R^S$. Then $T(f) + T(g) = f' + g'$, where $f'(s_0) = 0 = g'(s_0)$, and $f'(s) = f(s), g'(s) = g(s)$, for $s \neq s_0$.

Now

$$(f' + g')(s_0) = f'(s_0) + g'(s_0) = 0 + 0 = 0$$
$$(f' + g)(s) = f'(s) + g'(s) = f(s) + g(s) = (f + g)(s).$$

This proves $T(f) + T(g) = T(f + g)$.

Next let $f \in R^S$ and $x \in R$, then $xT(f) = xf'$, where $f'(s_0) = 0$, $f'(s) = f(s), s \neq s_0$. Now $(xf')(s_0) = xf'(s_0) = 0$.

$$(xf')(s) = xf'(s) = xf(s) = (xf)(s), \quad s \neq s_0.$$

Hence $xT(f) = T(xf)$. Thus T is linear.

This linear transformation T is not surjective, since any mapping $f \in R^S$ for which $f(s_0) \neq 0$ cannot be the image under T of a mapping belonging to R^S. Hence T is not an automorphism.

We have proved T is an endomorphism of R^S.

Note that T is also not injective. For if it were, the kernel of T would consist only of the zero vector θ of R^S. However if f is the mapping of $S \to R$ defined by $f(s_0) = 1$, $f(s) = 0$ for all $s \neq s_0$, then $T(f) = \theta$. Hence $f \in \ker T$. Since $f \neq \theta$, T is not injective.

Definition. A linear transformation T of $V \to W$ is said to be **invertible** if it has an inverse T^{-1}, that is, if there exists a mapping T^{-1} of $W \to V$ such that $TT^{-1} = 1_V$ and $T^{-1}T = 1_W$.

We know that T has the inverse T^{-1} if and only if it is bijective. Our next theorem proves the inverse T^{-1} is a linear transformation.

THEOREM 4

If T is an isomorphism of $V \to W$, then T^{-1} is an isomorphism of $W \to V$.

Proof: Since T is bijective we know T^{-1} exists and is bijective. We need to prove T^{-1} is a linear transformation of $W \to V$.

Let $\alpha', \beta' \in W$. Then $\alpha' T^{-1} = \alpha$ and $\beta' T^{-1} = \beta$, where α and β are the unique vectors in V for which $\alpha T = \alpha'$ and $\beta T = \beta'$. Thus $(\alpha + \beta)T = \alpha' + \beta'$, and therefore $(\alpha' + \beta')T^{-1} = \alpha + \beta = \alpha' T^{-1} + \beta' T^{-1}$. For $\alpha' \in W$, $\alpha' T^{-1} = \alpha$, where $\alpha T = \alpha'$. For $x \in F$, $x\alpha' = (x\alpha)T$. Hence $(x\alpha')T^{-1} = x\alpha = x(\alpha' T^{-1})$. This completes the proof that T^{-1} is linear.

THEOREM 5

Let $\alpha_1, \alpha_2, \ldots, \alpha_n$ be a basis for the finite-dimensional vector space V over F, and let $\beta_1, \beta_2, \ldots, \beta_n$ be arbitrary vectors in a vector space W over F. Then there exists a unique linear transformation T from $V \to W$ such that $\alpha_i T = \beta_i, i = 1, 2, \ldots, n$.

Proof: For any vector $\alpha \in V$,

$$\alpha = \sum_{i=1}^{n} x_i \alpha_i, \quad x_i \in F.$$

Define a mapping T of $V \to W$ by

$$\alpha T = \sum_{i=1}^{n} x_i \beta_i.$$

It is easy to show that T is a linear transformation, and clearly $\alpha_i T = \beta_i, i = 1, 2, \ldots, n$.

To prove the uniqueness of T, assume the existence of a second linear transformation T' with the property that $\alpha_i T' = \beta_i, i = 1, 2, \ldots, n$. Since T' is linear, then for

$$\alpha = \sum_{i=1}^{n} x_i \alpha_i,$$

we have

$$\alpha T' = \sum_{i=1}^{n} x_i \beta_i.$$

Thus $\alpha T' = \alpha T$, for every $\alpha \in V$. Hence $T' = T$.

Lemma 1. If T is an injective linear transformation of $V \to W$, then if $\alpha_1, \alpha_2, \ldots, \alpha_k$ are independent vectors of V, $\alpha_1 T, \alpha_2 T, \ldots, \alpha_k T$ are independent vectors of W.

Proof: Suppose $\sum_{i=1}^{k} x_i(\alpha_i T) = \bar{0}_W$. Since T is linear, this implies $\left(\sum_{i=1}^{k} x_i \alpha_i\right) T = \bar{0}_W$. Since T is injective, this implies $\sum_{i=1}^{k} x_i \alpha_i = \bar{0}_V$. The α are independent and therefore $x_i = 0, i = 1, 2, \ldots, k$.

Thus $\sum_{i=1}^{k} x_i(\alpha_i T) = \bar{0}_W$ implies each $x_i = 0$, and hence the $\alpha_i T$ are independent vectors.

Linear Transformations

THEOREM 6

If V is a finite dimensional vector space and if T is a linear transformation of $V \longrightarrow W$, then

$$\dim V = \dim(\ker T) + \dim(\operatorname{im} T).$$

Proof: Let $\dim V = n$. Let $\alpha_1, \alpha_2, \ldots, \alpha_j$ be a basis of the subspace $\ker T$ of V. Then $\dim(\ker T) = j$. This basis can be extended, by Lemma 9 of Chapter 2, to a basis $\alpha_1, \ldots, \alpha_j, \alpha_{j+1}, \ldots, \alpha_n$ of V. Hence for $\alpha \in V$, $\alpha = \sum_{i=1}^{n} x_i \alpha_i$, $x_i \in F$. Therefore $\alpha T = \sum_{i=1}^{n} x_i(\alpha_i T)$. Since $\alpha_i T = \bar{0}_W$, $i = 1, 2, \ldots, j$, we have

$$\alpha T = \sum_{i=j+1}^{n} x_i(\alpha_i T).$$

This proves that the vectors $\alpha_{j+1} T, \ldots, \alpha_n T$ span the subspace $\operatorname{im} T$ of W. We next show they are independent.

Assume

$$\sum_{i=j+1}^{n} y_i(\alpha_i T) = \bar{0}_W, \quad y_i \in F.$$

Then

$$\left(\sum_{i=j+1}^{n} (y_i \alpha_i) \right) T = \bar{0}_W.$$

Hence $\sum_{i=j+1}^{n} y_i \alpha_i \in \ker T$. Since $\alpha_1, \alpha_2, \ldots, \alpha_j$ is a basis of $\ker T$, we have

$$\sum_{i=j+1}^{n} y_i \alpha_i = \sum_{i=1}^{j} z_i \alpha_i, \quad z_i \in F.$$

Thus

$$\sum_{i=1}^{j} z_i \alpha_i - \sum_{i=j+1}^{n} y_i \alpha_i = \bar{0}_V.$$

Since $\alpha_1, \alpha_2, \ldots, \alpha_n$ are independent (they are a basis of V), it follows that all the $y_i = 0$ and all the $z_i = 0$. Thus $\sum_{i=j+1}^{n} y_i(\alpha_i T) = \bar{0}_W$ implies all the $y_i = 0$ and therefore the vectors $\alpha_i T$, $i = j + 1, \ldots, n$, are independent. Thus dim (im T) = $n - j$. Hence dim (ker T) + dim (im T) = $j + n - z = n = $ dim V.

Corollary. If two finite-dimensional vector spaces V and W are isomorphic, then dim V = dim W.

Proof: Let T be the isomorphism of $V \rightarrow W$. Then ker $T = \{\overline{0}_V\}$ and hence dim (ker T) = 0. Also im $T = W$. Therefore, by the theorem dim V = dim W.

THEOREM 7

Let V be a vector space over F of dimension n. Then V is isomorphic to the vector space $V_n(F)$.

Proof: Let $\alpha_1, \alpha_2, \ldots, \alpha_n$ be a basis of V. For any vector

$$\alpha = \sum_{i=1}^{n} x_i \alpha_i, \quad x_i \in F,$$

of V, define a mapping T of $V \rightarrow V_n(F)$ by

$$\alpha T = (x_1, x_2, \ldots, x_n).$$

It is a very simple exercise (left to the reader) to prove that T is a linear transformation and that it is bijective.

Exercise. Prove that the mapping T defined in Theorem 7 is an isomorphism.

Corollary. Two finite-dimensional vector spaces of the same dimension are isomorphic.

Proof: Let dim V = dim $W = n$. Let T be the isomorphism of $V \rightarrow V_n(F)$ and S the isomorphism of $W \rightarrow V_n(F)$. Then S^{-1} is an isomorphism of $V_n(F) \rightarrow W$. We know that the product TS^{-1} is a bijective mapping of $V \rightarrow W$. It is also linear, for $(\alpha + \beta)TS^{-1} = (\alpha T + \beta T)S^{-1} = (\alpha T)S^{-1} + (\beta T)S^{-1} = \alpha(TS^{-1}) + \beta(TS^{-1})$; and also $(x\alpha)TS^{-1} = ((x\alpha)T)S^{-1} = (x(\alpha T))S^{-1} = x((\alpha T)S^{-1}) = x(\alpha(TS^{-1}))$. Thus the mapping TS^{-1} is an isomorphism.

According to Theorem 7, every n-dimensional vector space over a field F is isomorphic to the vector space $V_n(F)$. This theorem then determines, up to isomorphism, all finite-dimensional vector spaces. Every finite-dimensional vector space can be regarded as a vector space of the type $V_n(F)$.

Observe that the two corollaries to Theorems 6 and 7 combine to prove that two finite-dimensional vector spaces are isomorphic if and only if they have the same dimension.

EXERCISES

1. U, V, and W are vector spaces over the same field. S is a linear transformation of $U \longrightarrow V$ and T is a linear transformation of $V \longrightarrow W$. Prove
   (a) $S \circ T$ is a linear transformation of $U \longrightarrow W$
   (b) if S and T are injective, then $S \circ T$ is injective
   (c) if S and T are isomorphisms, then $S \circ T$ is an isomorphism.

2. If S and T are endomorphisms of a vector space V, prove that $xS + yT$ is an endomorphism of V, where x and y are any scalars. When does $(S + T)(S - T) = S^2 - T^2$?

3. The mapping T of $V_3(R) \longrightarrow V_3(R)$, where R is the real field, is defined by

$$(x_1, x_2, x_3)T = (x_2, x_1 - x_2, -x_1).$$

(a) Verify that T is linear.
(b) Find ker T and im T and their dimensions.
(c) Find the image under T of the subspace spanned by the vectors $(1, 1, -1)$ and $(2, 2, 0)$.

2-2 NONSINGULAR TRANSFORMATIONS

The terms *nonsingular* and *singular* applied to linear transformations on vector spaces are so frequently used in the literature that they are defined here and used in the later chapters. The principal results in this section are some very important theorems on linear transformations from one finite-dimensional vector space to another finite-dimensional vector space, and we shall express them in these terms. However, for finite-dimensional vector spaces we shall see that nonsingular means the same as bijective, while singular will turn out to mean neither injective nor surjective. Thus when applied to finite-dimensional vector spaces these new terms introduce no really new ideas.

Definition. If V and W are arbitrary vector spaces (that is not necessarily finite-dimensional) a linear transformation T of $V \longrightarrow W$ is called **nonsingular** if T is injective; that is, if the kernel of T consists of only the zero vector of V. T is called **singular** if there exists a nonzero vector $\alpha \in V$ such that $\alpha T = \bar{0}_W$, that is, if T is not injective.

Although we shall be principally concerned with linear operators on a finite-dimensional vector space, we next prove one interesting theorem about nonsingular linear transformations in general.

THEOREM 8

A linear transformation T of $V \longrightarrow W$ is nonsingular if and only if there exists a mapping T' of im $T \longrightarrow V$ such that $TT' = 1_V$, where 1_V is the identity mapping on V. Moreover T' is a linear transformation.

Proof: Assume that T' exists. Then for $\alpha, \beta \in V$, $\alpha T = \beta T$ implies $\alpha TT' = \beta TT'$, that is $\alpha 1_V = \beta 1_V$ and hence $\alpha = \beta$. This proves T is injective and therefore nonsingular.

Conversely assume that T is nonsingular. Define a mapping T' of im $T \longrightarrow V$ by $\gamma T' = \alpha$, where $\gamma \in$ im T and α is the unique vector of V for which $\alpha T = \gamma$. Hence for any vector $\alpha \in V$, $\alpha TT' = \alpha$ and therefore $TT' = 1_V$.

We now show that T' is linear. Let $\gamma_1, \gamma_2 \in$ im T. Then there exist unique vectors $\alpha_1, \alpha_2 \in V$ such that $\gamma_1 = \alpha_1 T, \gamma_2 = \alpha_2 T$ and therefore $\gamma_1 T' = \alpha_1, \gamma_2 T' = \alpha_2$. Since $\gamma_1 + \gamma_2 = \alpha_1 T + \alpha_2 T = (\alpha_1 + \alpha_2) T$, we have $(\gamma_1 + \gamma_2) T' = \alpha_1 + \alpha_2 = \gamma_1 T' + \gamma_2 T'$. Also if c is any scalar, then $c\gamma_1 = c(\alpha_1 T) = (c\alpha_1)T$. Hence $(c\gamma_1)T' = c\alpha_1 = c(\gamma_1 T')$. Thus T' is linear.

We shall now turn our attention to finite-dimensional vector spaces.

THEOREM 9

If V and W are finite-dimensional vector spaces and dim $V =$ dim W, then a linear transformation T of $V \longrightarrow W$ is nonsingular if and only if T is an isomorphism.

Proof: If T is an isomorphism then T is injective and hence nonsingular.

Conversely, assume that T is nonsingular. Let $\alpha_1, \alpha_2, \ldots, \alpha_n$ be a basis for V. We first show that the vectors $\alpha_1 T, \ldots, \alpha_n T$ of W are linearly independent. For scalars $x_1, \ldots, x_n$ let $\bar{0}_W = x_1(\alpha_1 T) + \cdots + x_n(\alpha_n T) = (x_1 \alpha_1 + \cdots + x_n \alpha_n)T$. Since T is nonsingular this implies $x_1 \alpha_1 + \cdots + x_n \alpha_n = \bar{0}_V$ and hence that each scalar $x_i = 0$. The n vectors $\alpha_1 T, \ldots, \alpha_n T$ of W are therefore linearly independent. Since dim $W =$ dim $V = n$, it follows that these n vectors form a basis for W. Hence for any vector $\beta \in W$, $\beta = y_1(\alpha_1 T) + \cdots + y_n(\alpha_n T) = (y_1 \alpha_1 + \cdots + y_n \alpha_n)T$ where the y_i are scalars. Thus T is surjective and is therefore bijective and hence is an isomorphism.

It follows at once from this theorem that if V and W are finite-dimensional vector spaces of the same dimension, then a linear transformation T of $V \longrightarrow W$ is nonsingular if and only if it is bijective and hence if and only if T is invertible. Moreover the inverse T^{-1} must also be nonsingular.

The proof of the theorem also makes evident that if V and W are

Linear Transformations

finite-dimensional vector spaces of the same dimension, then a linear transformation T of $V \to W$ is nonsingular if and only if for any basis $\alpha_1, \ldots, \alpha_n$ of V the vector $\alpha_1 T, \ldots, \alpha_n T$ form a basis of W.

We point out that we can actually prove more than is stated in the theorem. We have the following corollary.

Corollary. Under the same conditions as obtain in the theorem, T is nonsingular if and only if T is surjective.

Proof: If T is nonsingular then, by the theorem it is an isomorphism, and hence surjective.

Conversely, assume T is surjective. Let β be any vector in W. Then there exists a vector $\alpha \in V$ such that $\beta = \alpha T$. If $\alpha_1, \alpha_2, \ldots, \alpha_n$ is a basis for V, then $\alpha = x_1 \alpha_1 + \cdots + x_n \alpha_n$, where the x_i are scalars. Hence $\beta = x_1(\alpha_1 T) + \cdots + x_n(\alpha_n T)$. The n vectors $\alpha_1 T, \ldots, \alpha_n T$ therefore span W and hence must be a basis of W. Thus T is nonsingular.

If dim V = dim W, where V and W are finite-dimensional vector spaces, then a singular linear transformation T of $V \to W$ is neither injective nor surjective.

In particular, then, for a linear operator T on a finite-dimensional vector space V (specialize the foregoing theorem to the case where $W = V$) we can say that (i) T is nonsingular if and only if it is an automorphism of V; (ii) T is nonsingular if and only if it is bijective; (iii) T is nonsingular if and only if it is invertible. Moreover, a singular linear operator on a finite-dimensional vector space is neither injective nor surjective. On the other hand, a linear operator on a finite-dimensional vector space is injective if and only if it is surjective.

THEOREM 10

The product $S \circ T$ (map composition) of two linear operators S and T on a finite-dimensional vector space V is a nonsingular linear operator on V if and only if both S and T are nonsingular.

Proof: Assume S and T are nonsingular. Then their inverses S^{-1}, T^{-1} exist. Now $T^{-1}S^{-1}(ST) = T^{-1}(S^{-1}S)T = T^{-1}1_V T = T^{-1}T = 1_V$, and similarly $(ST)T^{-1}S^{-1} = 1_V$. Hence ST is invertible and therefore nonsingular.

Conversely assume that ST is nonsingular. Then ST is invertible and has an inverse H. Then $H(ST) = 1_V = (ST)H$. If S were singular then $\alpha S = \bar{0}_V$ for some nonzero vector $\alpha \in V$, and hence $\alpha STH = \bar{0}_V$. However $\alpha STH = \alpha 1_V = \alpha$, and therefore $\alpha = \bar{0}_V$, a contradiction. Hence the first factor S of a nonsingular product ST is nonsingular. Hence S^{-1} exists. Since $STH = 1_V$, we get $TH = S^{-1}1_V = S^{-1}$. Thus the product

TH is nonsingular and therefore its first factor T must be nonsingular. Hence S and T are nonsingular.

Corollary. The product $T_1 T_2 \ldots T_n$ of any finite number n of linear operators on a finite-dimensional vector space is nonsingular if and only if each factor $T_i, i = 1, 2, \ldots, n$ is nonsingular.

Proof: The proof is by induction on n. The corollary is true by the theorem for $n = 2$. Let $n > 2$.

Then $S = T_1 T_2 \ldots T_n = (T_1 T_2 \ldots T_{n-1}) T_n$. Hence S is nonsingular if and only if the product $T_1 T_2 \ldots T_{n-1}$ and T_n are both nonsingular. By the induction hypothesis the product $T_1 T_2 \ldots T_{n-1}$ is nonsingular if and only if each $T_i, i = 1, 2, \ldots, n - 1$, is nonsingular. Thus S is nonsingular if and only if each $T_i, i = 1, 2, \ldots, n$, is nonsingular.

3-3 SETS OF LINEAR TRANSFORMATIONS

Let V and W be vector spaces over the same field F. We denote by Hom (V, W) the set of all linear transformations of $V \longrightarrow W$.

For $T_1, T_2 \in$ Hom (V, W), define an addition $T_1 + T_2$ by $\alpha(T_1 + T_2) = \alpha T_1 + \alpha T_2$ for all $\alpha \in V$. It is easy to verify that $T_1 + T_2 \in$ Hom (V, W). For $\alpha, \beta \in V$ have

$$(\alpha + \beta)(T_1 + T_2) = (\alpha + \beta) T_1 + (\alpha + \beta) T_2$$
$$= \alpha T_1 + \beta T_1 + \alpha T_2 + \beta T_2$$
$$= [\alpha T_1 + \alpha T_2] + [\beta T_1 + \beta T_2]$$
$$= \alpha(T_1 + T_2) + \beta(T_1 + T_2).$$

Also for any scalar x and any vector α, we have

$$x\alpha(T_1 + T_2) = (x\alpha) T_1 + (x\alpha) T_2$$
$$= x(\alpha T_1) + x(\alpha T_2)$$
$$= x[\alpha T_1 + \alpha T_2] = x[\alpha(T_1 + T_2)].$$

The reader should easily verify that addition of the linear transformation belonging to Hom (V, W) is both commutative and associative.

Define the **constant linear transformation** T_0 of $V \longrightarrow W$ by

$$\alpha T = \overline{0}_W, \quad \text{for all } \alpha \in V.$$

Clearly

$$T + T_0 = T_0 + T, \quad \text{for all } T \in \text{Hom}(V, W).$$

The fact that T_0 is linear can be quickly verified. For $T \in$ Hom (V, W) we define a mapping $-T$ of $V \longrightarrow W$ by

$$\alpha(-T) = -(\alpha T), \quad \text{for all } \alpha \in V.$$

Linear Transformations

Then again it can readily be proved that $-T \in \text{Hom}(V, W)$ and that

$$T + (-T) = T_0 = (-T) + T.$$

It now follows that under this associative binary operation of addition the set $\text{Hom}(V, W)$ is a commutative group with the neutral element T_0.

We next introduce a scalar multiplication in $\text{Hom}(V, W)$ and show that with this definition $\text{Hom}(V, W)$ becomes a vector space over the field F.

For $x \in F$ and $T \in \text{Hom}(V, W)$, define a mapping xT of $V \longrightarrow W$ by

$$\alpha(xT) = x(\alpha T), \quad \text{for all } \alpha \in V.$$

Let us verify that $xT \in \text{Hom}(V, W)$. For $\alpha, \beta \in V$, we have

$$(\alpha + \beta)(xT) = x[(\alpha + \beta)T] = x[\alpha T + \beta T]$$
$$= x(\alpha T) + x(\beta T) = \alpha(xT) + \beta(xT).$$

For $y \in F$, we have

$$(y\alpha)(xT) = x[(y\alpha)T] = x[y(\alpha T)] = xy(\alpha T) = yx(\alpha T) = y[\alpha(xT)].$$

These two results show that xT is a linear transformation of $V \longrightarrow W$; that is, $xT \in \text{Hom}(V, W)$.

We now claim that, with these definitions, $\text{Hom}(V, W)$ is a vector space over the same field F. Here xT is the definition of the scalar product of $x \in F$ and $T \in \text{Hom}(V, W)$. There is no difficulty in checking through all the axioms for a vector space and proving that **Hom (V, W) is a vector space over F.**

For example, let us prove that Axiom 3(c) in Chapter 2 is satisfied.

Let $x, y \in F$ and $T \in \text{Hom}(V, W)$. Now $x(yT)$ and $(xy)T$ both belong to $\text{Hom}(V, W)$ and we need to prove

$$x(yT) = (xy)T.$$

To do this we must prove that for every vector $\alpha \in V$,

$$\alpha[x(yT)] = \alpha[(xy)T].$$

This is done by keeping in mind the definition of the scalar product in $\text{Hom}(V, W)$.

$$\alpha[x(yT)] = x[\alpha(yT)] = x[y(\alpha T)]$$
$$= xy(\alpha T) = \alpha[(xy)T] \quad \text{for all } \alpha \in V \text{ and hence}$$
$$x(yT) = (xy)T.$$

If $W = V$, then for $T_1, T_2 \in \text{Hom}(V, V)$, we can use map composition to define a product $T_1 T_2$. We know such multiplication is associative, but not necessarily commutative. It is not hard to verify that $T_1 T_2 \in$ Hom (V,V); for $T_1, T_2 \in$ Hom (V,V) and $\alpha, \beta \in V$, we have

$$(\alpha + \beta) T_1 T_2 = [(\alpha + \beta) T_1] T_2 = [\alpha T_1 + \beta T_1] T_2$$
$$= (\alpha T_1) T_2 + (\beta T_1) T_2 = \alpha T_1 T_2 + \beta T_1 T_2$$

and

$$(x\alpha) T_1 T_2 = [(x\alpha) T_1] T_2 = [x(\alpha T_1)] T_2 = x[(\alpha T_1) T_2] = x(\alpha T_1 T_2).$$

It is also readily verified that for $T_1, T_2, T_3 \in$ Hom (V,V), the two distributive laws $T_1(T_2 + T_3) = T_1 T_2 + T_1 T_3$ and $(T_1 + T_2) T_3 = T_1 T_3 + T_2 T_3$ are satisfied.

Thus the vector space Hom (V,V) is seen to be a ring. Moreover, the mixed associative law

$$T_1(x T_2) = x(T_1 T_2) = (x T_1) T_2, \quad x \in F$$

is true.

We call such an algebraic system which combines the features of a vector space and a ring an **algebra over F**.

Let V be an n-dimensional vector space over a field F. We know that the multiplication of linear operators on V is associative and the product is a linear operator on V. If a linear operator on V is nonsingular then we know it has an inverse that is also a linear operator on V. We can conclude therefore that the set of nonsingular linear operators on V forms a noncommutative group and if V is n-dimensional then we shall denote this group by $L_n(F)$. It is called the **full linear group.**

3-4 AFFINE SPACES

The group concept is basic to all mathematics. As we have seen, a vector space is a commutative group with addition as the binary operation. We illustrate this concept with some examples and exercises and discuss some important nonlinear transformations. This in turn will lead to an introduction to affine spaces and some of their properties. This is a sort of extension of the notions of vector spaces and linear transformations, the intention being to increase the reader's comprehension of the geometrical aspects of a great deal of this study. Much of the work is left in the form of exercises for the reader, and this entire section should be regarded as an attempt to involve him in some very interesting generalizations. Generalizations have a way of bringing out the essential characteristics of related concepts.

Definition. A **translation** S on the vector space $V_n(F)$ is a transformation of $V_n(F) \longrightarrow V_n(F)$ of the form $\alpha S = \alpha + \tau$ for all $\alpha \in V_n(F)$, where τ is any fixed vector of $V_n(F)$.

Linear Transformations

(i) Prove S is a bijective mapping.
(ii) Prove S is a nonlinear transformation.
(iii) Prove that the product $S_1 \circ S_2$ (map composition) of two translations is a translation.
(iv) Prove that the inverse of the translation S, defined by $\alpha S = \alpha + \tau$, $\alpha \in V_n(F)$, is the translation S^{-1} defined by $\alpha S^{-1} = \alpha - \tau$.
(v) Prove that the set of translations on $V_n(F)$ is a commutative group with map composition as the binary operation.

Definition. An **affine transformation** g on the vector space $V_n(F)$ is a mapping of $V_n(F) \longrightarrow V_n(F)$ of the form

$$\alpha g = \alpha T + \tau, \quad \text{for all} \quad \alpha \in V_n(F),$$

where T is a linear operator on $V_n(F)$ and τ is any fixed vector of $V_n(F)$.

The affine transformations therefore include the linear transformations (take $\tau = \overline{0}_V$), as well as the translations (take T to be the identity map of $V_n(F)$).

(i) If $\alpha g_1 = \alpha T_1 + \tau_1, \alpha \in V_n(F)$, and $\alpha g_2 = \alpha T_2 + \tau_2, \alpha \in V_n(F)$, are two affine transformations, prove that $g_1 \circ g_2$ (map composition) is an affine transformation on $V_n(F)$.
(ii) An affine transformation is said to be **nonsingular** if its linear component T is nonsingular. Prove that the product (map composition) of two nonsingular affine transformations is a nonsingular affine transformation.
(iii) If g is a nonsingular affine transformation defined by $\alpha g = \alpha T + \tau$ for all $\alpha \in V_n(F)$, prove that the inverse g^{-1} of g is the nonsingular affine transformation given by $\alpha g^{-1} = \alpha T^{-1} - \tau T^{-1}$.
(iv) Prove that the set of all nonsingular affine transformations is a group with map composition as the binary operation. It is called the **affine group.**
(v) The group of translations on $V_n(F)$ is called a **subgroup** of the affine group. Explain what this means and define a subgroup of a group.
(vi) If g is a nonsingular affine transformation on $V_n(F)$ and S is a translation on $V_n(F)$, prove that the composite transformation $g \circ S \circ g^{-1}$ on $V_n(F)$ is a translation. This property is described by saying the group of translations is a **normal subgroup** of the affine group.

An affine transformation g on $V_3(R)$ is equivalent to a system of 3 equations of the form

$$y_i = a_{i1}x_i + a_{i2}x_2 + a_{i3}x_3 + b_i, \quad i = 1, 2, 3,$$

that express the coordinates (y_1, y_2, y_3) of the transformed vector αA in terms of the coordinates (x_1, x_2, x_3) of α and the coordinates (b_1, b_2, b_3) of the fixed vector τ.

An affine transformation is a linear transformation plus a translation. It is an affine transformation that is used in plane analytic geometry to reduce the general second-degree equation

$$ax^2 + by^2 + cxy + dx + ey + f = 0$$

to the form $a'x^2 + b'y^2 + c' = 0$.

An abstract affine space is defined as follows. Let V be a vector space over a field F whose characteristic is not 2.

Definition. A nonempty set A is called an **affine space** associated with the vector space V if a mapping f of $A \times A \longrightarrow V$ is defined for which
  (i) $f(a,c) = f(a,b) + f(b,c)$, for all $a,b,c \in A$.
  (ii) for any $a \in A$ and $\alpha \in V$, there exists a unique $b \in A$ such that $f(a,b) = \alpha$.

Let us denote an affine space by (A,f). Axiom (ii) is equivalent to requiring the existence of a mapping of $V \times A \longrightarrow A$ and an alternative definition of an affine space can be given in terms of this mapping.

Definition. The **dimension of an affine space** is the dimension of its associated vector space.

Two immediate consequences of the definition are:
1. $f(a,a) = \bar{0}_V$, for all $a \in A$.
   This follows from (i) by setting $c = b$.
2. $f(a,b) = -f(b,a)$, for all $a,b \in A$.
   This follows from (i) by setting $c = a$. (It is here where we need char $F \neq 2$.)

By Axiom (ii) we see that a vector $\alpha \in V$ determines a mapping of $A \longrightarrow A$ given by $a \longrightarrow b$, where b is the unique element of A for which $f(a,b) = \alpha$. The vectors of V are called **translations on A**. Thus a translation $\alpha \in V$ is a bijection of $A \to A$ and is often denoted by $a \to \alpha + a (= b)$.

Let (A,f) be an affine space associated with the vector space V over F and (A',f') an affine space associated with the vector space V' over the same field F.

Assume a mapping g of $A \longrightarrow A'$ is given. Then for any $a \in A$ we can define a mapping T_a of $V \longrightarrow V'$ by

(1) $\qquad \alpha T_a = f'(g(a), g(c)), \quad \alpha \in V,$

where c is the unique element of A for which $\alpha = f(a,c)$.

We next show that if the mapping T_a is linear for some $a \in A$, then $T_a = T_b$ for all $b \in A$.

Let T_a be linear and consider the mapping T_b defined by (1) for the point b of A.

Let β be any vector of V. Then for some unique $x \in A$, $\beta = f(b,x) = f(a,x) - f(a,b)$. Hence

$$\begin{aligned}
\beta T_a &= (f(a,x))T_a - (f(a,b))T_a \\
&= f'(g(a), g(x)) - f'(g(a), g(b)) \\
&= f'(g(b), g(x)) \\
&= (f(b,x))T_b \\
&= \beta T_b.
\end{aligned}$$

Hence $T_b = T_a$.

It therefore follows that the linear mapping T_a is independent of the choice of the point $a \in A$.

Definition. A mapping g of $A \longrightarrow A'$ is called an **affine mapping** if for some $a \in A$ the mapping T_a defined by (1) is linear.

Definition. In particular for $A' = A$, a bijective affine mapping g of $A \longrightarrow A$ is called an **affine transformation**.

Example 6. A vector space V over a field F can be regarded as an affine space (associated with itself) if we define a mapping f of $V \times V \longrightarrow V$ by

$$f(\alpha, \beta) = \beta - \alpha, \quad \text{for} \quad \alpha, \beta \in V.$$

This is easily verified. Now let g be an affine transformation of $V \longrightarrow V$. Then the mapping T_α, $\alpha \in V$, defined by (1) is linear and is the same mapping for all $\alpha \in V$. Let us take $\alpha = \bar{0}_V$ and designate the mapping simply by T. Then

$$\begin{aligned}
\alpha T &= f(g(\bar{0}_V), g(\alpha)) \\
&= g(\alpha) - g(\bar{0}_V).
\end{aligned}$$

Hence the affine transformation g has the form

$$g(\alpha) = \alpha T + \tau, \quad \text{for all} \quad \alpha \in V, \quad \tau \in V,$$

where T is a linear operator on V and τ is a fixed vector of V determined by $\tau = g(\bar{0}_V)$. This is the familiar form for an affine transformation on a vector space V. A translation in this context is a vector β such that if α is any vector of V then β maps α into $\alpha + \beta$. Thus the affine transformation g is described as a linear mapping T of $V \longrightarrow V$ followed by the translation τ.

Exercise. If we denote the mapping of $V \times A \longrightarrow A$ defined by axiom (ii) by h, prove that a translation is defined by $a \longrightarrow h(\alpha, a), a \in A$.

Prove also that $h(\bar{0}_V, a) = a$, for all $a \in A$ and that $h(\alpha + \beta, a) = h(\alpha, h(\beta, a))$. Show that if $f(a,b) = \alpha$ then $h(\alpha, a) = b$.

Exercise. Show that we can formulate an alternative set of axioms for an affine space by use of the mapping h as follows:

An **affine space** A over a field F is a nonempty set A for which there exists a vector space V over F and a mapping h of $V \times A \longrightarrow A$ for which
  (i) $h(\overline{0}_V, a) = a$, for all $a \in A$.
  (ii) $h(\alpha + \beta, a) = h(\alpha, h(\beta, a))$ for all $\alpha, \beta \in V$ and all $a \in A$.
  (iii) there exists a mapping of $A \times A \longrightarrow V$ such that $(a,b) \longrightarrow \alpha$ where $h(\alpha, a) = b$.

If we use the notation $h(\alpha, a) = \alpha + a$, write out these axioms in this notation.

Exercise. Prove that the inverse mapping g^{-1} of a nonsingular affine transformation g is an affine transformation and that if T is the linear transformation associated with g then T^{-1} is the associated linear transformation of g^{-1}.

Definition. A nonempty subset B of an affine space A is called an **affine subspace** of A if, for some $a \in B$, the set of vectors $f(a,b)$, for all $b \in B$, is a subspace of the associated vector space V of A.

Exercise. In this last definition prove that if B is an affine subspace of the affine space A, then any other point $c \in B$ would serve to define the same subspace. In other words, the definition is actually independent of the choice of the point $a \in B$.

Let $A_n(F)$ denote the affine group of all nonsingular affine transformations on the n-dimensional vector space V over F. Clearly the full linear group $L_n(F)$ and the group of translations on V are subgroups of $A_n(F)$. In fact, the latter is a commutative subgroup of the noncommutative group $A_n(F)$. If geometry is the study of properties of figures that are invariant under some group of transformations of a set of points, then affine geometry is the study of the invariant properties of figures in V under the affine group $A_n(F)$. Whereas in the vector geometry of V, under the full linear group, the origin 0 has a very special role, in affine geometry any two points (vectors) α and β can be mapped into one another by a translation. In contrast to this, under the full linear group only the origin can be mapped into the origin.

Let U be an affine subspace of the vector space V, regarded as an affine space with the mapping f of $V \times V \longrightarrow V$ defined by $f(\alpha, \beta) = \beta - \alpha$, $\alpha, \beta \in V$. Then for any vector $\tau \in U$, the set of vectors $\alpha - \tau$ for all $\alpha \in U$, by definition, forms a subspace U' of V. (In particular, then, every vector subspace of V is an affine subspace with $\tau = \overline{0}_V$.)

Let $\alpha, \beta \in U$. Consider the vector

$$\gamma = (1 - x)\alpha + x\beta, \quad x \in F.$$

γ is the set of all "points" on the line joining the point α to the point β of the affine space V. This set of points is called the **affine line** joining α and β.

Now $\alpha - \tau$ and $\beta - \tau$ are in the vector subspace U' and hence for all $x \in F$,

$$(1 - x)(\alpha - \tau) + x(\beta - \tau) = (1 - x)\alpha + x\beta - \tau$$

is in U' and thus the set of points $(1 - x)\alpha + x\beta$ belongs to U.

We have therefore proved that if an affine subspace U contains α and β then it contains all points on the affine line joining α and β.

We point out that the affine subspace U is nothing but a coset in the quotient vector space V/U' (see Chapter 4). An affine line (a one-dimensional affine subspace) in an n-dimensional affine space is simply a coset, under addition, in an $(n - 1)$-dimensional quotient space.

Exercise. By applying a nonsingular affine transformation to $\gamma = (1 - x)\alpha + x\beta$ prove that affine lines are mapped into affine lines under the affine group.

Exercise. Prove that an affine subspace of the affine space V can be defined as a subset U of V such that if $\alpha, \beta \in U$ then $(1 - x)\alpha + x\beta \in U$ for all $x \in F$; that is, U contains the entire line (the affine line) joining α and β. (We have already proved half of this theorem.)

Since, in particular, if an affine subspace contains α and β it contains all the "intermediate" points $(1 - x)\alpha + x\beta$ where $0 \leq x \leq 1$, an affine subspace is by definition a convex set.

Clearly an affine transformation maps affine subspaces into affine subspaces.

Exercise. Two subsets of an affine space V are called **parallel** if one is mapped into the other by a translation. Prove that an affine transformation maps parallel sets into parallel sets.

Exercise. If U is an affine subspace of V, is V/U an affine space?

EXERCISES

1. A plane shear parallel to the x-axis is a transformation of $V_2(R) \rightarrow V_2(R)$ of the form $(x, y) \rightarrow (x + ay, y)$ where a is a fixed scalar. Similarly $(x, y) \rightarrow (x, y + ax)$ is a plane shear parallel to the y-axis.

Describe the geometric effects of these transformations. Are they linear? Injective? Surjective?

2. Write out the transformations that are shears parallel to each of the three axes in $V_3(R)$ and describe their geometric effects.

3. Find the kernels and images, and their dimensions, of the following linear transformations T of $V_4(R) \longrightarrow V_4(R)$:
  (a) $(x, y, z, w)T = (x + y - z, 0, 0, w)$;
  (b) $(x, y, z, w)T = (x + z, y + w, y + z, 0)$;
  (c) $(x, y, z, w)T = (x - y, y, x + y, -y)$;
  (d) $(x, y, z, w)T = (2x, 0, 0, 0)$;
  (e) $(x, y, z, w)T = (-x, x - y - z - w, 0, x)$.

4. Let ϕ be a mapping of $M \longrightarrow M$, (M is the vector space defined in Example 4, Section 1 of Chapter 2) defined by $\phi(f) = xf(x)$, $f \in M$ $x \in [0, 1]$. Prove ϕ is linear.

5. V is an n-dimensional vector space with a basis $\alpha_1, \alpha_2, \ldots, \alpha_n$. T is an epimorphism of $V \longrightarrow W$, where W is a vector space.
  (a) Prove that the vectors $\alpha_i T$, $i = 1, 2, \ldots, n$, form a set of generators of W.
  (b) When do these vectors form a basis for W?

6. Give an example to show that the sum of two nonsingular linear transformations of $V_3(R) \longrightarrow V_3(R)$ is not necessarily nonsingular.

7. A linear operator (endomorphism) T on a vector space V has the property that $T^2 = 0$ (the zero operator) if and only if im $T \subset$ ker T.
  If $T^2 = 0$, prove that $I + T$ is an automorphism of V.

8. Prove that the mapping T defined in Theorem 7 is an isomorphism.

9. If T is a linear transformation $V \longrightarrow W$ of two vector spaces where V is a finite-dimensional vector space, prove that ker T and im T are finite-dimensional spaces.
  If in addition W is a finite-dimensional space, show
  (a) if T is injective, then dim $V \leq$ dim W;
  (b) if T is surjective, then dim $V \geq$ dim W;
  (c) if T is bijective, then dim $V =$ dim W.

10. Prove that two vector spaces of different dimensions are not isomorphic

11. Prove that two finite-dimensional vector spaces are isomorphic if and only if they have the same dimension.

12. Let T be the mapping of $V_3(F) \longrightarrow V_3(F)$, F the real field, defined by
$$(x, y, z)T + (x + y + z, x - y, -z).$$
  (a) Prove T is a linear transformation.
  (b) Prove T is an automorphism.
  (c) Find T^{-1} and T^{-2}.

13. Let S and T be the endomorphisms of the vector space $V_2(F)$, where F is the real field, defined by
$$(x, y)S = (2x + y, x - y), \quad (x, y)T - (x, x + 3y).$$
Exhibit in a similar form the following endomorphisms of $V_2(F)$: $S + T$, ST, TS, S, $-S$ and S^2.

14. Find a linear transformation of $V_2(F) \longrightarrow V_3(F)$ which is (a) injective (b) not injective.

Linear Transformations

15. T is the linear transformation of $V_3(F) \longrightarrow V_3(F)$ defined by $(x, y, z)T = (0, x + y, z)$. Find the kernel of T and the image of T and find their dimensions. If U is the subspace of $V_3(F)$ of all vectors of the form $(x, 0, z)$, find $T(U)$ and $T^{-1}(U)$. What are their dimensions?

16. Prove that multiplication in Hom (V, V) is associative and that it does satisfy the two distributive laws.

17. T is a linear transformation $V \longrightarrow W$ of two vector spaces and the subset B of V is a basis for V. What can be said about the subset $T(B)$ of W if T is (a) injective, (b) surjective, (c) bijective?

Chapter **4**

Quotient Spaces and Direct Sums

4-1 QUOTIENT SPACES

Let U be a subspace of a vector space V over a field F.

We shall introduce the notation V/U to designate the set whose elements have the form $\alpha + U, \beta + U, \gamma + U, \ldots$ where $\alpha, \beta, \gamma, \ldots$ are vectors of V. These elements of V/U are called **cosets of U in V**. We shall write U for $\bar{0}_V + U$. It must be understood that if $\alpha \in V$ then $\alpha + U$ stands for a single element of this new set V/U, and for each $\alpha \in V$ we can form such an element of V/U.

Example 1. Let U be the subspace of $V_3(R)$ spanned by the vector $(1,1,1)$; that is, U is the straight line through the origin and the point $(1,1,1)$. For any vector $(x,y,z) \in V_3(R)$ we can regard the coset $(x,y,z) + U$ as the set of vectors obtained by adding the vector (x,y,z) to each vector of U. This coset is therefore the set of all vectors on the line through the point (x,y,z) parallel to the line U. Hence the cosets of $V_3(R)/U$ are lines parallel to U.

Example 2. In the same way, if U is the subspace of $V_3(R)$ spanned by the vectors $(1,0,0)$ and $(0,1,0)$ then U is the set of all vectors in the xy-plane, and the cosets of $V_3(R)$ are the planes parallel to the xy-plane. For instance, the coset $(x,y,z) + U$ is the set of all vectors on the plane through the point (x,y,z) parallel to the xy-plane.

We intend to show how the set V/U can be made into a vector space over F by appropriate definitions of addition and scalar multiplication.

First of all, let us define two elements $\alpha + U$ and $\beta + U$ of V/U as being equal:

$$\alpha + U = \beta + U,$$

if and only if $\alpha - \beta \in U$. Since U is a subspace, we see that $\alpha - \beta \in U$ if and only if $\beta - \alpha \in U$.

Next we define an addition in the set V/U. If $\alpha + U$ and $\beta + U$ are cosets in V/U, we define

$$(\alpha + U) + (\beta + U) = \alpha + \beta + U.$$

Quotient Spaces and Direct Sums

Since addition of vectors in V is both associative and commutative, it follows easily that this addition of cosets is likewise associative and commutative. However, since we can have $\alpha + U = \beta + U$ without $\alpha = \beta$, it is necessary to prove that coset addition is well-defined. By this we mean the following.

Suppose $\alpha + U = \gamma + U$ and $\beta + U = \eta + U$. The definition of coset addition tells us that $(\alpha + U) + (\beta + U) = \alpha + \beta + U$ and that $(\gamma + U) + (\eta + U) = \gamma + \eta + U$, and since $(\alpha + U) + (\beta + U) = (\gamma + U) + (\eta + U)$, we therefore need to confirm that $\alpha + \beta + U = \gamma + \eta + U$. Now $\alpha + U = \gamma + U$ implies, by definition, that $\alpha - \gamma \in U$. Similarly from $\beta + U = \eta + U$ we get $\beta - \eta \in U$. Since U is a subspace, the vector $\alpha - \gamma + \beta - \eta = (\alpha + \beta) - (\gamma + \eta)$ is in U, and this implies $\alpha + \beta + U = \gamma + \eta + U$. Hence our coset addition is well-defined.

The coset $U = \bar{0}_V + U$ is the zero element of coset addition. For

$$U + \alpha + U = (\bar{0}_V + U) + (\alpha + U)$$
$$= (\bar{0}_V + \alpha) + U = \alpha + U.$$

Also, the additive inverse of the coset $\alpha + U$ is the coset $(-\alpha) + U = -\alpha + U$. For $(\alpha + U) + ((-\alpha) + U) = \alpha + (-\alpha) + U = \bar{0}_V + U = U$.

We can now make the statement that the set V/U of cosets of U in V is a commutative group under addition. This is part of the way to transforming it into a vector space.

The next definition is that of a scalar multiplication of cosets. For any $x \in F$, we define

$$x(\alpha + U) = x\alpha + U.$$

It is an easy exercise for the reader to confirm that all the axioms for scalar multiplication are satisfied and hence that this definition meets the requirements for scalar multiplication. However, again we must verify that our scalar multiplication of cosets is well-defined. Suppose then $\alpha + U = \beta + U$. For $x \in F$, we have therefore $x(\alpha + U) = x(\beta + U)$. By our definition, $x(\alpha + U) = x\alpha + U$ and $x(\beta + U) = x\beta + U$, and so we need to verify that $x\alpha + U = x\beta + U$. Since U is a subspace and since $\alpha - \beta \in U$, it follows that $x(\alpha - \beta) = x\alpha - x\beta \in U$. This, as we know, implies $x\alpha + U = x\beta + U$.

It follows now that, with these definitions, the set V/U of cosets is a vector space over the same scalar field F as that of the vector space V.

Definition. The vector space V/U is called the **quotient space** or **factor space of V by U**.

A subgroup of a group G is a subset of G that forms itself a group under the same binary operation as for G. Since a vector space V is a commutative group under vector addition, it is clear that a subspace U of V is a subgroup of the group V. From this standpoint we see that the cosets of V/U form under addition alone (that is, ignoring scalar multiplication) a commutative group. The coset U is the neutral element of the group and the coset $-\alpha + U$ is the inverse of the coset $\alpha + U$. In this way we define a quotient group V/U instead of a quotient space. Of course this commutative quotient group is nothing but the underlying group of the quotient vector space V/U. In fact the quotient vector space V/U is simply the commutative quotient group V/U with the scalar multiplication of its cosets defined, as before, by $x(\alpha + U) = x\alpha + U$.

A quotient space can be formed with any subspace of V. If $U = V$, then the only element of V/V is the zero vector V. If $U = \{\bar{0}_V\}$, then $V/\{\bar{0}_V\}$ and V are isomorphic vector spaces, the isomorphism being given by $\alpha + \{\bar{0}_V\} \rightarrow \alpha$, for $\alpha \in V$.

Example 3. Let U be the subspace of $V_3(R)$ that is spanned by the vector $(1,0,0)$. Then the quotient space $V_3(R)/U$ is spanned by the two distinct vectors $(0,1,0) + U$ and $(0,0,1) + U$. To see this, note that for any vector (x,y,z) of $V_3(R)$, we have $(x,y,z) = x(1,0,0) + y(0,1,0) + z(0,0,1)$, and therefore, since $x(1,0,0) \in U$,

$$(x,y,z) + U = U + y((0,1,0) + U) + z((0,0,1) + U)$$
$$= y(0,1,0) + U + z(0,0,1) + U$$
$$= (0,y,z) + U.$$

The vectors $(0,1,0) + U$ and $(0,0,1) + U$ are therefore also independent and hence they form a basis of V/U. Note that dim $V_3(R) = 3$, dim $U = 1$, and dim $V/U = 3 - 1 = 2$. We look into this fact very soon. (See Theorem 1.) Note also, for instance, that

$$(x_1, y, z) + U = (x_2, y, z) + U$$

since

$$(x_1, y, z) - (x_2, y, z) = (x_1 - x_2, 0, 0) \in U.$$

Example 4. The set P_1 of all polynomials over a field F of degree ≤ 1 forms a subspace of the vector space P_4 of all polynomials over F of degree ≤ 4. Form the quotient space P_4/P_1.

It is not hard to see that the set of vectors $x^2 + P_1$, $x^3 + P_1$, $x^4 + P_1$ spans P_4/P_1. Moreover they are clearly independent vectors. For let

$$c_1(x^2 + P_1) + c_2(x^3 + P_1) + c_3(x^4 + P_1) = P_1,$$

where $c_1, c_2, c_3 \in F$ and, of course, P_1 is the zero vector of P_4/P_1. This equation implies $c_1x^2 + c_2x^3 + c_3x^4 + P_1 = P_1$, hence $c_1x^2 + c_2x^3 + c_3x^4 \in P_1$. This is only possible if and only if $c_1 = c_2 = c_3 = 0$.

Quotient Spaces and Direct Sums

Hence the three vectors form a basis of P_4/P_1.

Again we point out that dim $P_4 = 5$, dim $P_1 = 2$ and dim $P_4/P_1 = 5 - 2 = 3$.

Let U be a subspace of the vector space V and form the quotient space V/U. Consider the mapping j of $V \to V/U$ defined by $\alpha j = \alpha + U$, $\alpha \in V$. We find

$$(\alpha + \beta)j = \alpha + \beta + U = (\alpha + U) + (\beta + U) = \alpha j + \beta j,$$

and

$$(x\alpha)j = x\alpha + U = x(\alpha + U) = x(\alpha j),$$

where $\alpha, \beta \in V$ and $x \in F$. This proves j is a linear transformation. j is surjective, for if $\alpha + U$ is any vector of V/U then $\alpha + U = \alpha j$, $\alpha \in V$. Note that ker $j = U$.

Definition. The epimorphism j is called the **natural** or **canonical homomorphism** of V onto the quotient space V/U.

We emphasize that im $j = V/U$.

THEOREM 1

If V is a finite-dimensional vector space and if U is a subspace of V, then

$$\dim V = \dim U + \dim V/U.$$

Proof: Consider the natural homomorphism j of $V \to V/U$. By Theorem 6, Chapter 3, we have dim $V = $ dim (ker j) + dim (im j). Since ker $j = U$ and im $j = V/U$, our theorem is proved.

Example 5. Let U be the subspace of $V_3(R)$ consisting of all vectors (x,y,z) whose first two components are equal, $x = y$. Geometrically, U is the plane $x = y$ and comprises all vectors that are points on this plane. U is therefore a two-dimensional subspace; for instance, a basis of U is $(1,1,0)$ and $(0,0,1)$. Hence by Theorem 1, the quotient space $V_3(R)/U$ is one-dimensional. Thus for any vector $(x,y,z) \notin U$, the coset $(x,y,z) + U$ serves as a basis for $V_3(R)/U$. If we choose the coset $(1,0,0) + U$ for our basis, then, for any coset $(x,y,z) + U$, we can determine a scalar t such that

$$(x,y,z) + U = t((1,0,0) + U).$$

This is equivalent to

$$(x,y,z) - t(1,0,0) = (x - t, y, z) \in U,$$

and hence to $x - t = y$. Thus $t = x - y$, and we have

$$(x,y,z) + U = (x - y)((1,0,0) + U).$$

Geometrically, the cosets of $V_3(R)/U$ are planes parallel to the plane $x = y$. The coset $(x,y,z) + U$ is a plane through the point (x,y,z) parallel to the plane $x = y$. Two cosets are equal

$$(x_1,y_1,z_1) + U = (x_2,y_2,z_2) + U$$

if and only if they are the same plane; that is, if and only if the vectors (x_1, y_1, z_1) and (x_2, y_2, z_2) are points on the same plane. For instance, $(4,2,3) + U = (2,0,0) + U$, since $(4,2,3) - (2,0,0) = (2,2,3) \in U$; or geometrically speaking, since $(4,2,3)$ and $(2,0,0)$ lie on the same plane $x - y = 2$.

4-2 EQUIVALENCE RELATIONS

If E is a nonempty set, any nonempty subset R of the cartesian product $E \times E$ is called a **relation on** E. For $x, y \in E$, $(x,y) \in R$ signifies that x stands in a certain relation to y.

For instance if Z is the ordered integral domain of integers, let R be the subset of all $(x,y) \in Z \times Z$ for which $x > y$. This is a relation on Z.

A mapping of $E \to E$ is a relation on E. The mapping associates to each x of E a unique y of E and this constitutes a subset R of elements (x,y) of $E \times E$. However not every relation R is a mapping, for R may contain elements (x,y) and (x,y') where $y' \neq y$. Such elements are excluded in a mapping.

We next define a special kind of relation, called an **equivalence relation**, that is of particularly common use.

Definition. An **equivalence relation** on a nonempty set E is a subset R of the cartesian product $E \times E$ which satisfies the three conditions:
  (i) $(x,x) \in R$, for all $x \in E$ (*reflexivity*),
  (ii) if $(x,y) \in R$ then $(y,x) \in R$, $x,y \in E$ (*symmetry*),
  (iii) if $(x,y) \in R$ and $(y,z) \in R$ then $(x,z) \in R$, $x, y, z \in E$ (*transitivity*).
By (ii) and (iii) we see that if $(x,y) \in R$ and $(x,z) \in R$ then $(y,z) \in R$.

Example 6. Let Z be the set of integers. An equivalence relation on Z is defined by the subset R of $Z \times Z$ for which $(x,y) \in R$ if and only if $x - y$ is divisible by 5, where x and y are integers. Clearly all three of the above requirements are satisfied for this subset R.

Example 7. If U is a proper subspace of a vector space V, then an equivalence relation on V is defined by the subset R of $V \times V$ for which $(\alpha, \beta) \in R$ if and only if $\alpha - b \in U$, where α and β are vectors of V. Again we can easily prove that the three conditions for R are satisfied.

The effect of an equivalence relation R on a set E is to partition E into disjoint nonempty subsets of E, called **R-equivalence classes**, whose union is E.

To see this, let $x \in E$. Form the subset E_x of E that contains all elements y of E for which $(x,y) \in R$. E_x is not empty, for by condition (i), $x \in E_x$. Thus each element of E is in an R-equivalence class. We next prove that each element of E is in only one R-equivalence class. First we show that if $y \in E_x$, then $E_y = E_x$. For if $z \in E_y$ then $(y,z) \in R$. Since $(x,y) \in R$, we have by condition (iii) that $(x,z) \in R$ and therefore $z \in E_x$. Hence $E_y \subset E_x$. Similarly we show $E_x \subset E_y$ and therefore $E_y = E_x$. On the other hand if $y \notin E_x$, then E_x and E_y are disjoint. For if $z \in E_x \cap E_y$, then $(x,z) \in R$ and $(y,z) \in R$ and this implies $(x,y) \in R$, that is $y \in E_x$. This contradiction proves $E_x \cap E_y$ is empty.

We have proved that the R-equivalence classes E_x are nonempty and disjoint and that their union is E. In other words every element of E is in one and only one R-equivalence class. The set of R-equivalence classes is called a **quotient set** of E.

The equivalence classes of Example 3 are called **residue classes modulo 5.** The equivalence classes of example 4 are called **cosets of V modulo U** and the quotient set in this case is the quotient vector space V/U.

The quotient space is therefore the vector space of cosets.

We conclude this section by asking and answering the following question: If $\beta_1, \beta_2, \ldots, \beta_n$, is a basis of the vector space V and if U is a subspace of V, do the vectors $\beta_1 + U, \beta_2 + U, \ldots \beta_n + U$ form a basis for the quotient space V/U?

Let V be a finite-dimensional vector space with the basis $\beta_1, \beta_2, \ldots, \beta_n$ and let U be a subspace of V. If j is the canonical epimorphism of $V \to V/U$, then for $\alpha \in V$, we have

$$\alpha = \sum_{i=1}^{n} x_i \beta_i$$

where the x_i are scalars, and

$$\alpha j = \left(\sum_{i=1}^{n} x_i \beta_i \right) + U = \sum_{i=1}^{n} x_i (\beta_i + U).$$

This proves that the vectors $\beta_i + U, i = 1, 2, \ldots, n$, form a set of generators of the quotient space V/U (they span V/U). If U is a proper subspace of V, then $1 \leq \dim U < n$, and by Theorem 1 we infer at once that $\dim V/U < n$, and so the n vectors $\beta_i + U$ cannot be linearly independent. They would not be a basis for V/U. If however U consists of the zero vector $\bar{0}_V$ only, then clearly $\dim U = 0$ and $\dim V/U = n$. This means ker j contains only the zero vector $\bar{0}_V$, and j is therefore injective. Since j is surjective, this means j is an isomorphism and in this case the n vectors $\beta_i + U$ are independent and therefore form a basis for V/U.

We can see this another way. If j is not an isomorphism then α is not injective and dim $U > 0$. Hence ker j contains a nonzero vector $\gamma = \sum_{i=1}^{n} y_i \beta_i$. We have

$$\gamma j = \left(\sum_{i=1}^{n} y_i \beta_i\right) + U = \sum_{i=1}^{n} y_i(\beta_i + U) = \bar{0}_{V/U} = U.$$

The significance of this last equation is that it is a linear combination of the n vectors $\beta_i + U$ whose coefficients y_i are not all zero and which equals the zero vector of V/U. This implies the n vectors $\beta_i + U$ are linearly dependent.

EXERCISES

1. An endomorphism P of V such that $PP = P$ is called a **projection** and an endomorphism J of V such that $JJ = 1_V$ is called an **involution**.
   If P is a projection of $V \to V$ prove that $J = 1_V - 2P$ is an involution of $V \to V$.

2. U is the subspace of $V_5(R)$ spanned by the vectors $(2,0,0,1,0)$, $(3,-5,0,0,0)$ and $(0,0,1,2,3)$. W is the subspace of $V_5(R)$ spanned by the vectors $(1,3,2,0,1)$, $(2,7,1,0,3)$, $(-3,0,0,2,1)$ and $(4,4,-1,-2,1)$.
   Find the dimensions of the vector spaces
   (a) U (b) W (c) $U + W$ (d) $U \cap W$ (e) $V_5(R)/W$.

3. U is the subspace of $V_5(R)$ spanned by the vectors $(1,2,3,4,5)$, $(5,4,3,2,1)$ and $(-8,-4,0,4,8)$. Find the dimensions of the vector spaces (a) U, (b) $V_4(R)/U$. Find bases for these two vector spaces.

4. Give an example of an endomorphism of $V_3(R)$ whose kernel is the plane spanned by the vectors $(1,0,1)$ and $(0,1,0)$ and whose image is the line spanned by the vector $(1,0,-1)$.

5. If T is the canonical epimorphism of $V \to V/U$, where U is a subspace of V, prove that the mapping of $V/U \to T(V)$ defined by $\alpha + U \to \alpha T$, $\alpha \in V$, is an isomorphism. What is the kernel of T?

6. Let $\alpha_1, \alpha_2, \ldots, \alpha_n$ be a basis of a vector space V and let U be a subspace of V.
   (a) Prove that the vectors $\alpha_1 + U, \ldots, \alpha_n + U$ form a set of generators of the quotient space V/U. (That is, these vectors span V/U.)
   (b) Show that in general the vectors $\alpha_i + U$ do not form a basis for V/U.
   (c) Cite a case (and prove) where they do.

7. If $\alpha_1 + U, \ldots, \alpha_k + U$ is a basis for V/U and if $\beta_1, \ldots, \beta_r$ is a basis for U, prove that $\alpha_1, \ldots, \alpha_k, \beta_1, \ldots, \beta_r$ is a basis for V.

8. G is a group whose binary operation is multiplication. Denote its

Quotient Spaces and Direct Sums

neutral element by 1 and the inverse of $x \in G$ by x^{-1}. H is a subgroup of G; that is, H is a subset of G that forms itself a group under multiplication. Call two elements x and y of G equivalent if and only if $x^{-1}y \in H$. Prove this is an equivalence relation on G.

9. If f is an epimorphism $V \to W$ of two vector spaces and if g is the canonical (natural) projection of $V \to V/\ker f$, prove that there exists a unique linear transformation h of $W \to V/\ker f$ such that $g = f \circ h$. This means the projection g factors into the composition of f and h.

Prove h is an isomorphism.

10. Let T be a linear transformation of $V \to W$ and let $U = \ker T$.
   (a) Prove V/U and im T are isomorphic vector spaces.
   (b) If j is the canonical epimorphism of $V \to V/U$ and if ι is the inclusion homomorphism of im $T \to W$, prove that T factors into

$$T = j \circ S \circ \iota$$

where $V \xrightarrow{j} V/U \xrightarrow{S} \text{im } T \xrightarrow{\iota} W$ and S is the isomorphism.

11. If U and W are subspaces of a vector space V,
   (a) Prove that the mapping T of $W \to (U + W)/U$ defined by $\alpha T = \alpha + U$, $\alpha \in W$, is an epimorphism.
   (b) Find ker T and prove that $W/(U \cap W)$ and $(U + W)/U$ are isomorphic.

4-3 DIRECT SUMS

Definition. A vector space V is said to be the **direct sum** of the two subspaces U and W of V if

1. $V = U + W$; that is, every vector of V is the sum of a vector of U and a vector of W.
2. $U \cap W = \bar{0}_V$; that is, the only vector in both U and W is the zero vector of V.

If V is the direct sum of the subspaces U and W, we write $V = U \oplus W$. This is known as an **(internal) direct sum**.

THEOREM 2

V is the direct sum of U and W, $V = U \oplus W$, if and only if every vector of V has a unique decomposition as the sum of a vector of U and a vector of W. (This means that if α is any vector of V, then there is a unique vector β of U and a unique vector γ of W such that $\alpha = \beta + \gamma$.)

Proof: Assume $V = U \oplus W$. Then $V = U + W$. For $\alpha \in V$, suppose $\alpha = \beta + \gamma = \beta' + \gamma'$ where $\beta, \beta' \in U$ and $\gamma, \gamma' \in W$. Then $\beta - \beta' = \gamma' - \gamma$, and $\beta - \beta' \in U$, $\gamma' - \gamma \in W$. Hence $\beta - \beta'$ and $\gamma' - \gamma$ belong to $U \cap W$. Since $U \cap W = \bar{0}_V$, $\beta' = \beta$ and $\gamma' = \gamma$. Thus the decomposition of α, as the sum of a vector of U and a vector of W, is unique.

Conversely assume each vector $\alpha \in V$ has such a unique decomposition. Then certainly $V = U + W$. Let $\beta \in U \cap W$. Then $\beta = \overline{0}_V + \beta, \overline{0}_V \in U, \beta \in W$. Also $\beta = \beta + \overline{0}_V$, and $\beta \in U, \overline{0}_V \in W$. This forces $\beta = \overline{0}_V$. Hence $U \cap W = \overline{0}_V$. Thus both (1) and (2) in the definition of a direct sum are satisfied. Hence $V = U \oplus W$.

Definition. Let V_1 and V_2 be two vector spaces over a field F. The (external) **direct sum** of V_1 and V_2 is the vector space V over F defined as follows:

Define V to be the set of all ordered pairs (α_1, α_2), $\alpha_1 \in V_1$, $\alpha_2 \in V_2$. Define an addition in V by $(\alpha_1, \alpha_2) + (\beta_1, \beta_2) = (\alpha_1 + \beta_1, \alpha_2 + \beta_2)$, where $\alpha_1, \beta_1 \in V_1$ and $\alpha_2, \beta_2 \in V_2$. Next define a scalar multiplication in V by $x(\alpha_1, \alpha_2) = (x\alpha_1, x\alpha_2)$, where $x \in F$. Verification of the axioms for a vector space is a very simple matter, and this proves V is a vector space over F.

If V is the external direct sum of the vector spaces V_1 and V_2, then we can always find two subspaces U and W of V such that
 (i) U and W are isomorphic to V_1 and V_2 respectively,
 (ii) V is the internal direct sum of U and W.

The set U of all ordered pairs $(\alpha_1, \overline{0}_{V_2})$, $\alpha_1 \in V_1$ forms a vector space over F, and this is also true of the set W of all ordered pairs $(\overline{0}_{V_1}, \alpha_2)$, $\alpha_2 \in V_2$. These statements are readily proved, again by verifying all the axioms for a vector space. Further, the mapping of $V_1 \to W$ defined by $\alpha_1 \to (\alpha_1, \overline{0}_{V_2})$ is an isomorphism of the two vector spaces. The reader can very easily check that it is indeed a bijective linear transformation. In the same way, we can show $\alpha_2 \to (\overline{0}_{V_1}, \alpha_2)$ is an isomorphism $V_2 \to W$ of the vector spaces V_2 and W. If we identify V_1 with U and V_2 with W, then we can write $V = U \oplus W$. In fact it is easy to see that $(\alpha_1, \alpha_2) = (\alpha_1, \overline{0}_{V_2}) + (\overline{0}_{V_1}, \alpha_2)$ is a unique decomposition of the vector $(\alpha_1, \alpha_2) \in V$ into the sum of a vector of U and a vector of W.

The concept of direct sum can be generalized to any finite number of subspaces.

Definition. A vector space V is said to be the **direct sum** $V = U_1 \oplus U_2 \oplus \cdots \oplus U_n$ of the n subspaces $U_1, U_2, \ldots, U_n$ of V if (1) $X = U_1 + U_2 + \cdots + U_n$ (2) the intersection of each U_i with the subspace generated by the remaining $U_j, j \neq i$, is $\overline{0}_V$.

Condition (2) in this last definition means that for each $i = 1, 2, \ldots, n$, the intersection of the subspace U_i with the subspace

$$\overline{U}_i = U_1 + \cdots + U_{i-1} + U_{i+1} + \cdots + U_n$$

contains only the zero vector. Here $\overline{U}_i$ is the subspace consisting of all sums of vectors from the subspaces $U_j, j \neq i$.

As in the case when $n = 2$, $V = U_1 \oplus \cdots \oplus U_n$ if and only if each vector $\alpha \in V$ can be written uniquely in the form

Quotient Spaces and Direct Sums

$$\alpha = \alpha_1 + \alpha_2 + \cdots + \alpha_n, \alpha_i \in U_i$$

for $i = 1, 2, \ldots, n$. This can be used as a definition of the direct sum of n subspaces.

Lemma 1. If V is any vector space and if $V = U \oplus W$, then V/U and W are isomorphic vector spaces.

Proof: Now every vector $\alpha \in V$ has a unique representation in the form $\alpha = \beta + \gamma, \beta \in U, \gamma \in W$.

Define a mapping of $V/U \to W$ by $\alpha + U \to \gamma$, where $\alpha = \beta + \gamma$. This mapping is easily shown to be an isomorphism.

Lemma 2. If V is a finite-dimensional vector space and if $V = U \oplus W$, then

$$\dim V = \dim U + \dim W.$$

Proof: By Lemma 1, $W \approx V/U$ and hence $\dim W = \dim V/U$ (See Corollary, Theorem 6, Chapter 3). Therefore by Theorem 1, $\dim V = \dim U + \dim W$.

Definition. An endomorphism P of a vector space V is called a **projection** if $PP = P$.

Lemma 3. If P is a projection of V then

$$V = \ker P \oplus \operatorname{im} P.$$

Proof: For any $\alpha \in V$, $\alpha = \alpha P + (\alpha - \alpha P)$. $\alpha P \in \operatorname{im} P$. Since $(\alpha - \alpha P)P = \alpha P - \alpha PP = \bar{0}_V$, $\alpha - \alpha P \in \ker P$. Hence every vector of V is the sum of a vector of $\operatorname{im} P$ and a vector of $\ker P$, that is $V = \operatorname{im} P + \ker P$. Let $\alpha \in \operatorname{im} P \cap \ker P$. Then $\alpha = \beta P$ for some $\beta \in V$ and $\alpha P = \bar{0}_V$. Hence $\alpha = \beta P = \beta PP = \alpha P = \bar{0}_V$. Therefore $\operatorname{im} P \cap \ker P = \bar{0}_V$. Thus $V = \operatorname{im} P \oplus \ker P$.

Example 8. If $V = U \oplus W$, then for each $\alpha \in V$ we have a unique representation of α in the form $\alpha = \beta + \gamma$, $\beta \in U$, $\gamma \in W$. It is easy to see that P_U defined by $\alpha P_U = \beta$ and P_W defined by $\alpha P_W = \gamma$ are projections. This can very simply be generalized to give n projections when V is the direct sum of n subspaces of V.

THEOREM 3

If U is a subspace of a finite-dimensional vector space V, then there exists a subspace W of V such that $V = U \oplus W$. (W is called a **complementary subspace** for U in V.)

Proof: Let $\alpha_1, \ldots, \alpha_k$ be a basis for U. By Lemma 9 we can complete this to a basis $\alpha_1, \ldots, \alpha_k, \alpha_{k+1}, \ldots, \alpha_n$ for V. Let W be the subspace of V spanned by the vectors $\alpha_{k+1}, \ldots, \alpha_n$. Then clearly $V = U + W$. Furthermore, $U \cap W = \bar{0}_V$. For if $\alpha \in U \cap W$, then $\alpha = a_1\alpha_1 +$

$\cdots + a_k \alpha_k$ and $\alpha = a_{k+1} \alpha_{k+1} + \cdots + a_n \alpha_n$. Hence $a_1 \alpha_1 + \cdots + a_k \alpha_k - a_{k+1} \alpha_{k+1} - \cdots - a_n \alpha_n = \bar{0}_V$. Since the α_i are independent vectors, all the $a_i = 0$ and hence $\alpha = \bar{0}_V$. Hence $U \cap W = \bar{0}_V$. This completes the proof that $V = U \oplus W$.

Theorem 3 is true if V is infinite dimensional, however the proof of this is omitted. Since the construction of W depends on the choice of a basis for U and on the particular completion of this to a basis for V, it is to be expected that W is not unique and indeed this is true. We do know, however, that W is isomorphic to the quotient space V/U.

Example 9. Let U be the subspace of $V_3(R)$ with the basis $(1,0,1)$ and $(-1,-1,0)$. Then $V_3(R) = U \oplus W$, where W is the subspace with the basis $(2,0,0)$. But this is true also if W is the subspace with, for instance, the basis $(0,1,0)$.

EXERCISES

1. If V is a finite-dimensional vector space and if $V = U \oplus W$, prove $\dim V = \dim U + \dim W$.

2. U is the subspace of $V_4(R)$ spanned by the vectors $(1,0,2,1)$ and $(0,-1,3,0)$. Find a subspace W of $V_4(R)$ for which $V_4(R) = U \oplus W$.

3. U and W are subspaces of the vector space V. Explain the difference between saying $V = U + W$ and $V = U \oplus W$.

4. V_1, V_2, V_3, V_4 are subspaces of a vector space V and $V = V_1 \oplus V_2 \oplus V_3 \oplus V_4$. Describe carefully what this signifies.

5. Find three subspaces V_1, V_2, V_3 of the vector space V of polynomials in x over a field F such that $V = V_1 \oplus V_2 \oplus V_3$.

6. If U is a subspace of a finite-dimensional vector space V, prove that the quotient space V/U is finite-dimensional.

4-4 EXACT SEQUENCES

While not actually essential to a comprehension of the material studied in this book, the notion of an exact sequence is introduced in this section. Exact sequences do provide a simple, graphic method (a sort of visual aid) of describing the elementary properties of linear transformations. They also serve to efficiently characterize the direct-sum property. Exact sequences are being used more and more in algebra and are a fundamental tool in homological algebra.

If $V \approx V_1 \oplus V_2$, then we know that V_1 and V_2 are isomorphic to two subspaces of V. This isomorphism is not dependent on any choice of bases in the vector spaces, and for this reason it is called a **canonical**

Quotient Spaces and Direct Sums

isomorphism. If two vector spaces are canonically isomorphic it is possible to identify them with one another. Thus if $V \approx V_1 \oplus V_2$, then we speak of V_1 and V_2 as being subspaces of V.

If V_1 and V_2 are subspaces of V and if $V = V_1 \oplus V_2$, we can define four canonical homomorphisms (linear transformations) as follows:

Let $\iota_1: V_1 \to V$ and $\iota_2: V_2 \to V$ be the inclusion mappings; that is, if $\alpha_1 \in V_1$ then $\alpha_1 \iota_1 = \alpha_1$, and if $\alpha_2 \in V_2$, then $\alpha_2 \iota_2 = \alpha_2$. This pair of homomorphisms are called the **canonical injections**. That ι_1 and ι_2 are linear mappings is easily verified.

If $\alpha \in V$, $\alpha = \alpha_1 + \alpha_2$, $\alpha_1 \in V_1$, $\alpha_2 \in V_2$, and this representation of α is unique. Define the projections $\pi_1: V \to V_1$ and $\pi_2: V \to V_2$ by $\alpha \pi_1 = \alpha_1$, $\alpha \pi_2 = \alpha_2$. One again quickly verifies that this pair of projections are linear. They are called the **canonical projections** of $V = V_1 \oplus V_2$ on the subspaces V_1 and V_2.

Moreover, the following relationships, involving map compositions among the four canonical homomorphisms, are again quickly proved:

$$\iota_1 \pi_1 = 1_{V_1}, \quad \iota_2 \pi_2 = 1_{V_2}$$
$$\pi_1 \iota_2 = 0, \quad \pi_2 \iota_1 = 0$$
$$\iota_1 \pi_1 + \iota_2 \pi_2 = 1_V$$
$$\pi_1 \circ \pi_1 = \pi_1, \quad \pi_2 \circ \pi_2 = \pi_2$$

Definition. Let U, V, and W be vector spaces and let S be a homomorphism (linear transformation) of $U \to V$ and T a homomorphism of $V \to W$. The sequence

(2) $$U \xrightarrow{S} V \xrightarrow{T} W$$

is said to be **exact at** V if im S = ker T.

In what follows we shall use the symbol 0 in sequences to signify a vector space that contains only the zero vector.

The following statements follow at once from the above definition of an exact sequence.

(A) If T is a homomorphism of $U \to V$, then T is injective if and only if the sequence

(3) $$0 \to U \xrightarrow{T} W$$

is exact at U. (For ker $T = 0$ if and only if the sequence is exact at U.)

(B) If T is a homomorphism of $V \to W$, then T is surjective if and only if the sequence

(4) $$V \xrightarrow{T} W \to 0$$

is exact at W. (For im $T = W$ if and only if there is exactness at W.)

(C) If T is a homomorphism of $V \to W$, then T is an isomorphism if and only if the sequence

(5) $$0 \to V \xrightarrow{T} W \to 0$$

is exact at both V and W. [This follows at once from (A) and (B).]

Definition. The sequence

(6) $$0 \to U \xrightarrow{S} V \xrightarrow{T} W \to 0$$

is called **exact** if it is exact at U, V, and W. This means therefore that (6) is exact if and only if S is injective, im S = ker T, and T is surjective.

Example 10. Let U be a subspace of the vector space V. Let ι be the inclusion homomorphism of $U \to V$ and j the canonical projection of $V \to V/U$. Hence if $\alpha \in U$, $\alpha\iota = \alpha$; and if $\alpha \in V$ then $\alpha j = \alpha + U$. Then the sequence

$$0 \to U \xrightarrow{\iota} V \xrightarrow{j} V/U \to 0$$

is exact at U, V, and V/U. For im $\iota = U =$ ker j, while ι is injective and j is surjective.

Definition. The exact sequence $V \xrightarrow{T} W \to 0$ is said to **split** if there exists a homomorphism S of $W \to V$ such that $ST = 1_W$.

Now $ST = 1_W$ implies S is injective and hence we get the exact sequence $0 \to W \xrightarrow{S} V$.

The exact sequence $0 \to W \xrightarrow{T} V$ is said to **split** if there exists a homomorphism S of $V \to W$ such that $TS = 1_W$. This implies S is surjective and hence the sequence $V \xrightarrow{S} W \to 0$ is exact.

The exact sequence

$$0 \to U \xrightarrow{S} V \xrightarrow{T} W \to 0$$

is said to **split** if it splits at both ends; that is if there exist homomorphisms H of $W \to V$ and K of $V \to U$ such that

$$HT = 1_W \quad \text{and} \quad SK = 1_U.$$

These equations imply H is injective and K is surjective. We also point out that, since T is surjective and ker $T =$ im S, $V/$im S and W are isomorphic.

Definition. An exact sequence that splits is said to be **split exact**.

Example 11. Let $V = U \oplus W$ and consider the sequence

(7) $$0 \to U \xrightarrow{\iota} V \xrightarrow{\pi} W \to 0$$

where ι is the inclusion homomorphism and π is the projective homo-

Quotient Spaces and Direct Sums

morphism. Then ι is injective and π is surjective; that is, the sequence is exact at U and W. We now show it is exact at V. According to (1) we have $\bar{0}_W = \alpha\iota\pi = \alpha\pi$ for all $\alpha \in U$. Hence $U \subset \ker \pi$. Now suppose $\alpha \in \ker \pi$, then $\alpha\pi = \bar{0}_W$. Since V is the direct sum of U and W, $\alpha = \beta + \gamma$, where $\beta \in U, \gamma \in W$. Hence $\beta = \beta\iota$ and $\gamma = \zeta\pi$ for some $\zeta \in V$. Thus $\alpha = \beta\iota + \zeta\pi$. Therefore, $\bar{0}_W = \alpha\pi = (\beta\iota)\pi + \zeta\pi\pi = \bar{0}_W + \zeta\pi = \gamma$. Hence $\alpha = \beta \in U$. We have proved $\ker \pi \subset U$. Therefore $\operatorname{im} \iota = U = \ker \pi$, and so the sequence is exact at V. The sequence is therefore exact.

Since $V = U \oplus W$, we see by (1) that there exist homomorphisms ι' of $W \to V$ and π' of $V \to U$ such that $\iota'\pi = 1_W$ and $\iota\pi' = 1_U$. Therefore the sequence is split exact.

We now prove the converse of this latter example.

Let U and W be subspaces of a vector space V. If the sequence

$$0 \to U \xrightarrow{S} V \xrightarrow{T} W \to 0$$

is split exact, then $V = U \oplus W$.

There exist therefore homomorphisms H of $W \to V$ and K of $V \to U$ such that $HT = 1_W$ and $SK = 1_U$. Moreover we see that H is injective and K is surjective.

Let α be any vector of V and write

$$\alpha = \alpha T + (\alpha - \alpha T).$$

Since $\alpha T \in W$, $(\alpha T)H = \alpha T$ and hence $(\alpha T)HT = \alpha TT$. But $(\alpha T)HT = \alpha T$ and we thus have $\alpha TT = \alpha T$. Therefore $(\alpha - \alpha T)T = \bar{0}_W$ and so $\alpha - \alpha T \in \ker T = \operatorname{im} S = U$. We have shown that α is the sum of a vector of U and a vector of W. This means $V = U + W$.

We next suppose $\beta \in U \cap W$. Then $\beta \in U$ implies $\beta T = \bar{0}_W$. $\beta \in W$ implies $\beta H = \beta$ and so $\beta HT = \beta T = \bar{0}_W$. Since $HT = 1_W$, this implies $\beta = \bar{0}_U$. Hence $U \cap W = \bar{0}_V$ and therefore $V = U \oplus W$. This completes the proof.

Furthermore, we can identify S and H as the canonical injections, and T and K as the canonical projections of this direct sum decomposition.

EXERCISES

1. Let $\alpha_1, \ldots, \alpha_n$ be a basis of a vector space V. If V_1 is the subspace spanned by the vectors $\alpha_1, \ldots, \alpha_k$, V_2 the subspace spanned by the vectors $\alpha_{k+1}, \ldots, \alpha_r$, and V_3 the subspace spanned by the vectors $\alpha_{r+1}, \ldots, \alpha_n$, prove $V = V_1 \oplus V_2 \oplus V_3$.

2. T is a linear transformation on a vector space V and U is the kernel of T. Show that all vectors of V which have the same image vector under T

belong to the same coset of V/U. Does this property characterize the cosets of V/U?

3. It was proved that if U is a subspace of a finite-dimensional vector space V, there exists a subspace W of V such that $V = U \oplus W$. Is W unique?

4. If U is a subspace of a finite-dimensional vector space V, prove that the quotient space V/U is finite-dimensional and that

$$\dim V/U = \dim V - \dim U.$$

5. If U is a subspace of a vector space V, prove that if U and V/U are finite-dimensional, then V is finite-dimensional.

6. V and W are vector spaces and T is a linear transformation of $V \to W$. U is a subspace of V and j is the natural epimorphism of $V \to V/U$. Prove there exists a linear transformation S of $V/U \to W$ such that $T = jS$ if and only if $U \subset \ker T$. Prove that S is surjective if T is.

7. U_1 and U_2 are subspaces of a finite-dimensional vector space V and $U_1 \cap U_2 = 0_V$. Prove there exists a subspace U_3 of V such that $V = U_1 \oplus U_2 \oplus U_3$.

8. U is a subspace of a vector space V. ι is the inclusion mapping of $U \to V$ and j is the natural epimorphism of $V \to V/U$. If T is a linear transformation of $V \to U$ such that $\iota T = 1_U$, prove (i) im $T = U$ (ii) $P = T\iota$ is a projection (i.e., $PP = P$).

9. If U is a subspace of a vector space V and if the quotient space V/U is finite-dimensional, prove there exists a subspace W of V such that $V = U \oplus W$. Prove that the natural epimorphism j of $V \to V/U$ induces an isomorphism of $W \to V/U$.

10. U is a subspace of a finite-dimensional vector space V. Let $\alpha_1, \ldots, \alpha_k$ be a basis of U and $\alpha_1, \ldots, \alpha_k, \alpha_{k+1}, \ldots, \alpha_n$ a basis of V. If j is the natural epimorphism of $V \to V/U$, prove $\alpha_{k+1}j, \ldots, \alpha_n j$ is a basis of V/U.

11. Let $V_1, \ldots, V_k$ be subspaces of a finite-dimensional vector space V. If $V = V_1 + \cdots + V_k$ and $\dim V = \dim V_1 + \cdots + \dim V_k$, prove that $V = V_1 \oplus \cdots \oplus V_k$.

12. U_1 is a subspace of a vector space V_1 and U_2 is a subspace of a vector space V_2. T is a given linear transformation of $V_1 \to V_2$. If j_1 and j_2 are the natural epimorphisms, prove there exists a unique linear mapping S such that the accompanying diagram is **commutative**—that is, $j_1 S = Tj_2$.

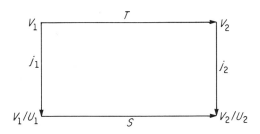

Quotient Spaces and Direct Sums

13. V and W are vector spaces and U is a subspace of V. Let ι be the inclusion mapping of $U \to V$. If T is a linear transformation of $W \to V$, prove there exists a linear transformation S of $W \to U$ such that $T = S\iota$ if and only if im $T \subset U$. Prove that S is injective if T is.

14. The sequence $U \xrightarrow{S} V \xrightarrow{T} W$ is exact. If
  (i) S is injective and
  (ii) T is surjective,
  (iii) there exist homomorphisms H of $W \to V$ and K of $V \to U$, such that

$$HT = 1_W \quad \text{and} \quad SK = 1_U,$$

prove V is isomorphic to $U \oplus W$.

15. U and W are subspaces of a finite-dimensional vector space V and $U \cap W = \bar{0}_V$. Prove that

$$\dim U + \dim W \leq \dim V.$$

16. If T is a surjective linear transformation of $V \to W$ and if U is a subspace of V such that ker $T \subset U$, prove

$$V/U \approx W/T(U).$$

17. U and W are subspaces of a finite-dimensional vector space V. Prove

$$\dim (U + W) = \dim U + \dim (W/U \cap W).$$

Chapter 5

Linear Transformations and Matrices

5-1 REPRESENTATION OF A LINEAR TRANSFORMATION BY A MATRIX

Suppose now that V and W are finite-dimensional vector spaces over the same field F. Let dim $V = m$ and dim $W = n$.

Let T be a linear transformation of $V \to W$. Choose bases $\alpha_1, \alpha_2, \ldots, \alpha_m$ for V and $\beta_1, \beta_2, \ldots, \beta_n$ for W. Then we get the following system of m equations:

$$(1) \qquad \alpha_i T = \sum_{j=1}^{n} a_{ij} \beta_j, \; a_{ij} \in F, \qquad i = 1, 2, \ldots, m,$$

which expresses the vectors $\alpha_i T$ of W in terms of the basis $\beta_1, \beta_2, \ldots, \beta_n$ of W.

The $m \times n$ rectangular array

$$(2) \qquad (a_{ij}) = \begin{bmatrix} a_{11} & a_{12} & \cdots & a_{1n} \\ a_{21} & a_{22} & \cdots & a_{2n} \\ \vdots & & & \vdots \\ a_{m1} & a_{m2} & \cdots & a_{mn} \end{bmatrix}$$

of elements of F (the coefficients of the basis vectors β_i of W), formed from the m equations (1), is called the $m \times n$ **matrix over** F of the linear transformation T, with respect to the α-basis for V and the β-basis for W.

In general such a rectangular array of m rows and n columns of elements of a field F is called an $m \times n$ **matrix over** F, regardless of whether it is interpreted as representing a linear transformation or not. We write the matrix as (a_{ij}), and this notation signifies that the element a_{ij} of F is the entry in the ith row and the jth column of the matrix (a_{ij}).

Of course any change in either or both of the bases of V and W results in a different matrix for T, since such a change would cause a change in the entries of the matrix of T.

If these bases are kept fixed, then each linear transformation T determines a unique $m \times n$ matrix and, conversely, to each $m \times n$ matrix (a_{ij}) over F there corresponds a unique linear transformation T determined by the system of equations (1). That T is unique follows from Theorem 5, Chapter 3. We describe this relationship as a **representation** of linear transformations by matrices. The linear transformation is said to be **represented** by a matrix. It must be kept in mind that here the central idea is the linear transformation itself and that matrices are introduced only for the possible convenience of their use in making computations with linear transformations. Otherwise things can get out of hand, and there is danger of the tail wagging the dog.

Now let ξ be any vector in V. Then in terms of the α-basis of V we have

(3) $$\xi = \sum_{i=1}^{m} x_i \alpha_i, \quad \text{where the } x_i \in F.$$

Let T be a linear transformation defined by (1). Then $\xi T \in W$, and in terms of the β-basis of W, we have

(4) $$\xi T = \sum_{i=1}^{n} y_i \beta_i, \quad \text{where the } y_i \in F.$$

Applying T to (3), we get

$$\xi T = \sum_{i=1}^{m} x_i (\alpha_i T) = \sum_{i=1}^{m} x_i \left(\sum_{j=1}^{n} a_{ij} \beta_j \right).$$

We write this last equation in the following form

(5) $$\xi T = \sum_{j=1}^{n} \left(\sum_{i=1}^{m} x_i a_{ij} \beta_j \right).$$

Since the vector ξT has a unique expression in terms of the basis vectors $\beta_1, \beta_2, \ldots, \beta_n$ of W, it follows at once from (4) and (5) that

(6) $$y_j = \sum_{i=1}^{m} x_i a_{ij}, \, j = 1, 2, \ldots, n.$$

This is a system of n linear equations relating the scalar components x_i of the vector ξ of V with the scalar components y_i of the transformed vector ξT of W. The system (6) is simply another form for the linear transformation T defined by (1). In fact, given the system (6) we can at once write down the system (1).

A linear transformation is therefore equivalent to a system (6) of linear equations. This justifies its title.

Note well that the rectangular array of the coefficients of the x_i in (6) is

(7)
$$\begin{bmatrix} a_{11} & a_{21} & a_{31} & \cdots & a_{m1} \\ a_{12} & a_{22} & a_{32} & \cdots & a_{m2} \\ \cdots & \cdots & \cdots & \cdots & \cdots \\ \cdots & \cdots & \cdots & \cdots & \cdots \\ a_{1n} & a_{2n} & a_{3n} & \cdots & a_{mn} \end{bmatrix}.$$

Comparison with the matrix (2) of T shows that the two matrices are not the same. The matrix (7) is an $n \times m$ matrix, obtained from the matrix (2), by interchanging the rows and columns of (2). The matrix (7) is called the **transpose** of the matrix (a_{ij}) in (2). The entry in the ith row and jth column of the transpose matrix is a_{ji}. This must be kept in mind when switching from the form (1) of the linear transformation to the form (6), and vice versa.

The form (6) for a linear transformation on a finite-dimensional vector space enables us to write down very easily such transformations at will.

Example 1. Let V be the vector space $V_4(R)$ and let W be the vector space $V_3(R)$, and let us choose the standard basis in each space. In virtue of (6), the system of linear equations

$$y_1 = 2x_1 - x_3 + x_4$$
$$y_2 = x_2 + x_3 - 3x_4$$
$$y_3 = 4x_1 + x_4$$

determines a linear transformation T of $V_4(R) \to V_3(R)$, defined by

$$(x_1, x_2, x_3, x_4)T = (y_1, y_2, y_3).$$

Let us find the matrix of T with respect to this choice of bases. We obtain

$$(1,0,0,0)T = (2,0,4) = 2(1,0,0) + 0(0,1,0) + 4(0,0,1)$$
$$(0,1,0,0)T = (0,1,0) = 0(1,0,0) + 1(0,1,0) + 0(0,0,1)$$
$$(0,0,1,0)T = (-1,1,0) = -1(1,0,0) + 1(0,1,0) + 0(0,0,1)$$
$$(0,0,0,1)T = (1,-3,1) = 1(1,0,0) - 3(0,1,0) + 1(0,0,1).$$

Hence the matrix representing T with respect to the standard bases is the 4×3 matrix

$$\begin{bmatrix} 2 & 0 & 4 \\ 0 & 1 & 0 \\ -1 & 1 & 0 \\ 1 & -3 & 1 \end{bmatrix}.$$

Linear Transformations and Matrices

The matrix of the system of linear equations is

$$\begin{bmatrix} 2 & 0 & -1 & 1 \\ 0 & 1 & 1 & -3 \\ 4 & 0 & 0 & 1 \end{bmatrix}$$

which we see is the transpose of the matrix of T.

NOTATION. We shall designate the **transpose** of the matrix A by the symbol A'.

5-2 OPERATIONS ON MATRICES

In this section we define addition and multiplication of matrices, and we assume that all matrices considered have their entries in some given field.

Two linear transformations S and T of $V \to W$ are equal (whether V and W are finite-dimensional or not), $S = T$, if and only if $\alpha S = \alpha T$, for **every** $\alpha \in V$. If dim $V = m$ and dim $W = n$, then with respect to given bases for V and W, let the matrices of S and T be (b_{ij}) and (a_{ij}) respectively. If $S = T$, then we see by (1) that $b_{ij} = a_{ij}$, for all $i = 1, 2, \ldots, m$ and all $j = 1, 2, \ldots, n$.

This leads us to define two $m \times n$ matrices (a_{ij}) and (b_{ij}) as equal

$$(a_{ij}) = (b_{ij})$$

if and only if $a_{ij} = b_{ij}$ for all i and j.

An $m \times n$ matrix all of whose entries are zero is called the $m \times n$ **zero matrix**.

The sum $S + T$ of two linear transformations of $V \to W$ was defined in Sect. 3-3 by

(8) $\qquad \alpha(S + T) = \alpha S + \alpha T, \alpha \in V.$

In particular, then, if dim $V = m$ and dim $W = n$, and if (b_{ij}) and (a_{ij}) are the $m \times n$ matrices (with respect to some choice of bases in V and W) respectively of S and T, then references to (1) and (8) show readily that the matrix of $S + T$ is $(b_{ij} + a_{ij})$.

In general, then, if (a_{ij}) and (b_{ij}) are $m \times n$ matrices over a field F, we define their **sum** by

(9) $\qquad (a_{ij}) + (b_{ij}) = (a_{ij} + b_{ij}).$

This sum is evidently an $m \times n$ matrix over F. Observe that the sum of two matrices is only defined for matrices that have the same number of rows and that have the same number of columns. As we have pointed out for the sum of two linear transformations, the addition of matrices is both commutative and associative. Of course the reader can verify this directly from (9).

If (a_{ij}) is an $m \times n$ matrix and if (0) is the $m \times n$ zero matrix, then

$$(a_{ij}) + (0) = (a_{ij}) = (0) + (a_{ij}).$$

Let us now specialize our theory to the case where $W = V$ and consider linear operators on V. We shall then be able to form the product of two such operators and this will lead us to a rule for multiplying appropriate matrices.

Let us assume that dim $V = n$. We choose a basis in V. With respect to this basis, every linear operator on V is represented by a unique $n \times n$ matrix. Let T and S be linear operators on V with matrices (a_{ij}) and (b_{ij}) respectively relative to this chosen basis of V. We want to determine the matrix of the product TS. We have

$$\alpha_i(TS) = (\alpha_i T)S = \sum_{j=1}^{n} a_{ij}(\alpha_j S)$$

$$= \sum_{j=1}^{n} a_{ij} \left(\sum_{k=1}^{n} b_{jk} \alpha_k \right) = \sum_{k=1}^{n} \left(\sum_{j=1}^{n} a_{ij} b_{jk} \right) \alpha_k.$$

Write

$$c_{ik} = \sum_{j=1}^{n} a_{ij} b_{jk},$$

then (c_{ik}) is the matrix of the linear operator TS, where c_{ik} stands for the entry in the ith row and kth column of the matrix (c_{ik}). This provides actually a very simple formula for computing the matrix of the product TS in terms of the entries of the matrices of T and S. We repeat:

If T is represented by the matrix (a_{ij}) and S by the matrix (b_{ij}) relative to some basis of V, then with respect to this basis TS is represented by the matrix (c_{ij}) where

$$c_{ij} = \sum_{r=1}^{n} a_{ir} b_{rj}.$$

Here c_{ij} is the entry in the ith row and jth column.

We use this last result to formulate a definition of the product of two matrices.

Definition. If (a_{ij}) and (b_{ij}) are two $n \times n$ matrices over a field F, we define the **product** of (a_{ij}) and (b_{ij}) to be the $n \times n$ matrix (c_{ij}), given by

(10) $$(a_{ij})(b_{ij}) = (c_{ij})$$

where

(11) $$c_{ij} = \sum_{k=1}^{n} a_{ik} b_{kj}.$$

As usual, c_{ij} denotes the entry in the product matrix that is located in the ith row and jth column. Notice how c_{ij} is obtained. Multiply the elements $a_{i1}, a_{i2}, \ldots, a_{in}$ in the ith row of (a_{ij}) respectively by the elements $a_{1j}, a_{2j}, \ldots, a_{nj}$ of the jth column of (b_{ij}), and add these n terms. This sum is c_{ij}.

If we form the product in the other order,

$$(b_{ij})(a_{ij}) = (d_{ij})$$

then we have

(12) $$d_{ij} = \sum_{k=1}^{n} b_{ik} a_{kj}.$$

A comparison of (12) with (11) shows at once that, in general, $d_{ij} \neq c_{ij}$. Hence we conclude

Lemma 1. *The multiplication of matrices is* not *in general commutative.*

However, the multiplication of matrices is associative. It requires only a frontal, though tedious assault to prove that

$$[(a_{ij})(b_{ij})](c_{ij}) = (a_{ij})[(b_{ij})(c_{ij})].$$

Exercise: Prove the multiplication of matrices is associative.

Example 2.

$$\begin{pmatrix} 2 & 1 \\ 0 & 1 \end{pmatrix} \begin{pmatrix} 1 & 0 \\ -2 & 1 \end{pmatrix} = \begin{pmatrix} 0 & 1 \\ -2 & 1 \end{pmatrix}$$

while

$$\begin{pmatrix} 1 & 0 \\ -2 & 1 \end{pmatrix} \begin{pmatrix} 2 & 1 \\ 0 & 1 \end{pmatrix} = \begin{pmatrix} 2 & 1 \\ -4 & -1 \end{pmatrix}.$$

On the other hand,

$$\begin{pmatrix} 0 & 0 \\ 1 & -1 \end{pmatrix} \begin{pmatrix} 2 & 0 \\ 3 & -1 \end{pmatrix} = \begin{pmatrix} 2 & 0 \\ 3 & -1 \end{pmatrix} \begin{pmatrix} 0 & 0 \\ 1 & -1 \end{pmatrix} = \begin{pmatrix} 0 & 0 \\ -1 & 1 \end{pmatrix}.$$

It is worth emphasizing that we have proved the following: let S and T be linear operators on a finite-dimensional vector space V and represented respectively by the matrices (a_{ij}) and (b_{ij}) with respect to some fixed basis for V. Then the linear operator ST on V is represented with respect to this same basis by the product $(b_{ij})(a_{ij})$ of these matrices.

The $n \times n$ matrix I_n defined by

$$I_n = \begin{bmatrix} 1 & 0 & 0 & \cdots & 0 & 0 \\ 0 & 1 & 0 & \cdots & 0 & 0 \\ 0 & 0 & 1 & \cdots & 0 & 0 \\ \cdots & \cdots & \cdots & \cdots & \cdots & \cdots \\ \cdots & \cdots & \cdots & \cdots & \cdots & \cdots \\ 0 & 0 & 0 & \cdots & 1 & 0 \\ 0 & 0 & 0 & \cdots & 0 & 1 \end{bmatrix}$$

is called the $n \times n$ **identity matrix**. All the entries on the principal diagonal of I_n are 1, and all other entries are 0.

It is a very easy exercise to prove that if $A = (a_{ij})$ is any $n \times n$ matrix, then

$$AI_n = A = I_n A.$$

Simply apply the formulas (10) and (11).

A matrix is said to be a **square matrix** if it has the same number of rows as columns. We have defined the product of two square matrices.

The definition of product however can be extended to the product of an $m \times n$ matrix (a_{ij}) and an $n \times r$ matrix (b_{ij}) as follows:

$$(a_{ij})(b_{ij}) = (c_{ij})$$

where

(13) $$c_{ij} = \sum_{k=1}^{n} a_{ik} b_{kj}, \quad \begin{matrix} i = 1, 2, \ldots, m \\ j = 1, 2, \ldots, r. \end{matrix}$$

Thus c_{ij} is obtained by multiplying the elements $a_{i1}, a_{i2}, \ldots, a_{in}$ of the ith row of (a_{ij}) respectively by the elements $b_{1j}, b_{2j}, \ldots, b_{nj}$ of the jth column of (b_{ij}) and then adding these n terms.

Notice that to form this product, the first matrix (a_{ij}) in the product must have as many columns as the second matrix (b_{ij}) has rows. It is quite evident that the product matrix has m rows and r columns. It is an $m \times r$ matrix. Observe that we cannot form the product in the other order $(b_{ij})(a_{ij})$. It is not defined, unless $r = m$.

Let $x \in F$ and let T be a linear transformation of $V \to W$. In Sec. 3-3 we defined the linear transformation xT of $V \to W$ as

$$\alpha(xT) = x(\alpha T), \quad \alpha \in V.$$

If we apply this definition to the linear transformation T, defined by

Linear Transformations and Matrices

(1), and form the linear transformation xT, we see that the matrix of xT is (xa_{ij}), where (a_{ij}) is the matrix of T. Clearly, the matrix (xa_{ij}) is obtained from the matrix (a_{ij}) by multiplying each entry of (a_{ij}) by x.

In general then we make the following definition.

Definition. If (a_{ij}) is a matrix over the field F and if $x \in F$, then we define

(14) $$x(a_{ij}) = (xa_{ij}).$$

Example 3.

$$\begin{bmatrix} 4 & 6 \\ 2 & 1 \\ 3 & 0 \end{bmatrix} \begin{bmatrix} 5 & 8 & 1 & -2 \\ -1 & 4 & 0 & 3 \end{bmatrix} = \begin{bmatrix} 14 & 56 & 4 & 10 \\ 9 & 20 & 2 & -1 \\ 15 & 24 & 3 & -6 \end{bmatrix}.$$

The product of a 3×2 matrix and a 2×4 matrix is a 3×4 matrix.

Example 4. If $X = (x_1, x_2, x_3)$ is any vector in $V_3(R)$ and if A is the matrix of a linear transformation of $V_3(R) \to V_2(R)$ (with respect to, say, the standard bases) then we can form the product matrix

$$XA = (x_1\, x_2\, x_3) \begin{bmatrix} a_{11} & a_{12} \\ a_{21} & a_{22} \\ a_{31} & a_{32} \end{bmatrix}$$

$$= (a_{11}x_1 + a_{21}x_2 + a_{31}x_3,\, a_{12}x_1 + a_{22}x_2 + a_{32}x_3).$$

We see that XA is a vector in $V_2(R)$.

Example 5. Let

$$A = \begin{bmatrix} 2 & 0 \\ 3 & -1 \end{bmatrix}, \quad B = \begin{bmatrix} 1 & -1 \\ 2 & 3 \end{bmatrix}, \quad C = \begin{bmatrix} 1 & 1 & 0 \\ 0 & -2 & -1 \end{bmatrix}.$$

Then

$$AB = \begin{bmatrix} 2 & -2 \\ 1 & -6 \end{bmatrix} \quad \text{and} \quad BC = \begin{bmatrix} 1 & 3 & 1 \\ 2 & -4 & -3 \end{bmatrix},$$

and we find that

$$(AB)C = A(BC) = \begin{bmatrix} 2 & 6 & 2 \\ 1 & 13 & 3 \end{bmatrix}.$$

Now suppose $\dim V = \dim W = n$, and let T be an isomorphism of

$V \to W$. For bases $\alpha_1, \alpha_2, \ldots, \alpha_n$ of V and $\beta_1, \beta_2, \ldots, \beta_n$ of W, we have

(15) $$\alpha_i T = \sum_{j=1}^{n} a_{ij} \beta_j, \quad i = 1, 2, \ldots, n.$$

and (a_{ij}) is the matrix of T with respect to these bases.

Since T is an isomorphism, it has an inverse T^{-1} which is also an isomorphism of $W \to V$. Hence

(16) $$\beta_i T^{-1} = \sum_{j=1}^{n} b_{ij} \alpha_j, \quad i = 1, 2, \ldots, n, \quad b_{ij} \in F,$$

and (b_{ij}) is the matrix of T^{-1} with respect to the chosen bases. Now

$$(\alpha_i T) T^{-1} = \alpha_i = \sum_{j=1}^{n} a_{ij} \left(\sum_{k=1}^{n} b_{jk} \alpha_k \right).$$

Hence

(17) $$\alpha_i = \sum_{k=1}^{n} \left(\sum_{j=1}^{n} a_{ij} b_{jk} \right) \alpha_k$$

for $i = 1, 2, \ldots, n$.

Since the α_i form a basis for V, they are independent vectors, and so from (17) we see that we must have

$$\sum_{j=1}^{n} a_{ij} b_{ji} = 1, \, i = 1, 2, \ldots, n$$

(18)

$$\sum_{j=1}^{n} a_{ij} b_{jk} = 0, \, k \neq i$$

Reference to (10) and (11), tells us that the equations (18) imply that

$$(a_{ij})(b_{ij}) = I_n,$$

the $n \times n$ identity matrix. Similarly, we can prove that

$$(b_{ij})(a_{ij}) = I_n.$$

We call the matrix (b_{ij}) the **inverse matrix** of the matrix (a_{ij}) and vice versa. The notation $(a_{ij})^{-1}$ is used for the inverse of (a_{ij}) and so we put

$$(b_{ij}) = (a_{ij})^{-1}.$$

Thus if (a_{ij}) is the matrix of the isomorphism T, then $(a_{ij})^{-1}$ is the matrix of its inverse T^{-1}, both matrices being formed with respect to the same bases.

Linear Transformations and Matrices

Since a linear transformation of $V \to W$ has an inverse if and only if the transformation is bijective, we can conclude that not all matrices have inverses.

Definition. An $n \times n$ matrix (a_{ij}) is said to be **invertible** or **nonsingular** if there exists a matrix (b_{ij}) such that

$$(a_{ij})(b_{ij}) = I_n = (b_{ij})(a_{ij})$$

and the matrix (b_{ij}) is called the **inverse** of the matrix (a_{ij}), and is denoted by $(a_{ij})^{-1}$.

Example 6. Let T be the linear transformation defined by (1). Suppose that the basis of V is changed from $\alpha_1, \alpha_2, \ldots, \alpha_n$ to another basis $\gamma_1, \gamma_2, \ldots, \gamma_n$, the basis of W being kept the same. Let P be the automorphism of V that transforms the α-basis into the γ-basis, so that

$$\gamma_i = \alpha_i P, \quad i = 1, 2, \ldots, m.$$

Let (p_{ij}) be the matrix of P relative to the α-basis. Then (p_{ij}) is an $m \times m$ matrix and we have

$$\gamma_i = \alpha_i P = \sum_{k=1}^{m} p_{ik} \alpha_k, \quad i = 1, 2, \ldots, m.$$

What effect does this change of basis in V have on the matrix representing T?

We have

$$\gamma_i T = \alpha_i PT = \sum_{k=1}^{m} p_{ik}(\alpha_k T), \quad i = 1, 2, \ldots, m.$$

Since

$$\alpha_k T = \sum_{j=1}^{n} a_{kj} \beta_j, \quad k = 1, \ldots, m,$$

it follows that

$$\gamma_i T = \sum_{k=1}^{m} p_{ik} \left(\sum_{j=1}^{n} a_{kj} \beta_j \right)$$

$$= \sum_{j=1}^{n} \left(\sum_{k=1}^{m} p_{ik} a_{kj} \right) \beta_j.$$

We see then that the new matrix representing T relative to the γ-basis of V and the β-basis of W is the product

$$(p_{ij})(a_{ij})$$

of the $m \times m$ matrix (p_{ij}) and the $m \times n$ matrix (a_{ij}). The product is therefore an $m \times n$ matrix.

Exercise. If (a_{ij}) is the matrix of a linear operator T on a finite-dimensional vector space V with respect to a chosen basis in V, find the matrices of T^2, $3T$, $2T + T^2$ with respect to this same basis.

5-3. CHANGE OF BASIS AND SIMILAR MATRICES

Let V be an n-dimensional vector space over the field F. Choose a basis $\alpha_1, \alpha_2, \ldots, \alpha_n$ for V. Let T be a linear operator on V and let (a_{ij}) be the matrix of T relative to this basis. If a change of basis is made in V from $\alpha_1, \alpha_2, \ldots, \alpha_n$ to a new basis $\beta_1, \beta_2, \ldots, \beta_n$, what is the matrix of T relative to this new basis?

We know (see Theorem 5, chapter 3) that there exists a unique linear operator P on V such that

$$\alpha_i P = \beta_i, \; i = 1, 2, \ldots, n.$$

Since the α_i and the β_i are bases of V, it is clear that P is an automorphism of V. Moreover P^{-1}, where $\beta_i P^{-1} = \alpha_i$, $i = 1, 2, \ldots, n$, is also an automorphism of V. Each $\beta_i \in V$ and is therefore a unique linear combination of the basic vectors α_i. Hence

$$\beta_i = \sum_{j=1}^{n} p_{ij} \alpha_j, \quad p_{ij} \in F, \quad i = 1, 2, \ldots, n.$$

Hence,

$$\alpha_i P = \sum_{j=1}^{n} p_{ij} \alpha_j, \quad i = 1, 2, \ldots, n,$$

and therefore (p_{ij}) is the matrix of the automorphism P with respect to the basis $\alpha_1, \alpha_2, \ldots, \alpha_n$.

We shall make use of two important results from the previous section to determine the matrix of T with respect to the new basis $\beta_1, \beta_2, \ldots, \beta_n$. These are:
1. If S and T are linear operators on V then the matrix of the linear operator ST on V is the product of the matrices of S and T in that order.
2. If T is a bijective linear operator (automorphism) on V then the matrix of T^{-1} is the inverse of the matrix of T.

The actual computation of the inverse of an invertible matrix will be given later. We do not need it here since we are concerned only with its existence and not its specific form.

Now PTP^{-1} is a linear operator on V. Let its matrix with respect to the basis $\alpha_1, \alpha_2, \ldots, \alpha_n$ be (c_{ij}). Then

$$(19) \qquad \alpha_i(PTP^{-1}) = \sum_{j=1}^{n} c_{ij}\alpha_j.$$

Applying P to this last equation, we get

$$\alpha_i PT = (\alpha_i P)T = \sum_{j=1}^{n} c_{ij}(\alpha_j P) = \sum_{j=1}^{n} c_{ij}\beta_j.$$

Since $\alpha_i P = \beta_i$, we obtain

$$(20) \qquad \beta_i T = \sum_{j=1}^{n} c_{ij}\beta_j, \ i = 1, 2, \ldots, n.$$

Equations (20) show that the matrix of T with respect to the new basis $\beta_1, \beta_2, \ldots, \beta_n$ is the matrix (c_{ij}).

Reference to (19) shows that (c_{ij}) is the matrix of PTP^{-1} with respect to the old basis $\alpha_1, \alpha_2, \ldots, \alpha_n$, so that

$$(c_{ij}) = (p_{ij})(a_{ij})(p_{ij})^{-1},$$

the product of the matrices of P, T, and P^{-1}, in that order.

Thus the matrix of T with respect to the new basis $\beta_1, \beta_2, \ldots, \beta_n$ is $(p_{ij})(a_{ij})(p_{ij})^{-1}$. We have proved

THEOREM 1

If (a_{ij}) is the matrix of a linear operator on a finite-dimensional vector space and if (p_{ij}) is the matrix of an automorphism defining a change of basis, then the matrix $(p_{ij})(a_{ij})(p_{ij})^{-1}$ is the matrix of the operator relative to the new basis.

Definition. Two linear operators S and T on a vector space V are called **similar** if there exists a nonsingular operator P such that $T = PSP^{-1}$.

Exercise. Prove that similarity is an equivalence relation on the set of linear operators on a vector space V.

Definition. Two matrices (a_{ij}) and (b_{ij}) are said to be **similar** if there exists a nonsingular matrix (p_{ij}) such that

$$(a_{ij}) = (p_{ij})(b_{ij})(p_{ij})^{-1}.$$

The matrices of a linear operator on a finite-dimensional vector space with respect to different bases are similar.

We stress the fact in Theorem 1 that the matrices (p_{ij}), (a_{ij}),

$(p_{ij})^{-1}$ are the matrices respectively of the linear operators P, T, and P^{-1} *all* with respect to the original α-basis for V.

Example 7. Let T be the linear operator on $V_2(R)$ defined by

$$(x_1, x_2) T = (x_1 + x_2, 2x_1 - x_2),$$

where (x_1, x_2) is any vector of $V_2(R)$ expressed in terms of the standard basis $\epsilon_1 = (1,0)$, $\epsilon_2 = (0.1)$; that is,

$$(x_1, x_2) = x_1 \epsilon_1 + x_2 \epsilon_2.$$

The matrix A of T with respect to the standard basis is read off from the equations

$$\epsilon_i T = (1,0) T = (1,2) = \epsilon_1 + 2\epsilon_2$$
$$\epsilon_2 T = (0,1) T = (1,-1) = \epsilon_1 - \epsilon_2.$$

We see that

$$A = \begin{pmatrix} 1 & 2 \\ 2 & -1 \end{pmatrix}.$$

Let us change the basis for $V_2(R)$ from ϵ_1, ϵ_2, to a basis

γ_1, γ_2 defined by

$$\gamma_1 = \epsilon_1 P = 2\epsilon_1 + \epsilon_2$$
$$\gamma_2 = \epsilon_2 P = 3\epsilon_1 + 2\epsilon_2.$$

Here P stands for the automorphism of $V_2(R)$ that changes the basis and the matrix of P with respect to the standard basis is read off from the last two equations to be $\begin{pmatrix} 2 & 1 \\ 3 & 2 \end{pmatrix}$. We also see, by applying the operator P^{-1} to these last two equations, that

$$\epsilon_1 = 2\epsilon_1 P^{-1} + \epsilon_2 P^{-1}$$
$$\epsilon_2 = 3\epsilon_1 P^{-1} + 2\epsilon_2 P^{-1},$$

and hence

$$\epsilon_1 P^{-1} = 2\epsilon_1 - \epsilon_2$$
$$\epsilon_2 P^{-1} = -3\epsilon_1 + 2\epsilon_2.$$

The matrix of the linear operator P^{-1} with respect to the standard basis is therefore $\begin{pmatrix} 2 & -1 \\ -3 & 2 \end{pmatrix}$. Multiplication of the matrices of P and P^{-1} confirms the fact that their product is the identity matrix $\begin{pmatrix} 1 & 0 \\ 0 & 1 \end{pmatrix}$. It now

Linear Transformations and Matrices

follows from Theorem 1 that the matrix of T with respect to this new basis is given by

$$\begin{pmatrix} 2 & 1 \\ 3 & 2 \end{pmatrix} \begin{pmatrix} 1 & 2 \\ 1 & -1 \end{pmatrix} \begin{pmatrix} 2 & -1 \\ -3 & 2 \end{pmatrix} = \begin{pmatrix} 3 & 3 \\ 5 & 4 \end{pmatrix} \begin{pmatrix} 2 & -1 \\ -3 & 2 \end{pmatrix} = \begin{pmatrix} -3 & 3 \\ -2 & 3 \end{pmatrix}.$$

This last result can be confirmed by a direct computation of $\gamma_1 T$ and $\gamma_2 T$ in terms of the new basis vectors γ_1, γ_2.

EXERCISES

1. Let $\alpha_1, \alpha_2, \alpha_3$ be a basis of $V_3(R)$ and β_1, β_2 a basis of $V_2(R)$, where R is the real field. A mapping T of $V_3(R) \to V_2(R)$ is defined by

$$(x_1, x_2, x_3) T = (x_1 + x_2 + x_3, \ 2x_1 - x_2)$$

where $(x_1, x_2, x_3) = x_1 \alpha_1 + x_2 \alpha_2 + x_3 \alpha_3$ and $(x_1 + x_2 + x_3, \ 2x_1 - x_2) = (x_1 + x_2 + x_3) \beta_1 + (2x_1 - x_2) \beta_2$.
  (a) Prove T is linear.
  (b) Find the matrix of T relative to these bases.
  (c) Find the matrix of T relative to the basis $\alpha_1 - \alpha_2, \ \alpha_1 + \alpha_2 + 3\alpha_3, \ 2\alpha_1 + \alpha_3$ of $V_3(R)$ and the basis $\beta_1 + 2\beta_2, \ 3\beta_1 - \beta_2$ of $V_2(R)$.

2. A mapping T of $V_4(R) \to V_4(R)$ is defined by

$$(x_1, x_2, x_3, x_4) T = (x_1 - x_4, \ 2x_3, \ x_2 + x_4, \ -x_1)$$

  (a) Prove T is an automorphism of $V_4(R)$.
  (b) Find the matrix of T relative to the standard basis of $V_4(R)$.
  (c) Find $(x_1, x_2, x_3, x_4) T^{-1}$ and find its matrix with respect to the standard basis.
  (d) Verify that the product of the two matrices is I_4.

3. If

$$A = \begin{bmatrix} 1 & 2 & 0 & 0 \\ 0 & -1 & 1 & 0 \\ 2 & 0 & 4 & 1 \\ 0 & 3 & 0 & 5 \end{bmatrix}$$

and

$$B = \begin{bmatrix} 0 & -1 & 2 & 0 \\ 6 & 1 & 0 & -1 \\ 4 & 0 & -3 & 1 \\ -2 & 1 & 0 & 0 \end{bmatrix}$$

show that

$$AB - BA = \begin{bmatrix} 8 & 0 & -5 & -4 \\ -8 & -9 & -4 & 7 \\ 16 & -12 & 4 & 2 \\ 10 & 13 & -1 & -3 \end{bmatrix}.$$

4. S and T are linear operators on $V_3(R)$ defined by

$$(x_1, x_2, x_3)S = (x_3, 2x_1 - x_2 + 3x_3, 0)$$
$$(x_1, x_2, x_3)T = (x_2 + x_3, -x_1, x_2).$$

(a) Define the operators $2S$, $S + T$, ST, TS, T^2 and T^{-1}.
(b) Find the matrices of these six operators relative to the standard basis of $V_3(R)$.

5. If

$$A = \begin{bmatrix} 2 & 0 & 1 \\ -1 & 3 & 6 \\ 4 & 1 & -2 \end{bmatrix}, \quad B = \begin{bmatrix} 1 & -1 & 2 \\ 5 & 4 & 1 \\ 3 & -4 & 7 \end{bmatrix}, \quad C = \begin{bmatrix} 1 & 2 \\ 4 & -1 \\ 3 & 0 \end{bmatrix},$$

find

$$A + B, AB, BA, 3A, 2A - B, 6A + 2B - 2A,$$
$$-B, (A + B)^2, (AB)C, A(BC), (AC)^2.$$

6. If

$$A = \begin{bmatrix} 1 & 0 & -2 \\ 1 & 3 & 4 \\ -1 & 2 & 3 \end{bmatrix}, \quad B = \begin{bmatrix} 3 & 0 & 0 \\ 2 & 1 & 5 \\ -1 & 0 & 4 \end{bmatrix},$$

find

(a) A^2B, $B^3 + AB$
(b) The transposes of AB, BA, $A + B$, $-2A$.

7. A linear transformation T of $V_3(R) \to V_4(R)$ is represented with respect to the standard bases by the matrix

$$\begin{bmatrix} 2 & -1 & 0 & 1 \\ 1 & 3 & 4 & 2 \\ 0 & -2 & 5 & 6 \end{bmatrix}$$

If (x_1, x_2, x_3) is any vector of $V_3(R)$, find the vector $(x_1, x_2, x_3)T$ of $V_4(R)$.

8. P_4 is the vector space of all polynomials of degree ≤ 4 in the variable x over the real field R. A mapping T of $P_4 \to V_3(R)$ is defined by

$$(a_0 + a_1 x + a_2 x^2 + a_3 x^3 + a_4 x^4)T = (a_0 + a_1 + a_4, a_3 - a_2, -a_0).$$

(a) Prove T is linear.

(b) Find the matrix of T relative to the basis $1, x, x^2, x^3, x^4$ of P_4 and the standard basis of $V_3(R)$.

(c) Prove that the matrix of T relative to the basis $1, 1 - x, x^2 + x, -x^3, x^4$ of P_4 and the standard basis of $V_3(R)$ is

$$\begin{bmatrix} 1 & 0 & -1 \\ 0 & 0 & 1 \\ 1 & -1 & 0 \\ 0 & -1 & 0 \\ 1 & 0 & 0 \end{bmatrix}.$$

9. Find two distinct linear transformations S and T of $V_3(R) \to V_3(R)$ for which both im S and im T are the subspace spanned by the vectors $(1, 1, 0)$ and $(1, 2, -1)$. If in addition it is required that ker S and ker T are the subspace spanned by $(1, 2, 3)$, find S and T.

10. A linear operator T on a vector space V is called an **involution** if $T \circ T = T^2 = I$. Show that, with respect to the standard basis of $V_2(R)$, the matrices

$$\begin{bmatrix} 1 & 0 \\ 0 & 1 \end{bmatrix}, \begin{bmatrix} -1 & 0 \\ 0 & -1 \end{bmatrix}, \begin{bmatrix} 3 & -4 \\ 2 & -3 \end{bmatrix} \quad \text{represent involutions on } V_2(R).$$

Find the matrices for all involutions on $V_2(R)$.

11. Let F be the real field. The matrix relative to the standard basis of an endomorphism T of $V_3(F)$ is

$$\begin{bmatrix} 2 & 0 & 0 \\ 0 & 1 & 2 \\ 0 & -1 & -2 \end{bmatrix}$$

Find the kernel of T and its dimension. Find the codomain (range) of T and its dimension.

12. An endomorphism T of $V_3(F)$, F the real field, is defined by $(x, y, z)T = (2x - y, z, x + y)$.

(a) Find the matrix A of T relative to the standard basis.

(b) Find the matrix B of T relative to the basis

$$(1, 1, 0), (1, 0, -1), (1, 1, 1).$$

(c) Find the matrix P for which $B = PAP^{-1}$.

13. Give an example of an endomorphism T of $V_3(R)$, R the real field, whose kernel is the plane spanned by the vectors $(1,0,1)$ and $(0,1,0)$ and whose image is the line spanned by the vector $(1, 0, -1)$.

14. Let T be the endomorphism of $V_3(F)$, F the real field, whose matrix with respect to the standard basis is

$$\begin{bmatrix} 0 & 2 & -2 \\ 1 & 1 & -4 \\ 0 & 1 & -1 \end{bmatrix}$$

(a) Find the kernel W_1 of T.
(b) Find the matrix of the endomorphism T^2.
(c) If I is the identity automorphism of $V_3(F)$, prove that the mapping $T^2 + I$ defined by $\alpha(T^2 + I) = \alpha T^2 + \alpha I = \alpha T^2 + \alpha$ is an endomorphism of $V_3(F)$.
(d) Find the kernel W_2 of the endomorphism $T_2 + I$.
(e) Prove $V = W_1 \oplus W_2$.

5-4 THE VECTOR SPACE Hom (V, W)

The discussion in Sec. 3-3 proved that if V and W are vector spaces over a field F then the set Hom (V, W) of linear transformations of $V \to W$ is a vector space over F.

Let us now assume that V and W are finite dimensional vector spaces and let dim $V = m$ and dim $W = n$. We want to determine the dimension of the vector space Hom (V, W). Choose bases $\alpha_1, \alpha_2, \ldots, \alpha_m$ for V and $\beta_1, \beta_2, \ldots, \beta_n$ for W.

Define $T_{ij} \in$ Hom (V, W) by

$$\alpha_i T_{ij} = \beta_j, \quad \begin{array}{l} i = 1, 2, \ldots, m \\ j = 1, 2, \ldots, n \end{array}$$

$$\alpha_k T_{ij} = 0, \quad k \neq i.$$

We are going to show that the mn vectors T_{ij} form a basis for the vector space Hom (V, W). Let $T \in$ Hom (V, W). Then

$$\alpha_i T = \sum_{j=1}^{n} a_{ij} \beta_j, \quad a_{ij} \in F.$$

Now

$$\alpha_i \left(\sum_{j=1}^{n} \sum_{i=1}^{m} a_{ij} T_{ij} \right) = \sum_{j=1}^{n} a_{ij} \beta_j = \alpha_i T$$

and this is true for each $i = 1, 2, \ldots, m$. Hence if α is any vector of V then $\alpha = \sum_{i=1}^{m} x_i \alpha_i$, $x_i \in F$, and

$$\alpha \left(\sum_{j=1}^{n} \sum_{i=1}^{m} a_{ij} T_{ij} \right) = \alpha T.$$

This proves

$$T = \sum_{j=1}^{n} \sum_{i=1}^{m} a_{ij} T_{ij}.$$

Therefore the T_{ij} span the vector space Hom (V, W). Moreover, their definition shows they are independent. They therefore form a basis for Hom (V, W), and hence dim $(V, W) = mn$.

Exercise. T is a linear transformation of $V \to W$ where V and W are finite-dimensional vector spaces with the bases $\alpha_1, \alpha_2, \ldots, \alpha_m$ and $\beta_1, \beta_2, \ldots, \beta_n$ respectively. The matrix representing T with respect to these bases is (a_{ij}). If the basis of V is changed by the automorphism P of V to a new basis and if the basis of W is changed by an automorphism Q of W to a new basis, prove that T is represented by a matrix

$$(p_{ij})(a_{ij})(q_{ij})^{-1}$$

relative to the new bases. Here (p_{ij}) and (q_{ij}) are the matrices of P and Q with respect to the old bases.

We have now seen that, with respect to two given bases for V and W, there is a one-to-one correspondence of the set Hom (V, W) of linear transformations of $V \to W$ with the set M of $m \times n$ matrices over F. Hom (V, W) is a vector space over F and, from the definitions (9) and (14), it follows that all the axioms for a vector space over F are satisfied by the set M. Moreover the correspondence of Hom (V, W) with M preserves addition and scalar multiplication. It could hardly be otherwise, since the definitions (9) and (14) for matrices were based on the definitions (8) and (13) for linear transformations. We have therefore the following isomorphism theorem:

THEOREM 2

If dim $V = m$, dim $W = n$, then the vector space Hom (V, W) is isomorphic to the vector space of all $m \times n$ matrices over the scalar field F.

In particular if $W = V$, this theorem extends to

THEOREM 3

If dim $V = n$, then the algebra Hom (V, V) is isomorphic to the algebra of all $n \times n$ matrices over the scalar field F.

Here isomorphism signifies an algebra isomorphism; that is, a bijective mapping that preserves multiplication as well as addition and scalar multiplication.

A matrix basis for the vector space of $m \times n$ matrices over F is the set of mn matrices, each of which has the following properties: Every entry except one is 0, and the one nonzero entry is 1. There are clearly mn distinct matrices of this form and any $m \times n$ matrix can be written as a linear combination of these mn matrices. Moreover they are independent; for a matrix is zero if and only if each of its entries is zero, while two matrices are equal if and only if their corresponding entries are equal. Hence these mn matrices form a basis for the vector space of $m \times n$ ma-

trices over F. This proves once again that
$$\dim \text{Hom}(V, W) = mn.$$

In virtue of the isomorphism theorem (Theorem 2) it is quite natural for linear transformations to pass on to matrices representing them some of their appropriate terminology, and vice versa. For example, the terms *nonsingular*, *invertible*, and *rank* are used for both linear transformations and matrices. Moreover, many theorems about linear transformations translate at once into analogous theorems about matrices.

EXERCISES

1. If $\dim V = m$, $\dim W = n$, find the dimensions of the following vector spaces
   (a) $\text{Hom}(W, V)$,
   (b) $\text{Hom}(V, V)$
   (c) $\text{Hom}(V, V_3(F))$,
   (d) $\text{Hom}(V, \text{Hom}(V, V))$,
   (e) $\text{Hom}(\text{Hom}(V, W) \text{Hom}(V, W))$

2. Let $\alpha_1, \alpha_2, \alpha_3, \alpha_4$ be a basis of a vector space V and let the matrix of a linear operator T on V be $A = (a_{ij})$, $1 \leq i \leq 4$, $1 \leq j \leq 4$, with respect to this basis. If a change of basis is made to the new basis $\alpha_1 + \alpha_2$, $\alpha_1 - \alpha_2 + \alpha_3$, $\alpha_3 + \alpha_4$, $\alpha_4 - \alpha_2$, find the matrix B of T with respect to this new basis. Determine the matrix P for the change of basis and verify that $B = PAP^{-1}$.

3. Given the matrices over the rational field

$$A = \begin{bmatrix} 2 & 1 & 0 & -3 \\ 0 & 1 & 0 & -1 \\ 4 & 3 & 0 & 0 \\ 1 & -2 & 5 & 0 \end{bmatrix}, \quad B = \begin{bmatrix} 2 & 1 & 0 & 1 \\ 0 & 1 & -2 & 0 \\ 1 & -1 & 1 & 0 \\ 2 & 0 & 1 & 1 \end{bmatrix},$$

$$C = \begin{bmatrix} 6 & 0 & -4 & 1 \\ 0 & 0 & 2 & -1 \\ 1 & 5 & 0 & 3 \\ 2 & -2 & 7 & 0 \end{bmatrix},$$

compute the matrices AB, AC, BC, $(AB)C$ and $A(BC)$.

4. A linear transformation T of $V_3(R) \to V_3(R)$ is defined in terms of the standard basis by
$$(x, y, z)T = (x - y - z, 2y + z, 3x).$$
Find the following matrices with respect to the standard basis for $V_3(R)$: (a) the matrix A of T, (b) the matrix A^{-1} of T^{-1}, (c) the matrix B of T^2, (d) the matrix C of $T^2 + 2T$.

Linear Transformations and Matrices

5. Compute the matrices (a) $(C - B)A^{-1}$ (b) $(A + I)^2$, where I is the 3×3 identity matrix. (c) $(A + A^{-1})^3$, where A, B, C are defined as in Exercise 4.

6. Hom (V, V) is the vector space of linear operators on the real vector space V. Determine which of the following subsets are subspaces of Hom (V, V):
  (a) The subset of all bijective linear operators on V.
  (b) The subset of all linear operators on V whose images are the same subspace U of V.

7. Find the matrix of the linear operator T on $V_3(R)$ defined in terms of the standard basis by

$$(x, y, z)T = (2x - z, x + y + z, z - y).$$

8. Prove that the linear operator T in Exercise 7 is an automorphism and find the matrix of T^{-1}. Next verify that the product of the matrices of T and T^{-1} is the identity matrix.

9. Let V be a four-dimensional vector space with the basis $\alpha_1, \alpha_2, \alpha_3, \alpha_4$. An endomorphism T of V is defined by

$$(x_1\alpha_1 + x_2\alpha_2 + x_3\alpha_3 + x_4\alpha_4)T = (2x_1 + x_2)\alpha_1 - 3x_3\alpha_2 + x_4\alpha_3 + x_2\alpha_4,$$

where the x_i are scalars. Find the matrix of T with respect to this basis. Is T an automorphism?

10. V is a three-dimensional vector space and $\alpha_1, \alpha_2, \alpha_3$ is a basis of V. A linear operator T on V is defined by

$$(x_1\alpha_1 + x_2\alpha_2 + x_3\alpha_3)T = (2x_1 + x_2)\alpha_1 - 3x_3\alpha_2 + x_2\alpha_3,$$

where x_1, x_2, x_3 are scalars. Find the matrix of T with respect to this basis. Is T an automorphism?

11. Find a basis for the vector space of all 3×3 matrices over a field F. What is the dimension of this space? Find a proper subspace of this space.

12. Prove that the sets of all matrices over a field of the form

(i) $\begin{bmatrix} a & b \\ -b & c \end{bmatrix}$ (ii) $\begin{bmatrix} -a & a \\ b & c \end{bmatrix}$

are subspaces of the vector space of all 2×2 matrices over this field. Find the dimensions of this vector space and the two subspaces.

13. An $n \times n$ matrix (a_{ij}) over a field F is called **symmetric** if $a_{ij} = a_{ji}$ for all $i, j = 1, 2, \ldots, n$. Prove the set of $n \times n$ symmetric matrices form a subspace of the space of all $n \times n$ matrices over F. Find bases for this space and this subspace.

14. Prove that similarity is an equivalence relation on the set Hom (V, V).

Chapter **6**

Inner-Product Vector Spaces and Dual Spaces

6-1 THE INNER PRODUCT

The definition of a vector space includes no means of ascribing either a length to a vector or to defining a distance between two vectors. Some further requirement is needed if we wish to do this. We introduce in this chapter the usual instrument for this purpose, an **inner product.** An inner product is a kind of multiplication of vectors that yields, however, not a vector but a scalar, and with this additional property we shall see that we can begin to make measurements in a vector space. We emphasize that a vector space does not inherently have an inner product. In order for the space to have an inner product, one has to be defined on it. A finite-dimensional vector space, for instance, can have many inner products.

Definition. An **inner product** in a vector space over the REAL field F is a mapping (function) f of the cartesian product $V \times V \to F$, denoted by $(\alpha,\beta)f = \alpha \cdot \beta \in F$ for all α and β of V, which has the following properties:.

(I) (*Symmetry*) $\alpha \cdot \beta = \beta \cdot \alpha$ for all $\alpha,\beta \in V$.
(II) (*Bilinearity*)
 (a) $(x\alpha) \cdot \beta = x(\alpha \cdot \beta) = \alpha \cdot (x\beta)$ for all $x \in F$ and all $\alpha,\beta \in V$.
 (b) $(\alpha + \beta) \cdot \gamma = \alpha \cdot \gamma + \beta \cdot \gamma$ and
  $\gamma \cdot (\alpha + \beta) = \gamma \cdot \alpha + \gamma \cdot \beta$ for all $\alpha,\beta,\gamma \in V$.
(III) (*Positive Definiteness*) $\alpha \cdot \alpha \geq 0$ for all $\alpha \in V$, and $\alpha \cdot \alpha = 0$ if and only if $\alpha = \overline{0}_V$.

Thus an inner product in V is a function which assigns to each pair α and β of vectors of V a real number, denoted by $\alpha \cdot \beta$, and satisfying the above axioms. If an inner product is defined in a vector space V, then V is called an **inner-product vector space.**

It follows at once from Axiom II(b) that $\overline{0}_V \cdot \alpha = 0 = \alpha \cdot \overline{0}_V$ for all $\alpha \in V$.

Inner-Product Vector Spaces and Dual Spaces

It also follows easily by induction that the properties assumed in Axiom II(b) can be generalized to

$$(\alpha_1 + \alpha_2 + \cdots + \alpha_n) \cdot \gamma = \alpha_1 \cdot \gamma + \alpha_2 \cdot \gamma + \cdots + \alpha_n \cdot \gamma$$
$$\gamma \cdot (\alpha_1 + \alpha_2 + \cdots + \alpha_n) = \gamma \cdot \alpha_1 + \gamma \cdot \alpha_2 + \cdots + \gamma \cdot \alpha_n.$$

Throughout this section, unless otherwise stated, the field F of scalars is taken to be the field of real numbers.

THEOREM 1

Let V be an n-dimensional vector space over F with a basis $\alpha_1, \alpha_2, \ldots, \alpha_n$. Then for $\alpha, \beta \in V$,

$$\alpha \cdot \beta = \sum_{i=1}^{n} a_i b_i,$$

where $\alpha = \sum_{1}^{n} a_i \alpha_i$ and $\beta = \sum_{i}^{n} b_i \alpha_i$ is an inner product in V.

Proof: Axioms I and II are easily seen to be satisfied. Since $\alpha \cdot \alpha = \sum_{1}^{n} a_i^2$ and since the a_i are real numbers, Axiom III is satisfied, since $\sum_{i=1}^{n} a_i^2 = 0$ if and only if each $a_i = 0$; that is, if and only if $\alpha = \bar{0}_V$.

In Theorem 1 we have defined an inner product in a finite-dimensional vector space in terms of a given basis of the space. Since a finite-dimensional vector space has many different bases, it follows that we can have many different inner products in such a space.

If we choose the standard basis $(1, 0, \ldots, 0)$, $(0, 1, 0, \ldots, 0)$, $(0, 0, \ldots, 0, 1)$ as the basis of the vector space $V_n(R)$, then, for any vectors α and β of $V_n(R)$, let $\alpha = (x_1, x_2, \ldots, x_n)$ and $\beta = (y_1, y_2, \ldots, y_n)$ in terms of this basis. An inner product in $V_n(R)$ is defined by

$$\alpha \cdot \beta = x_1 y_1 + x_2 y_2 + \cdots + x_n y_n.$$

This particular inner product, defined in terms of the standard basis for $V_n(R)$, is known as the **standard inner product of $V_n(R)$**.

Example 1. Let $\alpha = (x_1, x_2)$ and $\beta = (y_1, y_2)$ be vectors of $V_2(R)$ expressed in terms of the standard basis for $V_2(R)$. Let us define an inner product in $V_2(R)$ by

$$\alpha \cdot \beta = x_1 y_1 - x_1 y_2 - x_2 y_1 + 3 x_2 y_2.$$

The axioms are readily verified for this definition of an inner product. We note that $\alpha \cdot \alpha = (x_1 x_2)^2 + 2x_2^2$ and so $\alpha \cdot \alpha = 0$ if and only if $x_1 = 0$ and $x_2 = 0$, that is, if and only if $\alpha = \bar{0}_V$.

Example 2. A very interesting example of an inner product occurs in analysis. The set V of all continuous real functions defined on some real interval $[a,b]$, is an infinite-dimensional vector space over the real field. For $f, g \in V$, define

$$f \cdot g = \int_a^b f(x)g(x)\,d(x).$$

The axioms for an inner product can be shown to hold for this definition.

Definition. The **length** or **norm of a vector** α of an inner product vector space V is denoted by $|\alpha|$ and defined by

$$|\alpha| = +\sqrt{\alpha \cdot \alpha}.$$

Since $\alpha \cdot \alpha \geq 0$ this definition is meaningful. Thus $|\alpha| \geq 0$ and $|\alpha| = 0$ if and only if $\alpha = \bar{0}_V$.

Clearly if x is a real number, then

$$|x\alpha| = \sqrt{x\alpha \cdot x\alpha} = |x|\,|\alpha|.$$

Here $|\alpha|$ is the norm of the vector α, and $|x|$ stands for the absolute value of the real number x. In particular, $|-\alpha| = |\alpha|$.

Definition. The **distance** between two vectors α and β of an inner product vector space is defined to be

$$|\alpha - \beta| = \sqrt{(\alpha - \beta) \cdot (\alpha - \beta)} = \sqrt{\alpha \cdot \alpha - 2\alpha \cdot \beta + \beta \cdot \beta}.$$

This definition of a distance function d of $V \times V \to F$,

$$d(\alpha, b) = |\alpha - \beta|, \quad \alpha, \beta \in V,$$

makes V what is called a **metric space**.

Example 3. In $V_3(F)$, F the real field, the vectors $\alpha = (2, -3, 6)$ and $\beta = (-1, 0, 4)$ have lengths $|\alpha| = 7, |\beta| = \sqrt{17}$. The distance between them is $|\alpha - \beta| = |(3, -3, 2)| = \sqrt{22}$.

THEOREM 2 (Schwarz Inequality)

For any pair of vectors α and β, of an inner-product vector space

$$|\alpha \cdot \beta| \leq |\alpha|\,|\beta|$$

Proof: If either α or β is the zero vector, the statement is merely $0 \leq 0$, and so is true. Now assume α and β are nonzero vectors. For any scalars x, y

$$0 \le |x\alpha + y\beta|^2 = x^2|\alpha|^2 + 2xy(\alpha \cdot \beta) + y^2|\beta|^2.$$

Putting $x = |\beta|$, $y = |\alpha|$ and canceling $2|\alpha||\beta| \ne 0$, we get $-(\alpha \cdot \beta) \le |\alpha||\beta|$. Letting $x = |\beta|$, $y = -|\alpha|$ we find in the same way that

$$\alpha \cdot \beta \le |\alpha||\beta|.$$

Thus $|\alpha \cdot \beta| \le |\alpha||\beta|$.

Using the Schwarz inequality we can prove the following inequalities:

(A) If α and β are any vectors, then

$$|\alpha + \beta| \le |\alpha| + |\beta|$$

(B) If α, β, γ are any three vectors, then

$$|\alpha - \beta| \le |\alpha - \gamma| + |\gamma - \beta| \quad \text{(Triangle Inequality.)}$$

To prove (A), we have

$$|\alpha + \beta|^2 = |\alpha|^2 + 2\alpha \cdot \beta + |\beta|^2 \le |\alpha|^2 + 2|\alpha||\beta| + |\beta|^2$$
$$= (|\alpha| + |\beta|)^2.$$

Hence $|\alpha + \beta| \le |\alpha| + |\beta|$.

(B) follows from (A) at once, for

$$\alpha - \beta = (\alpha - \gamma) + (\gamma - \beta)$$

and therefore

$$|\alpha - \beta| = |(\alpha - \gamma) + (\gamma - \beta)| \le |\alpha - \gamma| + |\gamma - \beta|.$$

When V is a vector space with an inner product over the real field, we have defined the *distance* $d(\alpha, \beta)$ between two vectors to be $d(\alpha, \beta) = |\alpha - \beta|$. This distance function satisfies the four axioms for such a function: (1) $|\alpha - \beta| > 0$ if $\alpha \ne \beta$; (2) $|\alpha - \beta| = 0$ if and only if $\alpha = \beta$; (3) $|\alpha - \beta| = |\beta - \alpha|$; (4) the triangle inequality, $|\alpha - \beta| \le |\alpha - \gamma| + |\gamma - \beta|$. Thus V becomes a **metric space**. The inner product led to a definition of the length $|\alpha|$ of a vector. $|\alpha|$ is also called the **norm** of α and a linear space with a norm is called a **normed linear space**. Two very important normed linear spaces are Banach space and Hilbert space. They are complete normed linear spaces. A metric space is called **complete** if every Cauchy sequence in the space converges to a limit in the space. The real numbers, but not the rationals alone, form a complete linear space.

A great deal of the material developed in this chapter can be generalized to theorems in an infinite-dimensional linear space, of which one of the outstanding examples is Hilbert space.

EXERCISES

1. If α and β are vectors in an inner product vector space, prove that $|\alpha \cdot \beta| = |\alpha||\beta|$ if and only if α and β are linearly dependent.

2. If α and β are vectors in an inner product vector space, prove $2(\alpha \cdot \beta) = |\alpha + \beta|^2 - |\alpha|^2 - |\beta|^2$.

3. If $\alpha = (x, y)$ and $\beta = (u, v)$ are vectors in $V_2(F)$, determine whether each of the following definitions of an inner product is acceptable.
    (a) $\alpha \cdot \beta = (x + y)(u + v)$
    (b) $\alpha \cdot \beta = (x + y)(u + v) + vy$
    (c) $\alpha \cdot \beta = (x + u)(y + v)$
    (d) $\alpha \cdot \beta = (x + u)(y + v) + uv$

6-2 ORTHONORMAL BASIS

Throughout this section we shall assume V to be an inner-product vector space.

Definition. The **angle** θ, $0 \leq \theta \leq \pi$, between two nonzero vectors α and β is defined by

$$\cos \theta = \frac{\alpha \cdot \beta}{|\alpha||\beta|}$$

By the Schwarz inequality $|\alpha \cdot \beta| \leq |\alpha||\beta|$, and hence $-|\alpha||\beta| \leq \alpha \cdot \beta \leq |\alpha||\beta|$, so that $-1 \leq \dfrac{\alpha \cdot \beta}{|\alpha||\beta|} \leq 1$. This justifies our definition.

Definition. Two vectors α and β are said to be **orthogonal** if $\theta = 90°$. If α and β are orthogonal vectors we write $\alpha \perp \beta$.

Thus $\alpha \perp \beta$ if and only if $\alpha \cdot \beta = 0$.

Definition. A vector α is **normal** if $|\alpha| = 1$, that is if the length of α is 1.

If we multiply a nonzero vector α by the scalar $1/|\alpha|$, we get a vector $\alpha/|\alpha|$ of unit length. For

$$\left|\frac{\alpha}{|\alpha|}\right| = \frac{|\alpha|}{|\alpha|} = 1.$$

The vector α is said to have been **normalized**.

Definition. A basis for a vector space V is called **orthogonal** if each pair of distinct vectors of the basis is orthogonal. An orthogonal basis is called **orthonormal** if each vector of the basis is normal.

Lemma 1. If U is a subspace of a vector space V over F, then the set $U^\perp$ of all vectors of V that are orthogonal to every vector of U, is a subspace of V.

Proof: For $\alpha, \beta \in U^\perp$ and $\gamma \in U$, we have $\alpha \cdot \gamma = 0$, $\beta \cdot \gamma = 0$ and hence $(\alpha + \beta) \cdot \gamma = 0$. Thus $\alpha + \beta \in U^\perp$. Also if $x \in F$, then for $\alpha \in U^\perp$, $\gamma \in U$, $(x\alpha) \cdot \gamma = x(\alpha \cdot \gamma) = 0$. Hence $x\alpha \in U^\perp$. By Theorem 2 of Chapter 2 this proves $U^\perp$ is a subspace.

Definition. The subspace $U^\perp$ is called the **orthogonal complement** of U.

THEOREM 3

There exists an orthogonal basis for any finite dimensional inner-product vector space.

Proof: Let $\alpha_1, \alpha_2, \ldots, \alpha_n$ be a basis for the finite-dimensional vector space V. We want to find an orthogonal basis $\beta_1, \beta_2, \ldots, \beta_n$ for V. For convenience, define

$$c_j^i = \frac{\beta_i \cdot \alpha_j}{|\beta_i|^2}, \quad i, j = 1, 2, \ldots, n.$$

We start by taking $\beta_1 = \alpha_1$. Hence $L(\beta_1) = L(\alpha_1)$. Next choose the scalar x so that $(\alpha_2 - x\beta_1) \cdot \beta_1 = 0$. We find $x = c_2^1$. Take $\beta_2 = \alpha_2 - c_2^1 \beta_1$. Then $\beta_2 \cdot \beta_1 = 0$ and $L(\beta_1, \beta_2) = L(\alpha_1, \alpha_2)$. Now choose scalars x_1 and x_2 so that the vector $\alpha_2 - x_1 \beta_1 - x_2 \beta_2$ is orthogonal to β_1 and to β_2. We find $x_1 = c_3^1, x_2 = c_3^2$. Then take $\beta_3 = \alpha_3 - c_3^1 \beta_1 - c_3^2 \beta_2$. Then $\beta_1, \beta_2, \beta_3$ are orthogonal vectors, and clearly $L(\beta_1, \beta_2, \beta_3) = L(\alpha_1, \alpha_2, \alpha_3)$.

Now use induction on k and assume $\beta_1, \beta_2, \ldots, \beta_k$ are orthogonal vectors given by

(1) $\quad \beta_i = \alpha_i - c_i^1 b_1 - c_i^2 \beta_2 - \cdots - c_i^{i-1} \beta_i, i = 1, 2, \ldots, k$

and such that $L(\alpha_1, \ldots, \alpha_k) = L(\beta_1, \ldots, \beta_k)$.

Determine scalars $x_1, x_2, \ldots, x_k$ such that

$$\gamma = \alpha_{k+1} - x_1 \beta_1 - x_2 \beta_2 - \cdots - x_k \beta_k$$

is orthogonal to each of the vectors $\beta_1, \beta_2, \ldots, \beta_k$. Then for each j from 1 to k, we get

$$\beta_j \cdot \gamma = \beta_j \cdot \alpha_{k+1} - x_j (\beta_j \cdot \beta_j) = 0,$$

that is,
$$x_j = c^j_{k+1}, \quad j = 1, 2, \ldots, k.$$
Hence $\beta_{k+1} = \alpha_{k+1} - c^1_{k+1}\beta_1 - c^2_{k+1}\beta_2 - \cdots - c^k_{k+1}\beta_k$ is a vector orthogonal to $\beta_1, \beta_2, \ldots, \beta_k$. Moreover, it is clear that $L(\beta_1, \beta_2, \ldots \beta_{k+1}) = L(\alpha_1, \alpha_2, \ldots, \alpha_{k+1})$. Our induction is completed and proves that $\beta_1, \beta_2, \ldots, \beta_n$, where each β_i, $i = 1, 2, \ldots, n$, is given by (1), is an orthogonal set of vectors which span V—that is, $L(\beta_1, \ldots, \beta_n) = L(\alpha_1, \ldots \alpha_n)$. Hence $\beta_1, \beta_2, \ldots, \beta_n$ form an orthogonal basis for V.

Corollary 1. There exists an orthonormal basis for a finite-dimensional vector space V.

Proof: Simply normalize each vector $\beta_1, \beta_2, \ldots, \beta_n$ of an orthogonal basis for V and we obtain an orthonormal basis $\dfrac{\beta_1}{|\beta_1|}, \dfrac{\beta_2}{|\beta_2|}, \ldots, \dfrac{\beta_n}{|\beta_n|}$ for V.

Corollary 2. With respect to the inner product defined in Theorem 1 the given basis $\alpha_1, \alpha_2, \ldots, \alpha_n$ is an orthonormal basis of V.

Proof: This follows from the formula in Theorem 1 for the inner product. We see at once that
$$\alpha_i \cdot \alpha_j = 0, \quad j \neq i$$
$$= 1, \quad j = i.$$

Theorem 3 shows that from any finite linearly independent set of vectors $\alpha_1, \alpha_2, \ldots, \alpha_n$ we can derive a set $\beta_1, \beta_2, \ldots, \beta_n$ of orthogonal vectors. This process is known as the *Gram-Schmidt method*, and we are said to have **orthogonalized** the vectors $\alpha_1, \alpha_2, \ldots, \alpha_n$.

Example 4. Orthogonalize the vectors $(1, 2, 3)$, $(1, 0, 1)$ and $(2, 1, 0)$ of the vector space $V_3(R)$ with the standard inner product.

Choose $\beta_1 = (1, 0, 1)$. Let $\beta_2 = (2, 1, 0) - c(1, 0, 1) = (2 - c, 1, -c)$. In order that $\beta_1 \cdot \beta_2 = 0$ we see that $2 - c - c = 0$; that is, $c = 1$. Hence take $\beta_2 = (1, 1, -1)$. Now let $\beta_3 = (1, 2, 3) - a(1, 0, 1) - b(1, 1, -1) = (1 - a - b, 2 - b, 3 - a + b)$. We require that $\beta_3 \cdot \beta_1 = 1 - a - b + 3 - a + b = 0$, which means $a = 2$. Also $\beta_3 \cdot \beta_2 = 1 - a - b + 2 - b - 3 + a - b = 0$; that is, $b = 0$. Hence $\beta_3 = (-1, 2, 1)$. Thus the vectors $(1, 0, 1), (1, 1, -1), (-1, 2, 1)$ are orthogonal. If they are normalized we obtain the unit vectors $\dfrac{1}{\sqrt{2}}(1, 0, 1)$, $\dfrac{1}{\sqrt{3}}(1, 1, -1)$, $\dfrac{1}{\sqrt{6}}(-1, 2, 1)$.

If $\beta_1, \beta_2, \ldots, \beta_n$ is an orthogonal basis for V then for every vector $\alpha \in V$ we have

$$\alpha = \sum_{i=1}^{n} x_i \alpha_i, \quad x_i \in F.$$

Hence

$$\alpha \cdot \alpha_j = \left(\sum_{i=1}^{n} x_i \alpha_i\right) \cdot \alpha_j = x_j (\alpha_j \cdot \alpha_j)$$

and so

$$x_j = \frac{\alpha \cdot \alpha_j}{|\alpha_j|^2}, \quad j = 1, 2, \ldots, n.$$

Hence with respect to an orthogonal basis every vector α can be expressed uniquely by

$$\alpha = \sum_{i=1}^{n} \frac{\alpha \cdot \alpha_i}{|\alpha_i|^2} \alpha_i.$$

If the basis is orthonormal, we have

(2) $$\alpha = \sum_{i=1}^{n} (\alpha \cdot \alpha_i) \alpha_i.$$

The reader should note well this last equation, which gives the form of a vector α with respect to an orthonormal basis. It will be used frequently in subsequent work.

Let $\alpha_1, \alpha_2, \ldots, \alpha_n$ be an orthonormal basis for a vector space V. Let $\alpha, \beta \in V$. Then

$$\alpha = \sum_{1}^{n} a_i \alpha_i, \quad \beta = \sum_{1}^{n} b_i \alpha_i.$$

Therefore

$$\alpha \cdot \beta = \sum_{1}^{n} a_i b_i.$$

Thus we can see that, relative to an orthonormal basis, the inner product has this unique and particularly simple form.

THEOREM 4

If U is a subspace of a finite-dimensional vector space V, then $V = U \oplus U^\perp$.

Proof: Choose an orthonormal basis $\beta_1, \ldots, \beta_k$ for U. Let $\alpha \in V$. Then

$$\alpha = \sum_1^k (\alpha \cdot \beta_i)\beta_i + \left(\alpha - \sum_1^k (\alpha \cdot \beta_i)\beta_i\right).$$

Consider the vector

$$\gamma = \alpha - \sum_{i=1}^k (\alpha \cdot \beta_i)\beta_i.$$

Now $\gamma \cdot \beta_j = (\alpha \cdot \beta_j) - (\alpha \cdot \beta_j)(\beta_j \cdot \beta_j) = 0$, for $j = 1, 2, \ldots, k$. Hence is orthogonal to each vector of the basis $\beta_1, \beta_2, \ldots, \beta_k$ for U and hence $\gamma \in U^\perp$. Thus each vector $\alpha \in V$ is the sum of a vector from U and vector from $U^\perp$. Moreover $U \cap U^\perp = \overline{0}_V$. Hence $V = U \oplus U^\perp$.

Corollary. *If $\alpha_1, \alpha_2, \ldots, \alpha_k$ is an orthonormal basis for a subspace U of a vector space V of dimension n, then this basis can be enlarged to a orthonormal basis for V.*

Proof: Let $U^\perp$ be the orthogonal complement of U. By Theorem Corollary 1, $U^\perp$ has an orthonormal basis $\beta_1, \beta_2, \ldots, \beta_h$. The vectors $\alpha_1, \ldots, \alpha_k, \beta_1, \ldots, \beta_h$ form an orthonormal set. Moreover, by our theorem, $k + h = n$, and hence these vectors form a basis for V.

If $V = U \oplus U^\perp$ then every $\alpha \in V$ has a unique representation in the form $\alpha = \beta + \gamma$, $\beta \in U$, $\gamma \in U^\perp$. The linear transformation P of $V \to U$ defined by $\alpha P = \beta$ is called an **orthogonal projection** and β called the **orthogonal projection of α on U.**

Example 5. Let U be the one-dimensional subspace of $V_3(R)$ spanned by the vector $(1,1,1)$. The vectors of U therefore have the form (t,t,t), $t \in R$. Assume the standard inner product in $V_3(R)$. The $U^\perp$ is the subspace of $V_3(R)$ consisting of all vectors (x,y,z), for which $tx + ty + tz = t(x + y + z) = 0$. Since t is an arbitrary real number this last equation forces $x + y + z = 0$. Hence the vectors (x,y,z) of $U^\perp$ are those for which $x + y + z = 0$. We see that two linearly independent vectors of this form are $(1,-1,0)$ and $(0,1,-1)$. Any vector in $U^\perp$ is a linear combination of these two vectors and so dim $U^\perp = 2$. Since $U \cap U^\perp$ is the zero vector and since the vectors $(1,1,1)$, $(1,-1,0)$ and $(0,1,-1)$ span $V_3(R)$, we confirm the fact that $V_3(R) = U \oplus U^\perp$. Geometrically the subspace U is the straight line through the origin and the point $(1,1,1)$, while the subspace $U^\perp$ is the plane $x + y + z = 0$. The line is perpendicular to the plane.

EXERCISES

1. If we propose in $V_2(F)$, F the real field, an inner product defined by $\alpha \cdot \beta = (x_1 + y_1)(x_2 + y_2)$ where $\alpha = (x_1, y_1)$ and $\beta = (x_2, y_2)$, test all the axioms to ascertain if this proposal is acceptable.

2. If U is a finite-dimensional subspace of an inner product vector space V, prove $(U^\perp)^\perp = U$. (Recall that $V = U \oplus U^\perp$.)

3. Starting with the independent vectors $\alpha_1 = (0, 1, 1)$, $\alpha_2 = (-1, 0, 0)$, $\alpha_3 = (2, 1, 0)$ construct an orthogonal basis $\beta_1, \beta_2, \beta_3$ for $V_3(F)$, where F is the real field.

4. If V is an inner-product vector space, prove that if $|\alpha \cdot \beta| = |\alpha||\beta|$ then the vectors α and β of V are dependent.

5. Let F be the real field and let $V_3(F)$ be the vector space with the standard or usual inner product. If U is the subspace of $V_3(F)$ spanned by the vector $(1, 2, 0)$, find the subspace U and verify that $V_3(F) = U \oplus U^\perp$. Express the vector $(2, 3, 4)$ as the unique sum of a vector of U and a vector of $U^\perp$.

6. Let V be an inner-product vector space. If $V = U \oplus W$, show that the mapping P defined by $\alpha P = \beta$ where $\alpha = \beta + \gamma \in V$, $\beta \in U$, $\gamma \in W$ is well-defined and is linear. The vector β is called the **projection of α on U along W**. If $W = U^\perp$, β is called the **orthogonal projection of α on U**. Prove that for any subspace of V, a vector of V has an orthogonal projection on the subspace.

7. If $\alpha_1, \ldots, \alpha_k$ are orthonormal vectors of an inner-product vector space V, prove the orthogonal projection of a vector $\alpha \in V$ on $U = L(\alpha_1, \alpha_2, \ldots, \alpha_k)$ is given by

$$\sum_1^k (\alpha \cdot \alpha_i)\alpha_i.$$

8. Prove that for every finite dimensional real vector space an inner product exists.

9. $V_3(F)$ is a vector space over the real field with the usual inner product. If U is the subspace spanned by the vectors $(1, 0, 2)$ and $(-1, 1, 1)$, find the orthogonal projection of the vector $(2, 1, 0)$ on U.

10. Let V be the vector space of continuous real functions on the closed interval $[0, 1]$ with the inner product

$$f \cdot g = \int_0^1 f(x)g(x)dx, \quad f, g \in V.$$

Find two orthonormal vectors of V that span the same subspace of V as the vectors f and g, where $xf = 1$, $xg = x$ for all $x \in [0, 1]$.

11. U is the subspace of $V_4(R)$ spanned by the vectors $(2, 0, -1, 3)$, $(5, -1, -3, 7)$ and $(1, 1, 0, 2)$. W is the subspace spanned by the vectors $(3, -1, -2, 4)$ and $(0, 0, 1, -1)$. Find the dimensions of the following vector spaces: (a) U (b) $U^\perp$ (c) W (d) $W^\perp$ (e) $U^\perp \cap W^\perp$ (f) $U \cap W$ (g) $U^\perp + W^\perp$ (h) $U + W$.

12. In Example 11, find bases for the following vector spaces: (a) U (b) $V_4(R)/U^\perp$ (c) $V_4(R)/U$ (d) $V_4(R)/W^\perp$.

13. In Example 11, find $(V_4(F)/W)^\perp$ and determine its dimension. Does $V_4(R)/W^\perp = (V_4(R)/W)^\perp$?

14. U is a subspace of the finite-dimensional vector space V. Prove that the vector spaces V/U and $U^\perp$ are isomorphic and find the isomorphism.

15. If S is a nonempty subset of a vector space V, $S^\perp$ is the set of all vectors of V that are orthogonal to every vector of S. Prove $S^\perp$ is a subspace of V.

16. S is a nonempty subset of V and $[S]$ is the subspace of V generated by S.
  (a) Prove $[S] \subset (S^\perp)^\perp$.
  (b) If V is finite-dimensional, prove that $[S] = (S^\perp)^\perp$.

17. Consider the set of functions

$$1, \sin x, \cos x, \sin 2x, \cos 2x, \ldots, \sin nx, \cos nx$$

defined on the interval $[-\pi,\pi]$. A linear combination with real coefficients

$$P(x) = \frac{a_0}{2} + a_1 \cos x + b_1 \sin x + \cdots + a_n \cos nx + b_n \sin nx$$

is called a **trigonometric polynomial of degree n**.

  (a) Prove that the set of all such polynomials of degree n is a $(2n + 1)$ dimensional real vector space V.
  (b) If $P, Q \in V$, prove that

$$P \cdot Q = \int_{-\pi}^{\pi} P(x) Q(x)\, dx$$

defines an inner product in V.

  (c) Show that the set of functions

$$1, \sin x, \cos x, \ldots, \sin nx, \cos nx$$

form an orthogonal basis of V.

  (d) Normalize these basis vectors.
  (e) Use equation (2) to find the (Fourier) coefficients a_k and b_k of P above.

6-3 ISOMETRIES

Definition. Let V and W be inner-product vector spaces over the real field R. A linear transformation T of $V \to W$ is called an **isometry** if (i) im $T = W$, (ii) $\alpha T \cdot \beta T = \alpha \cdot \beta$, for all $\alpha, \beta \in V$.

Thus an isometry is a surjective linear transformation that preserves the inner product. An isometry will therefore preserve all derivatives of the inner product, such as length, angle, distance, and, in particular orthogonality. This is easy to prove as follows:

(a) Length is preserved, since
$$\alpha \cdot \alpha = \alpha T \cdot \alpha T$$
for all $\alpha \in V$, and hence $|\alpha| = |\alpha T|$.

(b) Distance is preserved, since
$$(\alpha - \beta) \cdot (\alpha - \beta) = (\alpha - \beta)T \cdot (\alpha - \beta)T$$
$$= (\alpha T - \beta T) \cdot (\alpha T - \beta T),$$
and hence $|\alpha - \beta| = |\alpha T - \beta T|$ for all $\alpha, \beta \in V$.

(c) Angle is preserved, since
$$\cos \theta = \frac{\alpha \cdot \beta}{|\alpha||\beta|} = \frac{\alpha T \cdot \beta T}{|\alpha T||\beta T|}, \quad \alpha \neq \overline{0}_V, \quad \beta \neq \overline{0}_V.$$

(d) Orthogonality is preserved, since $\alpha T \cdot \beta T = 0$ if and only if $\alpha \cdot \beta = 0$.

Definition. If T is an isometry of $V \to W$, then the vector spaces V and W are said to be **isometric**.

THEOREM 5

Let V and W be inner product vector spaces. A linear transformation T of $V \to W$ is an isometry if and only if T is an isomorphism that preserves the inner product.

Proof: If T is an isomorphism and preserves the inner product, then T is certainly surjective and hence is, by definition, an isometry.

Conversely, assume that T is an isometry. Then T is surjective. We need therefore show only that T is injective. To do this use Theorem 3 of Chapter 3. Let α be any vector in ker T. Then $\alpha \cdot \alpha = \alpha T \cdot \alpha T = 0$. Hence $|\alpha| = 0$ and therefore $\alpha = \overline{0}_V$. Thus T is injective and hence is an isomorphism.

Corollary. If T is an isometry, then T^{-1} is an isometry.

Proof: Since T is an isomorphism, we know T^{-1} is an isomorphism of $W \to V$. For $\alpha', \beta' \in W$, let $\alpha = \alpha' T^{-1}, \beta = \beta' T^{-1}$. Hence
$$(\alpha' T^{-1}) \cdot (\beta' T^{-1}) = \alpha \cdot \beta = (\alpha T) \cdot (\beta T) = \alpha' \cdot \beta'.$$
and therefore T^{-1} is an isometry.

THEOREM 6

If V and W are n-dimensional vector spaces with inner products, then there exists an isometry T of $V \to W$.

Proof: Let $\alpha_1, \ldots, \alpha_n$ and $\alpha'_1, \ldots, \alpha'_n$ be orthonormal bases for V and W respectively. Then by Theorem 5, Chapter 3, there exists an isomorphism T of $V \to W$ such that $\alpha_i T = \alpha'_i, i = 1, 2, \ldots, n$.

If α and β are any vectors in V, then

$$\alpha = \sum_1^n a_i \alpha_i, \quad \beta = \sum_1^n b_i \alpha_i, \quad \text{and} \quad \alpha T = \sum_1^n a_i \alpha_i',$$

$$\beta T = \sum_1^n b_i \alpha_i'.$$

Now

$$\alpha \cdot \beta = \sum_1^n a_i b_i$$

and

$$(\alpha T) \cdot (\beta T) = \sum_1^n a_i b_i.$$

Hence $\alpha \cdot \beta = (\alpha T) \cdot (\beta T)$ and therefore T is an isometry.

Theorem 6 demonstrates that all n-dimensional vector spaces with inner products are essentially the same, that is such a vector space unique up to isometry.

Definition. A linear operator on a real inner-product vector space that is an isometry is called an **orthogonal operator** or **orthogonal transformation**.

An orthogonal operator therefore preserves lengths, distances, and angles, and transforms orthogonal vectors into orthogonal vectors.

In the light of Theorem 5 we can say that an orthogonal operator on an inner-product vector space V is an automorphism of V that preserves the inner product. By the corollary to Theorem 5 we know also that the inverse of an orthogonal operator is an orthogonal operator. Obviously the identity operator 1_V is orthogonal.

Now suppose S and T are orthogonal operators on V. Then the product ST is an automorphism of V. (Prove this!) Moreover

$$\alpha(ST) \cdot \beta(ST) = (\alpha S)T \cdot (\beta S)T = \alpha S \cdot \beta S = \alpha \cdot \beta$$

for all $\alpha, \beta \in V$. Hence the product ST is an orthogonal operator. Therefore the orthogonal operators on V form a group under multiplication (map composition is associative). It is called the **orthogonal group of linear operators** on V.

EXERCISES

1. If T is a linear transformation $V \to W$ of two vector spaces which preserves lengths of vectors—that is, $|\alpha T| = |\alpha|$ for every $\alpha \in V$—is necessarily an isometry? Prove your answer.

2. F is the real field. An endomorphism T of $V_3(F)$ is defined by $(1,0,0)T = (1,0,0)$, $(0,1,0)T = \left(0, \frac{1}{2}, \frac{\sqrt{3}}{2}\right)$, $(0,0,1)T = \left(0, \frac{-\sqrt{3}}{2}, \frac{1}{2}\right)$.

(a) Find $(x, y, z)T$, where (x, y, z) is an arbitrary vector.
(b) Prove T is an isometry.

3. If $\alpha_1, \alpha_2, \ldots, \alpha_n$ and $\beta_1, \beta_2, \ldots \beta_n$ are two orthonormal bases of an inner-product vector space V, prove there is a unique orthogonal operator T on V such that

$$\alpha_\iota T = \beta_\iota, \quad \iota = 1, 2, \ldots, n.$$

4 ORTHOGONAL OPERATORS ON FINITE-DIMENSIONAL INNER-PRODUCT VECTOR SPACES

[We shall assume throughout this section that V is a real finite-dimensional inner-product vector space.]

Examples of orthogonal transformations in the euclidean plane or space are the rotations of the rectangular axes into new sets of rectangular axes. In the case of the plane, we have already introduced an example of an orthogonal transformation with the equations

$$x' = x \cos \theta - y \sin \theta$$
$$y' = x \sin \theta + y \cos \theta$$

which represents a rotation of the rectangular axes through an angle θ. This transformation $(x, y) \to (x', y')$ can also be regarded as mapping the point (x, y) of the plane into the point (x', y') of the plane. It preserves the distance between points. Moreover, it is a linear transformation, for the equations are linear, and its matrix is $\begin{bmatrix} \cos \theta & \sin \theta \\ -\sin \theta & \cos \theta \end{bmatrix}$. Note that the sum of the squares of the entries in any row (or column) is 1, and the sum of the products of corresponding entries of two rows (or columns) is 0. These are characteristic features of an "orthogonal matrix."

In three-dimensional space with the standard inner product, the rigid motions (rotations) are orthogonal transformations and they form a group called the **euclidean group.**

Our first result is

THEOREM 7

A linear operator T on V is orthogonal if and only if it maps any orthonormal basis of V into an orthonormal basis of V.

Proof: Assume that T is orthogonal and let $\alpha_1, \alpha_2, \ldots, \alpha_n$ be an orthonormal basis of V. Since T is an automorphism, it is nonsingular and therefore $\alpha_1 T, \ldots, \alpha_n T$ is also a basis of V. Furthermore, $|\alpha_i T| =$

$|\alpha_i| = 1$, $i = 1, 2, \ldots, n$, and $\alpha_i T \cdot \alpha_j T = \alpha_i \cdot \alpha_j = 0$, $i \neq j$. Hence $\alpha_1 T, \ldots, \alpha_n T$ is an orthonormal basis of V.

Conversely, assume that T transforms the orthonormal basis $\alpha_1, \alpha_2, \ldots, \alpha_n$ of V into an orthonormal basis $\alpha_1 T, \alpha_2 T, \ldots, \alpha_n T$ of V. Then T is surjective and hence is nonsingular, and so T must be an automorphism of V. (See the corollary to Theorem 9, Chapter 3.) Let $\alpha, \beta \in V$. Then $\alpha = x_1 \alpha_1 + \cdots + x_n \alpha_n$ and $\beta = y_1 \alpha_1 + \cdots + y_n \alpha_n$ for scalars x_i and y_i, where $i = 1, 2, \ldots, n$. Hence $\alpha \cdot \beta = x_1 y_1 + \cdots x_n y_n$. Now

$$\alpha T \cdot \beta T = [x_1(\alpha_1 T) + \cdots + x_n(\alpha_n T)] \cdot [y_1(\alpha_1 T) + \cdots + y_n(\alpha_n T)]$$
$$= x_1 y_1 + x_2 y_2 + \cdots + x_n y_n,$$

since the $\alpha_i T$, $i = 1, 2, \ldots, n$, are an orthonormal basis of V. Hence $\alpha T \cdot \beta T = \alpha \cdot \beta$ and therefore T is orthogonal.

THEOREM 8

A linear operator T on V is orthogonal if and only if it preserves the length of a vector.

Proof: If T is orthogonal then T is an isometry and so preserves lengths.

Conversely, assume that T preserves lengths. Let $\alpha \in \ker T$. Then $|\alpha| = |\alpha T| = 0$ and hence $\alpha = \overline{0}_V$. This proves that T is injective. Hence T is nonsingular and therefore T is an automorphism. (Theorem of Chapter 3.)

We next show that T preserves the inner product. Let $\alpha, \beta \in V$. Then

$$|\alpha T + \beta T|^2 = |\alpha T|^2 + 2(\alpha T \cdot \beta T) + |\beta T|^2$$
$$= |\alpha|^2 + 2(\alpha T \cdot \beta T) + |\beta|^2.$$

Also

$$|\alpha T + \beta T|^2 = |(\alpha + \beta) T|^2 = |\alpha + \beta|^2$$
$$= |\alpha|^2 + 2(\alpha \cdot \beta) + |\beta|^2.$$

Hence $\alpha \cdot \beta = \alpha T \cdot \beta T$. This completes the proof that T is orthogonal.

Next we determine the form of the matrix of an orthogonal linear operator T on V with respect to an orthonormal basis $\alpha_1, \alpha_2, \ldots, \alpha_n$ of V.

Let

$$\alpha_i T = \sum_{j=1}^{n} a_{ij} \alpha_j, \quad i = 1, 2, \ldots, n.$$

Since the $\alpha_i T$ also form an orthonormal basis, we have

$$|\alpha_i T|^2 = 1 = |\sum_{j=1}^{n} a_{ij} \alpha_j|^2 = \sum_{j=1}^{n} a_{ij}^2,$$

Inner-Product Vector Spaces and Dual Spaces

$$\alpha_i T \cdot \alpha_k T = 0 = \left(\sum_{j=1}^{n} a_{ij}\alpha_j\right) \cdot \left(\sum_{j=1}^{n} a_{kj}\alpha_j\right) = \sum_{j=1}^{n} a_{ij}a_{kj},$$

where $i \neq k$.

Thus the matrix $A = (a_{ij})$ of T with respect to an orthonormal basis of V has the properties:
  (i) The sum of the squares of the entries in any row of A is 1.
  (ii) The sum of the product of the corresponding entries of two distinct rows of A is 0.

If we interpret the rows of A to be vectors in the vector space $V_n(R)$, with the usual inner product, then the row vectors of A form an orthonormal basis for $V_n(R)$. A is called an **orthogonal matrix**.

We emphasize that a matrix is orthogonal if and only if it represents an orthogonal transformation *with respect to an orthonormal basis*. In virtue of (i) and (ii) above, we can also define an orthogonal matrix as a matrix that possesses these two properties (i) and (ii).

If A is an orthogonal matrix then it is easy to see that these two properties imply

$$AA^t = A^tA = I,$$

where I is the $n \times n$ identity matrix and A^t is the transpose of A. Hence $A^t = A^{-1}$. This property characterizes an orthogonal matrix. For if $AA^t = I$ then equating the entries on both sides of this matrix equation yields the two properties (i) and (ii) and these imply that the $\alpha_i T$ above form an orthonormal basis of V and hence prove T is an orthogonal linear operator.

It is easy to prove that the two properties (i) and (ii) above are true for the columns of an orthogonal matrix.

We can use this last result to generalize the concept of an orthogonal matrix over the real field to that of an orthogonal matrix over an arbitrary field.

Definition. A square matrix A over an arbitrary field F is said to be **orthogonal** if and only if $AA^t = I$; that is $A^t = A^{-1}$.

This definition disassociates the concept of an orthogonal matrix from any relationship to the inner product of a vector space.

If A is an orthogonal $n \times n$ matrix then the matrix equation $AA^t = I$ yields the following n^2 equations in the entries of A.

$$\sum_{j=1}^{n} a_{ij}a_{kj} = 0, \quad i \neq k$$
$$= 1, \quad i = k.$$

Given any nonzero vector α_1 of the inner-product vector space $V_n(R)$, we can always construct an orthogonal matrix with $\alpha_1/|\alpha_1| =$

$(x_1, \ldots, x_n)$ as its first row. The vector α_1 is the basis of a subspace U of $V_n(R)$ and this basis, as we have seen, can be enlarged to a basis $\alpha_1, \alpha_2, \ldots, \alpha_n$ for $V_n(R)$. Starting with α_1, and retaining it as a basis vector for $V_n(R)$, we can construct an orthogonal basis for $V_n(R)$. Finally normalizing this orthogonal basis we get a set of vectors headed by $\alpha_1/|\alpha_1|$ with which we construct an orthogonal $n \times n$ matrix, by using these vectors for its rows (or columns). This matrix is not unique, since there are many ways of extending the given vector α_1 to a basis for all of $V_n(R)$.

We shall soon be putting orthogonal matrices into active service.

EXERCISES

1. In the rotation of rectangular axes in three dimensions the equations

$$x'_1 = a_1 x_1 + b_1 x_2 + c_1 x_3$$
$$x'_2 = a_2 x_1 + b_2 x_2 + c_2 x_3$$
$$x'_3 = a_3 x_1 + b_3 x_2 + c_3 x_3$$

determine the coordinates (x'_1, x'_2, x'_3) of a given point P, with respect to the axes $0x'_1, 0x'_2, 0x'_3$, in terms of the coordinates (x_1, x_2, x_3) of P with respect to axes $0x, 0y, 0z$, which have the same disposition and the same origin. For $i = 1,2,3$, (a_i, b_i, c_i) are the direction cosines of $0x'_i$ with respect to $0x, 0y, 0z$ respectively.

Prove that

$$\begin{bmatrix} a_1 & b_1 & c_1 \\ a_2 & b_2 & c_2 \\ a_3 & b_3 & c_3 \end{bmatrix}$$

is an orthogonal matrix and find its inverse. The equations therefore represent an orthogonal transformation.

2. $V_3(R)$ is the real vector space with the standard inner product. Given the vector $\alpha_1 = (-3, 0, 4)$, construct an orthogonal matrix with the components of the unit vector $\alpha_1/|\alpha_1|$ as its first row.

6-5 DUAL SPACES

The notion of a dual space is one of very great importance in the theory of metric spaces in analysis. (As we have seen, an inner-product vector space is made into a metric space by defining the distance d between two vectors as $d(\alpha, \beta) = |\alpha - \beta|$.) We mention here merely the example of Hilbert space, which is an infinite-dimensional inner-product vector space in which every Cauchy sequence converges to a limit in the space. Hilbert space is self-adjoint—that is, it is identical with its dual. This self-duality is one of the most important and distinctive features of this metric space.

Hilbert space is a basic concept in analysis and has widespread applications.

However, the definition of a dual space does not involve the use of an inner product.

Definition. If V is a vector space over a field F, then a linear mapping f of $V \to F$ is called a **linear form** or **linear functional** on V.

This means $(\alpha + \beta)f = \alpha f + \beta f$, for all $\alpha, \beta \in V$, and $(x\alpha)f = x(\alpha f)$ for all $\alpha \in V$ and $x \in F$. The values of a linear form are scalars.

As usual we can introduce an addition $f + g$ of linear forms on V by defining $f + g$ as

$$\alpha(f + g) = \alpha f + \alpha g, \text{ for all } \alpha \in V.$$

Also a scalar multiplication xf, $x \in F$, is defined by $\alpha(xf) = x(\alpha f)$, $\alpha \in V$.

The set of all linear forms on V is the vector space Hom (V, F). However it is customary to denote the vector space of linear forms on V by V^*. V^* **is a vector space over** F.

Definition. V^* is called the **dual space** of V.

From now on we shall deal exclusively with the dual of a finite-dimensional vector space.

A field F is clearly a one-dimensional vector space over itself. If dim $V = n$ then, as we have seen earlier, dim $V^* =$ dim Hom $(V, F) = n \cdot 1 = n$.

THEOREM 9

Let V be an n-dimensional vector space with a basis $\alpha_1, \alpha_2, \ldots, \alpha_n$. If $c_1, c_2, \ldots, c_n$ are any n scalars, then there is a unique linear form f for which $\alpha_i f = c_i$, $i = 1, 2, \ldots, n$. This form f is given by

(3) $\quad (x_1\alpha_1 + \cdots + x_n\alpha_n)f = x_1 c_1 + \cdots + x_n c_n.$

Proof: Let f be the mapping of $V \to F$ defined by (3), where $\alpha = x_1\alpha_1 + \cdots + x_n\alpha_n$ is an arbitrary vector of V. If $\beta = y_1\alpha_1 + \cdots + y_n\alpha_n$, then

$$(\alpha + \beta)f = (x_1 + y_1)c_1 + \cdots + (x_n + y_n)c_n$$
$$= \alpha f + \beta f.$$

Also if $y \in F$, then

$$(y\alpha)f = y(x_1 c_1 + \cdots + x_n c_n) = y(\alpha f).$$

Hence $f \in V^*$. It is unique, since a linear transformation f of $V \to F$

is uniquely determined by its values on the basis vectors. For
$$(x_1\alpha_1 + \cdots + x_n\alpha_n)f = x_1(\alpha_1 f) + \cdots + x_n(\alpha_n f).$$

THEOREM 10

Let $\alpha_1, \alpha_2, \ldots, \alpha_n$ be a basis of the vector space V over F. The unique linear forms defined by
$$\alpha_j f_i = 0, \quad j \neq i$$
$$= 1, \quad j = i,$$
$i = 1, 2, \ldots, n$, form a basis for V^*.

Proof: Let $\alpha = x_1\alpha_1 + \cdots + x_n\alpha_n$, $x_i \in F$, and let $f \in V^*$. Then
$$\alpha f = x_1(\alpha_1 f) + \cdots + x_n(\alpha_n f)$$
and
$$\alpha f_i = x_i, \; i = 1, 2, \ldots, n.$$
Hence
$$\alpha f = (\alpha f_1)(\alpha_1 f) + \cdots + (\alpha f_n)(\alpha_n f)$$
$$= \alpha[(\alpha_1 f)f_1 + \cdots + (\alpha_n f)f_n]$$
for every $\alpha \in V$. Therefore
$$f = (\alpha_1 f)f_1 + \cdots + (\alpha_n f)f_n.$$
The $\alpha_i f$ are scalars and so we have shown that every linear form is a linear combination of the f_i, $i = 1, 2, \ldots, n$. Hence $f_1, f_2, \ldots, f_n$ span V^*.

Moreover, the f_i are independent. Assume $x_1 f_1 + \cdots + x_n f_n = 0$, where the $x_i \in F$ and 0 denotes the zero linear form (the one that maps every vector into zero). For each of the basis vectors α_i of V we therefore have
$$\alpha_i(x_1 f_1 + \cdots + x_n f_n) = 0, \; i = 1, 2, \ldots, n.$$
$$x_1(\alpha_i f_1) + \cdots + x_n(\alpha_i f_n) = 0.$$
Since $\alpha_i f_j = 0$, $j \neq i$, this proves $x_i = 0$, $i = 1, 2, \ldots, n$. Therefore the f_i are independent and form a basis for V^*.

Definition. The basis $f_1, \ldots, f_n$ of V^* defined in the last theorem is called the **dual basis** to the basis $\alpha_1, \ldots, \alpha_n$ of V.

Corollary. If dim $V = n$, then dim $V^* = n$.

Let $\alpha_1, \alpha_2, \ldots, \alpha_n$ be a basis of V and let $f_1, f_2, \ldots, f_n$ be the dual basis of V^*. Then if $f \in V^*$,

(4) $\quad f = (\alpha_1 f)f_1 + (\alpha_2 f)f_2 + \cdots + (\alpha_n f)f_n.$

Remember this form (4). It will be useful to us later.

Inner-Product Vector Spaces and Dual Spaces

THEOREM 11

Let $\alpha_1, \ldots, \alpha_n$ be a basis of V and let $f_1, \ldots, f_n$ be the dual basis of V^*. The mapping T of $V \to V^*$ defined by

$$(x_1 \alpha_1 + \cdots + x_n \alpha_n)T = x_1 f_1 + \cdots + x_n f_n$$

is an isomorphism of $V \to V^*$.

Proof: The proof is virtually immediate and is left to the reader.

THEOREM 12

Let V be a finite-dimensional inner-product vector space over the real field R. Then for every linear form $f \in V^*$, there exists a unique $\alpha \in V$ such that $\gamma f = \alpha \cdot \gamma$ for all $\gamma \in V$.

Proof: f is a linear mapping of $V \to R$ and so im f is a subspace of R. The only subspaces of R are 0 and R. If im $f = 0$, we take the vector α to be $\alpha = \overline{0}_V$, and clearly this α is unique for this case, where f is the zero linear form.

Suppose then im $f = R$. If dim $V = n$, then dim (ker f) $= n - 1$. This follows from Theorem 6, Chapter 3. Let $\alpha_1, \ldots, \alpha_{n-1}$ be an orthonormal basis for ker f and extend it to an orthonormal basis for V. (Theorem 3, Corollary 2.)

If $\gamma \in V$, then

$$\gamma = \sum_{i=1}^{n} (\gamma \cdot \alpha_i) \alpha_i$$

and hence

$$\gamma f = \sum_{i=1}^{n} (\gamma \cdot \alpha_i)(\alpha_i f).$$

For $i = 1, 2, \ldots, n - 1$, $\alpha_i \in$ ker f, and therefore $\gamma f = (\gamma \cdot \alpha_n)(\alpha_n f) = ((\alpha_n f) \alpha_n) \cdot \gamma$, since $\alpha_n f$ is a scalar. Hence $\alpha = (\alpha_n f) \alpha_n$ is the promised vector for which $\gamma f = \alpha \cdot \gamma$, for all $\gamma \in V$.

This α is unique. For if $\alpha \cdot \gamma = \beta \cdot \gamma$ for all $\gamma \in V$, then $(\alpha - \beta) \cdot \gamma = 0$ for all $\gamma \in V$. If we put $\gamma = \alpha - \beta$ in this equation, it follows at once that $\alpha - \beta = \overline{0}_V$ and $\beta = \alpha$.

The last theorem provides a specific isomorphism of $V^* \to V$.

Corollary. If V is a finite-dimensional inner-product vector space, then the mapping $f \to \alpha$, where $f \in V^*$, $\alpha \in V$, defined by $\gamma f = \alpha \cdot \gamma$, for all $\gamma \in V$, is an isomorphism of $V^* \to V$.

Proof: The details are left as an exercise (Exercise 2).

Thus we see again that if V is a finite-dimensional vector space, then dim $V^* =$ dim V.

The obvious difference between a finite-dimensional vector space and an infinite-dimensional vector space is that no finite set of vectors spans the latter. However, there is another basic difference in their theories, for if V is an infinite-dimensional vector space then it can be proved that $\dim V^* > \dim V$.

If V is a finite-dimensional inner-product vector space, let us next see how we can define an inner product in V^* that is induced by the inner product in V.

The choice of an inner product on V determines a linear transformation T of $V \to V^*$ defined by $\alpha T = f$, where $\beta f = \beta \cdot \alpha$ for all $\beta \in V$ and where $f \in V^*$. By the above corollary it follows that T is an isomorphism. Conversely, a linear transformation T of $V \to V^*$ determines an inner product on V defined by

$$\beta \cdot \alpha = \beta f, \text{ where } f = \alpha T.$$

Now if V is a finite-dimensional inner-product vector space, let T be the isomorphism determined by this inner product. For $f, g \in V^*$, define an inner product on V^* by

$$f \cdot g = (fT^{-1})g = (fT^{-1}) \cdot (gT^{-1}).$$

It can be easily verified that this is actually an inner product. It is called the inner product on V^* induced by the inner product on V.

The Bidual of V

The linear forms on V^* form again a vector space over F denoted by V^{**} and called the **bidual** of V. These linear forms are linear mappings of $V^* \to F$. If $\dim V = n$, then $\dim V^* = n$, and hence $\dim V^{**} = n$.

Assume $\dim V = n$ and let $\alpha_1, \alpha_2, \ldots, \alpha_n$ be a basis for V. Consider the mapping T of $V \to V^{**}$ defined by $\alpha T = \phi_\alpha$, $\alpha \in V$, where ϕ_α is the linear form on V^* given by

$$f\phi_\alpha = \alpha f, \text{ for all } f \in V^*.$$

If $\phi_\alpha = \phi_\beta$, then $f\phi_\alpha = f\phi_\beta$ for all $f \in V^*$. Hence $\alpha f = \beta f$ and therefore $(\alpha - \beta)f = 0$ for all $f \in V^*$. Now $\alpha - \beta = x_1\alpha_1 + \cdots + x_n\alpha_n$ and so $(\alpha - \beta)f_i = x_i$. But $(\alpha - \beta)f_i = 0$ for all $i = 1, 2, \ldots, n$. Hence $x_i = 0$, $i = 1, 2, \ldots, n$, and thus $\alpha = \beta$. This proves T is injective. Since $\dim V = \dim V^{**}$, T must be an isomorphism and hence every linear form on V^* can be expressed in the form ϕ_α for some $\alpha \in V$.

If $\alpha_1, \ldots, \alpha_n$ is a basis for V, then it can be proved easily that $\phi_{\alpha_1}, \ldots, \phi_{\alpha_n}$ is a basis for V^{**}. (See Exercise 3.)

We observe that the isomorphism T of $V \to V^{**}$ is *not* dependent on a choice of bases for V and V^{**}. Such an isomorphism is called a **canonical isomorphism**.

Inner-Product Vector Spaces and Dual Spaces

The Transpose of a Linear Mapping

Let V and W be two vector spaces over the same field F, and let T be a linear transformation of $V \to W$. If $g \in W^*$ then the composite mapping $T \circ g \in V^*$, for $V \xrightarrow{T} W \xrightarrow{g} F$, exists.

Using this composite mapping we can see that T induces a linear transformation T' of $W^* \to V^*$ defined by

$$(5) \qquad g T' = T \circ g, \; g \in W^*.$$

The mapping T' is called the **transpose** of T. It is very simple to prove that T' is actually a linear mapping; for $g, g' \in W^*$, $x \in F$, we have

$$(g + g')T' = T \circ (g + g') = T \circ g + T \circ g'$$
$$= gT' + g'T'.$$
$$(xg)T' = T \circ (xg) = x(T \circ g) = x(gT').$$

Note that the left-hand side of (5) is *not* a composite mapping, it is the functional notation.

Now suppose V and W are finite-dimensional vector spaces. Let dim $V = m$ and let $\alpha_1, \alpha_2, \ldots, \alpha_m$ be a basis of V. Denote by $f_1, f_2, \ldots, f_m$ the dual basis of V^*. Let dim $W = n$ and let $\beta_1, \beta_2, \ldots, \beta_n$ be a basis of W. Denote by $g_1, g_2, \ldots, g_m$ the dual basis of W^*.

Let T be a linear transformation of $V \to W$ and let T be represented by the $m \times n$ matrix (a_{ij}) relative to the α-basis of V and the β-basis of W. We wish to determine the matrix of the transpose linear transformation T' of $W^* \to V^*$ with respect to the dual bases of V^* and W^*.

From (5) we have

$$g_i T' = T \circ g_i, \; i = 1, 2, \ldots, n.$$

Now make use of (4) and we obtain

$$g_i T' = (\alpha_1 T \circ g_i) f_1 + (\alpha_2 T \circ g_i) f_2 + \cdots + (\alpha_m T \circ g_i) f_m$$

where

$$\alpha_k T \circ g_i = \left(\sum_{j=1}^{n} a_{kj} \beta_j \right) g_i$$
$$= \sum_{j=1}^{n} a_{kj} (\beta_j g_i)$$

Since the g_i form the dual basis to the β_i, $\beta_j g_i = 0$, if $j \neq i$, and $\beta_i g_i = 1$. Hence $\alpha_k T g_i = a_{ki}$, and therefore

$$g_i T' = a_{1i} f_1 + a_{2i} f_2 + \cdots + a_{mi} f_m, \; i = 1, 2, \ldots, n.$$

Thus the matrix of T' with respect to the dual bases is

$$\begin{bmatrix} a_{11} & a_{21} & \cdots & a_{m1} \\ a_{12} & a_{22} & \cdots & a_{m2} \\ \cdots\cdots\cdots\cdots\cdots \\ a_{1n} & a_{2n} & \cdots & a_{mn} \end{bmatrix}$$

This is an $n \times m$ matrix that is the transpose of the $m \times n$ matrix (a_{ij}) of T. This serves to explain the origin of the name "transpose linear transformation" for T'.

The following two illustrative examples are intended to help the reader to better comprehend the relationship between a finite-dimensional vector space and its dual space. However we also make use of them to explain the meanings of the terms: well-defined mapping and canonical isomorphism. The second example is involved with quotient spaces and constitutes a review of this very important idea. The reader is referred to Chapter 4 for the introduction to a quotient space.

Example 6. If U is a subspace of a finite-dimensional vector space V, then there exists a subspace of V^* that is isomorphic to U^*.

Let $\alpha_1, \ldots, \alpha_k$ be a basis of U. Extend this basis to a basis $\alpha_1, \ldots, \alpha_k, \alpha_{k+1}, \ldots, \alpha_n$ of V. Let W denote the set of all $f \in V^*$ such that $\alpha_i f$, $i = 1, 2, \ldots, k$, are arbitrary, while all $\alpha_i f = 0$ for $i = k + 1, \ldots, n$. It is virtually obvious that W is a subspace of V^*. If $f \in U^*$, the mapping of $U^* \to W$ given by $f \to f'$, where $\alpha_i f' = \alpha_i f$, $i = 1, 2, \ldots, k$ and $\alpha_i f' = 0$, $i = k + 1, \ldots, n$, is an isomorphism. (The proof is easy and direct and is left to the reader.)

This isomorphism of $U^* \to W$ is not actually dependent on the choice of basis for U, for if some other basis is chosen, its vectors are linear combinations of the α_i. Hence the subspace W would remain the same subspace of V^*. For this reason this isomorphism of $U^* \to W$ is said to be **canonical**. It is customary and convenient to identify W with U^* and write $W = U^*$, when the two spaces are canonically isomorphic.

Example 7. Again suppose U is a subspace of a finite-dimensional vector space V. Following the convention above we shall regard U^* as a subspace of V^*.

We are going to prove that the following quotient spaces are isomorphic

$$V^*/U^* \approx (V/U)^*.$$

Let j be the canonical epimorphism of $V \to V/U$ given by $\alpha j = \alpha + U$, $\alpha \in V$. Let $\phi \in (V/U)^*$. Then $V \xrightarrow{j} V/U \xrightarrow{\phi} F$ shows that the composite linear mapping $j \circ \phi \in V^*$. Let us define a mapping T of $V^*/U^* \to (V/U)^*$ by $(f + U^*)T = \phi$, where $f \in V^*$ and ϕ is defined by $f = j\phi$.

Now two elements $f + U^*$ and $g + U^*$ can be equal, $f + U^* = g + U^*$, without $f = g$. In fact they are equal, as we know, if and only if $f - g \in U^*$. Hence it is necessary to show that if $f + U^* = g + U^*$, then T maps them both into the same element of $(V/U)^*$. Let $(f + U^*)T = \phi$ and $(g + U^*)T = \psi$ and suppose $f + U^* = g + U^*$. Now $f - g = j\phi - j\psi \in U^*$. We consider the two cases:

1. If $\alpha_i \in U$, $\alpha_i j = \alpha_i + U = U$. Hence
$\alpha_i(j\phi - j\psi) = \alpha_i j(\phi - \psi) = U(\phi - \psi) = 0$, since U is the zero vector of V/U.
2. If $\alpha_i \notin U$, then $\alpha_i j(\phi - \psi) = 0$, since $j(\phi - \psi) \in U^*$.

Hence $\alpha j(\phi - \psi) = 0$ for all $\alpha \in V$. Therefore $\phi - \psi = 0$ (the zero linear form on V/U), and so $\phi = \psi$. All of this proves that the mapping T is **well-defined**.

The rest of the proof is now concerned with showing that T is an isomorphism. By applying the two tests for linearity, the reader can easily show that T is a linear transformation. Moreover, T is surjective, since for every $\phi \in (V/U)^*$, $f = j\phi \in V^*$ and therefore $(f + U^*)T = \phi$. Also T is injective, since $\phi = 0$ implies $f = 0$ and therefore $f \in U^*$. Thus ker T contains only the zero vector U^* of V^*/U^* and so T is injective (Theorem 3, Chapter 3). Thus T is an isomorphism.

The isomorphism T is not dependent on any choice of bases and hence it is a canonical isomorphism. We therefore write

$$V^*/U^* = (V/U)^*.$$

EXERCISES

1. Let P_3 be the vector space of all polynomials over the real field R (that is, with real coefficients) of degree ≤ 3. Define a mapping f of $P_3 \to R$ by

$$(a_0 + a_1 x + a_2 x^2 + a_3 x^3)f = a_0 + a_1 + a_2 + a_3.$$

(a) Prove $f \in P_3^*$, the dual space of P_3.
(b) Find the dimension of P_3^*.
(c) Find the basis of P_3^* that is the dual basis to the basis $1, x, x^2, x^3$ of P_3.
(d) Find an isomorphism of $P_3 \to P_3^*$.
(e) Express the linear form f of part (a) in terms of this dual basis.

2. Prove that the mapping $f \to \alpha$, defined in the corollary to Theorem 13 is an isomorphism of $V^* \to V$.

3. Prove that if $\alpha_1, \ldots, \alpha_n$ is a basis of V then $\phi_{\alpha_1}, \ldots, \phi_{\alpha_n}$ is a basis for V^{**}. The ϕ_α are the linear forms on V^* defined in the paragraph above on the bidual of V.

4. U and W are subspaces of a finite-dimensional vector space V and $V = U \oplus W$. Use the result in the first illustrative example to prove

(a) $V^* = U^* \oplus W^*$. (This is a canonical isomorphism.)

(b) If V and V^* have inner products defined in terms of bases by the usual formula, then $(U^\perp)^* = (U^*)^\perp$. (This is a canonical isomorphism.)

5. If V is a finite-dimensional vector space over F, prove that the mapping ψ of $V \times V^* \to F$ defined by $(\alpha, f)\psi = \alpha f, \alpha \in V, f \in V^*$, is bilinear.

6. Let S be a nonempty subset of a finite-dimensional vector space V, and denote by $S_\perp$ the set of all $f \in V^*$ such that $\alpha f = 0$, for every $\alpha \in S$. Prove that $S_\perp$ is a subspace of V^*.

7. U and W are subspaces of a finite-dimensional vector space V and $V = U \oplus W$. Let $\pi_1: V \to U$ and $\pi_2: V \to W$ be the canonical projections. For $f \in V^*$ prove that $f = f_1 + f_2$, where $\alpha f_1 = \alpha \pi_2 f$ and $\alpha f_2 = \alpha \pi_1 f$, $\alpha \in V$. Then show that this proves $V^* = U_\perp \oplus W_\perp$.

8. If T is a linear transformation $V \to W$ of two vector spaces, prove that im T is finite-dimensional if and only if $V/\ker T$ is finite-dimensional. (If U is a subspace of V, the dimension of V/U is called the **codimension** of U.) [Hint: consider the mapping of $V/\ker T \to \text{im } T$ defined by $\alpha + \ker T \to \alpha T$, $\alpha \in V$.]

9. V is a vector space over the field F. A mapping f of $V \times V \to F$ is said to be **bilinear** if $(\alpha, \beta) \to (\alpha, \beta)f$ is linear in α and linear in β, where α and β are arbitrary vectors of V.

(a) Write out the equations that express this.

(b) Defining the sum of two bilinear functions in the usual way, and defining a "scalar product" xf, $x \in F$, by $(\alpha, \beta)(xf) = (x\alpha, \beta)f = (\alpha, x\beta)f$, prove the set of all bilinear functions of $V \times V \to F$ is a vector space over F.

(c) If V is an n-dimensional vector space, what is the dimension of this vector space of bilinear functions?

10. Define a **trilinear function** of $V \times V \times V \to F$. Show, in the same way, that the trilinear functions form a vector space over F. This generalizes to **multilinear** functions defined on an r-fold cartesian product $V \times V \times \cdots \times V \to F$.

11. Let P_4 be the vector space of all polynomials in x of degree ≤ 4 with real coefficients. Prove that an inner product is defined in P_4 by

$$\int_{-1}^{1} f(x)g(x)\,dx, \quad f(x), \ g(x) \in P_4.$$

12. Find a basis for P_4 and also an orthonormal basis for P_4.

13. Let S be a subset of a finite-dimensional vector space V. The annihilator S^0 of S is defined to be the subset of all $f \in V^*$ such that

$$\alpha f = 0, \quad \text{for all} \quad \alpha \in S.$$

If U is a subspace of V^*, prove

(a) U^0 is a subspace of V^*

(b) $\dim U + \dim U^0 = \dim V = \dim V^*$

(c) $(U + W)^0 = U^0 \cap W^0$ and $(U \cap W)^0 = U^0 + W^0$, where W is also a subspace of V.

Chapter **7**

Matrices and Determinants

7-1 PERMUTATIONS

In defining a determinant, we shall need to make use of some elementary properties of a permutation group on a finite set. It is the purpose of this section to provide a brief introduction to a group of permutations. The group concept, as we have pointed out, is a very important one in mathematics.

However, since we shall not be developing this theory but only applying it once (to determinants), the reader actually needs only to understand the statements of results and can with impunity omit all the proofs.

Let S be a finite set of n elements. A **permutation** of S is a bijective mapping of S into itself. Let f be a permutation of S. If we designate the elements of S by $1, 2, \ldots, n$, then $(1)f, (2)f, \ldots, (n)f$ will be the same symbols, but not necessarily in the same order. The effect of a permutation is, in general, to rearrange the symbols $1, 2, \ldots, n$ in some different order.

A permutation f on this same set S is usually written as

$$f = \begin{bmatrix} 1 & 2 & 3 & \cdots\cdots & n \\ j_1 & j_2 & j_3 & \cdots\cdots & j_n \end{bmatrix}$$

where $j_1, \ldots, j_n$ are the n symbols $1, 2, \ldots, n$ in some arrangement. In this notation, $j_i = (i)f$, $i = 1, 2, \ldots, n$, and j_i is the symbol into which f maps the symbol i.

It is easy to determine the number of distinct permutations of a set of n elements. Let f be a permutation. Starting with 1, there are n distinct elements into which f can map 1. For each choice of $(1)f$ there are $n-1$ elements into which f can map the symbol 2, and for each choice of $(1)f$ and $(2)f$ there are $n-2$ elements into which f can map the symbol 3. Continuing in this way, it is not hard to see that there are $n(n-1)(n-2)\cdots 3 \cdot 2 \cdot 1 = n!$ (factorial n) different permutations on a set of n elements.

For example,

$$\begin{pmatrix} 1 & 2 & 3 \\ 1 & 2 & 3 \end{pmatrix}, \begin{pmatrix} 1 & 2 & 3 \\ 1 & 3 & 2 \end{pmatrix}, \begin{pmatrix} 1 & 2 & 3 \\ 2 & 1 & 3 \end{pmatrix}, \begin{pmatrix} 1 & 2 & 3 \\ 2 & 3 & 1 \end{pmatrix}, \begin{pmatrix} 1 & 2 & 3 \\ 3 & 1 & 2 \end{pmatrix}, \begin{pmatrix} 1 & 2 & 3 \\ 3 & 2 & 1 \end{pmatrix}$$

are the six distinct permutations on a set of three elements.

The product $f \circ g$ of two permutations f and g of a set S is defined by map composition, that is, by $(s)fg = (sf)g$, $s \in S$. As we have seen in Chapter 1, map composition is always an associative binary operation. The identity permutation I is the permutation for which $(s)I = s$, for all $s \in S$. It is the neutral element of this binary operation. Since each permutation f is a bijective mapping of S into itself, we know that it has an inverse f^{-1}, and since this inverse is also bijective it is a permutation. All of this expresses that fact that the set of permutations of the elements of a set S is a group. In particular, if S is a set of n elements the group of permutations on S is called the **symmetric group of degree** n. We denote this group by S_n. It contains $n!$ distinct permutations. Since in general if f and g are permutations, $f \circ g \neq g \circ f$, the group of permutations on a set is not commutative.

Example 1. Let $f = \begin{pmatrix} 1 & 2 & 3 & 4 & 5 \\ 5 & 3 & 2 & 4 & 1 \end{pmatrix}$ and $g = \begin{pmatrix} 1 & 2 & 3 & 4 & 5 \\ 4 & 5 & 2 & 1 & 3 \end{pmatrix}$ be permutations in the group S_5. Then $f \circ g = \begin{pmatrix} 1 & 2 & 3 & 4 & 5 \\ 3 & 2 & 5 & 1 & 4 \end{pmatrix}$ and $g \circ f = \begin{pmatrix} 1 & 2 & 3 & 4 & 5 \\ 4 & 1 & 3 & 5 & 2 \end{pmatrix}$. To find $f \circ g$ we start with f. Now f maps 1 into 5. Next, go to g and we see that g maps 5 into 3. Hence the product $f \circ g = fg$ maps 1 into 3. Next we see that f maps 2 into 3 and g maps 3 into 2. Hence fg maps 2 into 2. We continue in this way through all five symbols. Similarly we can compute the product gf. We note that $fg \neq gf$.

A **cycle** on m elements of a set S is a permutation of S that permutes these m elements among themselves while leaving all other elements of S unchanged.

For example, the permutation $\begin{pmatrix} 1 & 2 & 3 & 4 & 5 \\ 3 & 2 & 5 & 1 & 4 \end{pmatrix}$ in S_5 is a cycle on the four elements 1, 3, 4, 5 of the set S of the five elements 1, 2, 3, 4, 5. We note that we can write this cycle in the form (1354), where this notation signifies that 1 maps into 3, 3 into 5, 5 into 4, and 4 into 1. The absence of 2 indicates that 2 is mapped into itself. A cycle is known as a *circular permutation* and we can picture this cycle as shown on next page.

The permutations f and g in S_5, in the example above, are not cycles but we can write them as products of cycles. $f = (15)(23)$ and $g = (14)(253)$. However the products fg and gf are cycles and it is not hard

to see that

$$fg = (15)(23)(14)(253) = (1354).$$

Another example of writing a permutation as the product of cycles is

$$\begin{pmatrix} 1\,2\,3\,4\,5\,6\,7\,8\,9 \\ 9\,7\,2\,6\,5\,1\,3\,8\,4 \end{pmatrix} = (1946)(273).$$

Here the permutation is a product of a four-cycle and a three-cycle, and we note that they are disjoint cycles—that is they have no symbol in common. In the light of the next lemma we see that this is not accidental.

Lemma 1. Every permutation is the unique product of disjoint cycles.

Let

$$f = \begin{pmatrix} 1 & 2 & 3 & \ldots n \\ j_1 & j_2 & j_3 & \ldots j_n \end{pmatrix}.$$

Start with 1. If $j_1 = 1$, skip 1, since it is mapped into itself and pass on to 2. If however $j_1 \neq 1$, start a cycle with $(1\ j_1 \ldots)$ and write after j_1 the symbol into which f maps j_1. If this is 1, then close the cycle and write it as $(1\ j_1)$. If not, the cycle continues. In this way we write $(1\ j_1, \ldots, j_r)$, say, $(j_r)f = 1$. We next carry out this same procedure with the remaining symbols, if any, until we have exhausted them. A little practice will soon convince the reader that this yields a unique product for f in terms of disjoint cycles. We give another example:

$$\begin{pmatrix} 1\,2\,3\,4\,5\,6\,7\,8\,9 \\ 5\,7\,3\,9\,8\,4\,1\,2\,6 \end{pmatrix} = (15827)(496).$$

We omit 3, since it is mapped into itself.

A two-cycle is called a **transposition**, and it is not hard to show that any permutation can be written as a product of transpositions. We first write the permutation as a product of cycles and then express each cycle as a product of transpositions. For example, if $(j_1\ j_2\ j_3\ \ldots j_n)$ is a cycle, then we see that one way of writing it as a product of transpositions is

$$(j_1\,j_2\ldots j_n) = (j_1\,j_2)(j_1\,j_3)\ldots(j_1\,j_n).$$

For example, $(23675) = (23)(26)(27)(25)$. In the product on the right, the first transposition maps $2 \to 3$, while the other transpositions do not act on 3; hence we start a cycle with $(23\ldots)$. The first transposition maps $3 \to 2$ and the second one maps $2 \to 6$, while the last two transpositions do not act on 6. Thus $3 \to 6$, and so the cycle continues with $(236\ldots)$. It is easy to see that we reach in this way the cycle (23675).

Consider the product

$$P = \prod_{i<j} (x_i - x_j), \quad i,j = 1,2,\ldots,n.$$

It is quite easy to see that any transposition of the n symbols $1, 2, \ldots,$ forming the subscripts of x in P will change the sign of P. In fact, a permutation of the n symbols has the effect on P of either leaving P fixed, or of merely changing its sign.

For example, for $n = 4$ consider the effect on

$$P = (x_1 - x_2)(x_1 - x_3)(x_1 - x_4)(x_2 - x_3)(x_2 - x_4)(x_3 - x_4)$$

of the permutation (124). Applying it to P we get

$$(x_2 - x_4)(x_2 - x_3)(x_2 - x_1)(x_4 - x_3)(x_4 - x_1)(x_3 - x_1) = (-1)^4 P = P.$$

Observe that $(124) = (12)(14)$, a product of two transpositions. On the other hand, the permutation $(1243) = (12)(14)(13)$, a product of three transpositions, changes the sign of P. For if (1243) is applied to P we get

$$(x_2 - x_4)(x_2 - x_1)(x_2 - x_3)(x_4 - x_1)(x_4 - x_3)(x_1 - x_3) = (-1)^3 P = -P.$$

A permutation can be expressed as a product of transpositions in many ways. For example, $(1243) = (12)(14)(13)$ and $(1243) = (4312) = (43)(41)(42)$. Moreover the reader will have no difficulty convincing himself that we can also write

$$(1243) = (12)(14)(13)(24)(24).$$

The permutation (1243), however, will always be the product of an odd number of transpositions.

If applying a permutation to P either gives us back P or merely changes the sign of P, it would then follow that any permutation, when written as a product of transpositions, must be always the product of either an even number of transpositions (an **even permutation**) or an odd number of transpositions (an **odd permutation**). This follows since a transposition always changes the sign of P. Moreover, the product of two even permutations is an even permutation, while the product of two odd ones is even and the product of an even and an odd one is an odd permutation.

Matrices and Determinants

The inverse of the permutation $\begin{pmatrix} 1 & 2 & \cdots & n \\ j_1 & j_2 & & j_n \end{pmatrix}$ is the permutation $\begin{pmatrix} j_1 & j_2 & \cdots & j_n \\ 1 & 2 & & n \end{pmatrix}$, since their product is the identity permutation $\begin{pmatrix} 1 & 2 & \cdots n \\ 1 & 2 & \cdots n \end{pmatrix}$. Any transposition is its own inverse. If a permutation f is written $f = t_1 t_2 \ldots t_k$, where the t_i are transpositions, then we see that $f^{-1} = t_k t_{k-1} \ldots t_1$. Moreover, the inverse of a permutation is even or odd according to whether the permutation itself is even or odd. This follows at once from the fact that the product of a permutation and its inverse is the identity permutation, that is, an even permutation.

We conclude, therefore, that the set A_n of even permutations forms a group. Since A_n is a subset of the group S_n, we call A_n a subgroup of S_n. The group A_n is very important in other branches of algebra. It has a name. It is called the **alternating group of degree n**.

EXERCISES

1. Prove that the effect of the transposition (rs), $r < s$, on the product P is to change the signs of $2(s - r) - 1$ of its factors. Since $2(s - r) - 1$ is an odd integer, the effect of the transposition is to change P to $-P$.

2. Write out the 24 permutations on the set $S = \{1, 2, 3, 4\}$. Express them all as products of transpositions and find their inverses.

3. If $\sigma = (42135)$, prove $\sigma^{-1} = (53124)$.

4. Prove $\sigma = \begin{pmatrix} 1 & 2 & 3 & 4 & 5 & 6 & 7 & 8 \\ 3 & 7 & 1 & 4 & 8 & 2 & 6 & 5 \end{pmatrix} = (13)(276)(58)$.
Is σ an even or odd permutation?

5. Write the inverse of $\sigma = \begin{pmatrix} 1 & 2 & 3 & 4 & 5 & 6 & 7 \\ 2 & 7 & 5 & 6 & 3 & 4 & 1 \end{pmatrix}$ as a product of transpositions.

6. Prove that an n-cycle is an even permutation if n is odd and an odd permutation if n is even.

7-2 RANK OF A MATRIX

Let $A = (a_{ij})$ be an $m \times n$ matrix over some field F. The **row space** of A is defined to be the subspace of $V_n(F)$ spanned by the m rows of A regarded as vectors of $V_n(F)$, and the **column space** of A is defined to be the subspace of $V_m(F)$ spanned by the n columns of A regarded as vectors of $V_m(F)$.

Definition. The dimension of the row space of a matrix A is called the **row rank** of A, and the dimension of the column space of A is called the **column rank** of A.

We remind the reader that the transpose matrix A^t of the $m \times n$ matrix $A = (a_{ij})$ is the matrix whose entry in the ith row and jth column is a_{ji}, $i = 1, 2, \ldots, m$, $j = 1, 2, \ldots, n$. A^t is therefore an $n \times m$ matrix obtained from A by writing the columns of A as the rows of A^t.

THEOREM 1

The row rank of A equals the column rank of A.

Proof: Let r be the row rank of A and let c be the column rank of A. Clearly rearranging the rows of A does not change r. Moreover, any linear dependence

$$\sum_{i=1}^{n} x_i a^i = 0$$

among the column vectors of A determines a solution of the system of equations.

(1) $\quad a_{i1} x_1 + \cdots + a_{in} x_n = 0, \quad i = 1, 2, \ldots, n.$

Thus rearranging rows of A merely has the effect of rearranging the equations (1). Hence the column rank c is not changed by any rearrangement of the rows of A.

Let us assume the first r rows of A are linearly independent. Then the matrix

$$A^* = (a_{ij}), \quad i = 1, 2, \ldots, r; \quad j = 1, 2, \ldots, n$$

has the same row rank as A. It also has the same column rank as A, since the system of equations

(2) $\quad a_{i1} x_1 + \cdots + a_{in} x_n = 0, \quad i = 1, 2, \ldots, r$

is equivalent to the system (1). This is to say that every solution of (1) is obviously a solution of (2). Furthermore, if $\alpha_1, \ldots, \alpha_r$ is a basis of the row space of A, then the remaining rows α_i, $i = r+1, \ldots, m$, are linear combinations of $\alpha_1, \ldots, \alpha_r$. Hence any solution of (2) is a solution of (1).

Thus the column rank of A^* is also c. Since the column vectors of A^* belong to $V_r(F)$, an r-dimensional vector space, $c \leq r$.

This same argument applied to the transpose A^t of the matrix A shows that $r \leq c$. Hence $r = c$.

In the light of this result we have

Definition: The **rank of a matrix** A over a field F is the row (or column) rank of A.

Definition. Let T be a linear transformation of a finite-dimensional vector space V into a finite-dimensional vector space W. The **rank of T** is defined as the dimension of the range im T of T.

If we choose bases in V and in W and let A be the matrix of the linear transformations T with respect to these bases, then the rank of T is equal to the rank of the matrix A. For if $\alpha_1, \ldots, \alpha_m$ is a basis for V, then for $\zeta \in V$,

$$\zeta = \sum_{i=1}^{m} x_i \alpha_i \quad \text{and} \quad \zeta T = \sum_{i=1}^{m} x_i T(\alpha_i).$$

Thus any vector in im T is a linear combination of the row vectors of A.

EXERCISES

1. Find the ranks of the two matrices

(a) $\begin{bmatrix} 1 & 2 & -3 & 4 \\ 3 & 0 & -2 & -1 \\ -3 & -6 & 9 & -12 \\ 4 & -4 & 6 & -8 \end{bmatrix}$ (b) $\begin{bmatrix} 2 & 6 & -4 & 14 \\ 3 & 9 & -6 & 21 \\ 1 & 3 & -2 & 7 \end{bmatrix}$

2. Find the rank of the linear transformation T of $V_3(R) \to V_4(R)$ defined by

$$(x, y, z)T = (x + y + z, \ z, \ -y, \ 2x)$$

with respect to the standard bases of both spaces.

7-3 ELEMENTARY MATRICES

In Chapter 5 we explained the isomorphism existing between the ring of endomorphisms of an n-dimensional vector space V over F and the ring of $n \times n$ matrices over F. With respect to a fixed basis of V, each endomorphism (linear operator on V) corresponds to a unique $n \times n$ matrix. The isomorphism means that products of endomorphisms correspond to the products of their matrices, and, if an endomorphism has an inverse (that is, it is an automorphism) then its matrix has an inverse. Since a linear transformation of $V \to V$, where V is a finite-dimensional vector space, is nonsingular if and only if it has an inverse, we are led quite naturally to the following repetitions of two definitions given earlier (Chapter 5).

Definition. An $n \times n$ matrix A over F is called **invertible** or **nonsingular** if there exists a matrix A^{-1} such that $AA^{-1} = A^{-1}A = I$, where

I is the $n \times n$ identity matrix with each entry on the principal diagonal equal to 1 and with all other entries equal to 0.

Definition. The matrix A^{-1} is called the **inverse** of the matrix A.

Since the multiplication of matrices is associative, if a matrix A has an inverse, then the inverse is unique. For suppose A has the two inverses B and C. Then $BAC = (BA)C = IC = C$ and $BAC = B(AC) = BI = B$, where I is the identity matrix. Hence $B = C$.

If we translate Theorem 10 of Chapter 3 into terms of matrices it reads as follows:

Lemma 2. If A and B are $n \times n$ matrices over F, then their product AB is a nonsingular matrix if and only if both A and B are nonsingular matrices. If AB is nonsingular, then its inverse matrix is $B^{-1}A^{-1}$.

Proof: We need only prove the form for the inverse. This follows at once by direct multiplication. For

$$ABB^{-1}A^{-1} = A(BB^{-1})A^{-1} = AIA^{-1} = AA^{-1} = I,$$

and similarly $B^{-1}A^{-1}AB = I$.

Corollary. If $A_1, A_2, \ldots, A_k$ are $n \times n$ matrices over F, then their product is a nonsingular matrix if and only if each factor A_i, i, 1, 2, $\ldots$, k is a nonsingular matrix. Moreover,

$$(A_1 A_2 \ldots A_k)^{-1} = A_k^{-1} A_{k-1}^{-1} \ldots A_2^{-1} A_1^{-1}.$$

Proof: Use induction on k as in Theorem 10 of Chapter 3. The second part of the corollary is proved by direct multiplication.

Definition. The **elementary row operations** on a matrix over a field F are:
1. The interchange of any two rows of the matrix.
2. The multiplication of a row of the matrix by a nonzero element c of the field F.
3. The addition to any row of the matrix of c times the corresponding elements of any other row, where $c \in F$.

These elementary row operations have inverses which are also elementary row operations of the same types. Clearly we have

1'. The inverse of (1) is the same as (1).
2'. The inverse of (2) is obtained by using $1/c = c^{-1}$ in place of c.
3'. The inverse of (3) is obtained by using $-c$ in place of c.

Similar definitions can be given for elementary column operations.

Definition. An $m \times n$ matrix A is said to be **row-equivalent** to an $m \times n$ matrix B if A can be obtained from B by a finite succession of elementary row operations.

Definition. An **elementary $n \times n$ matrix** E is any matrix obtained from the $n \times n$ identity matrix I by a single elementary row operation.

It is very easy to see that the elementary matrix obtained by using the inverse of the elementary row operation used to form E, is the inverse E^{-1} of E, since $E^{-1}E = EE^{-1} = I$. This proves

Lemma 3. An elementary matrix is nonsingular.

THEOREM 2

Two matrices are row-equivalent if and only if they have the same row space.

Proof: This clearly follows from the nature of the three elementary row operations, since they merely produce new row vectors that are linear combinations of the old ones. Moreover, the effect of an elementary row operation is undone by its inverse, and it therefore follows that each set of row vectors is expressible in terms of the other set. This means the two row spaces are the same.

THEOREM 3

An $n \times n$ matrix A has rank n, if and only if it is row-equivalent to the $n \times n$ identity matrix I.

Proof: If A and I are row-equivalent, then they have the same row space. Hence the dimension of the row space of A is n; that is, the rank of A is n.

Conversely, assume the rank of A is n. Then the row space of A is $V_n(F)$, and this is the row space of I. Hence A and I are row-equivalent.

THEOREM 4

An $n \times n$ matrix A over F is nonsingular if and only if its row vectors form a basis for $V_n(F)$.

Proof: This follows at once from the fact that A is the matrix of a linear transformation of $V_n(F) \to V_n(F)$ which carries the standard basis $(1,0,\ldots,0), \ldots, (0,0,\ldots,0,1)$ into the n row vectors of A. This linear transformation, and hence A, is nonsingular if and only if these row vectors form a basis for $V_n(F)$.

Corollary. An $n \times n$ matrix A is nonsingular if and only if it is row equivalent to I.

Lemma 4. Let A be an $n \times n$ matrix over F and let E be an elementary matrix. Then the matrix EA is the matrix obtained from A by performing the same elementary row operation on A as is used on I to produce E.

Proof: This follows at once for each of the three elementary row operations by writing out the matrices and using the rule for forming the product of two matrices.

THEOREM 5

A square matrix A is nonsingular if and only if it is equal to the product of elementary matrices.

Proof: If A is nonsingular, then A^{-1} is nonsingular, and hence A^{-1} is row equivalent to I. By Lemma 4, this means

$$E_1 E_2 \ldots E_k A^{-1} = I,$$

where $E_1, \ldots, E_k$ are elementary matrices. Hence

$$A = E_1 E_2 \ldots E_k.$$

Conversely, let A be the product of elementary matrices, $A = E_1 E_2 \ldots E_k$. Since an elementary matrix is nonsingular, we see that the product $E_k^{-1} \ldots E_2^{-1} E_1^{-1}$ exists and that it is the inverse of A. Hence A is nonsingular.

THEOREM 6

Two matrices A, B are row-equivalent, if and only if $B = PA$, where P is some nonsingular matrix.

Proof: If they are row-equivalent, then by Lemma 4, we have $B = E_1 E_2 \ldots E_n A$, where the E_i are elementary matrices. Since each of them is nonsingular, then $P = E_1 E_2 \ldots E_n$ is nonsingular. For $E_n^{-1} \ldots E_2^{-1} E_1^{-1} = P^{-1}$. Hence $B = PA$, where P is nonsingular.

Conversely, assume $B = PA$, where P is nonsingular. Then, by Theorem 5, $P = E_1 E_2 \ldots E_n$, where the E_i are elementary matrices. Hence $B = E_1 E_2 \ldots E_k A$ and therefore, by Lemma 4, B is row-equivalent to A.

Results analogous to Lemma 4 and Theorems 2–6 hold for elementary column operations. If E' is an $n \times n$ matrix formed by performing an elementary column operation on the $n \times n$ identity matrix and if A is an $n \times n$ matrix, then AE' is the matrix obtained from A by performing the same elementary column operation on A. Moreover, two matrices are column-equivalent if and only if they have the same column space; that is, the same rank. We can show similarly that two matrices A and B are column-equivalent if and only if there exists a nonsingular matrix Q such that $B = AQ$.

It therefore follows that if A and B are two matrices such that $B = PAQ$, where P and Q are nonsingular matrices, then A and B have the same rank. Two matrices A and B for which $B = PAQ$, where P

and Q are nonsingular, are called **equivalent**. Hence equivalent matrices have the same rank.

A matrix B is therefore equivalent to a matrix A if it is possible to derive B from A by performing on A a finite number of elementary row operations and/or a finite number of elementary column operations.

Elementary matrices can be used advantageously to determine the inverse of a nonsingular square matrix A.

By Theorem 3, A is row-equivalent to the identity matrix I, and therefore by Lemma 4,

$$E_r E_{r-1} \ldots E_2 E_1 A = I,$$

where the E_i are elementary matrices.

Multiplying on the right by A^{-1}, we get

$$E_r E_{r-1} \ldots E_2 E_1 I = A^{-1}.$$

This last matrix equation shows that the same elementary row operations performed on I (as were performed on A to obtain I) will yield the inverse of A. This is an efficient way of finding A^{-1}, particularly when the order of A is 4 or more.

If A is singular then a succession of elementary row operations on A will not of course yield I, but will yield a singular matrix that is row-equivalent to A.

Example 2. Consider $A = \begin{bmatrix} 2 & 6 \\ -1 & -2 \end{bmatrix}$.

(a) Multiply its first row by $\frac{1}{2}$: $\begin{bmatrix} 1 & 3 \\ -1 & -2 \end{bmatrix}$

(b) Add the new first row to the second row: $\begin{bmatrix} 1 & 3 \\ 0 & 1 \end{bmatrix}$

(c) Multiply the new second row by 3 and subtract from this new first row: $\begin{bmatrix} 1 & 0 \\ 0 & 1 \end{bmatrix}$

If we perform these same elementary row operations, in the same order, on $\begin{bmatrix} 1 & 0 \\ 0 & 1 \end{bmatrix}$ we get in succession the matrices

$$\begin{bmatrix} \frac{1}{2} & 0 \\ 0 & 1 \end{bmatrix}, \begin{bmatrix} \frac{1}{2} & 0 \\ \frac{1}{2} & 1 \end{bmatrix}, \text{ and } \begin{bmatrix} -1 & -3 \\ \frac{1}{2} & 1 \end{bmatrix}.$$

This last matrix $\begin{bmatrix} -1 & -3 \\ \frac{1}{2} & 1 \end{bmatrix}$ is the inverse of $\begin{bmatrix} 2 & 6 \\ -1 & -2 \end{bmatrix}$.

In terms of the elementary matrices corresponding to these elementary row operations, we can write

$$\begin{bmatrix} 1 & -3 \\ 0 & 1 \end{bmatrix} \begin{bmatrix} 1 & 0 \\ 1 & 1 \end{bmatrix} \begin{bmatrix} 1/2 & 0 \\ 0 & 1 \end{bmatrix} \begin{bmatrix} 2 & 6 \\ -1 & -2 \end{bmatrix} = \begin{bmatrix} 1 & 0 \\ 0 & 1 \end{bmatrix}$$

$$\begin{bmatrix} 1 & -3 \\ 0 & 1 \end{bmatrix} \begin{bmatrix} 1 & 0 \\ 1 & 1 \end{bmatrix} \begin{bmatrix} 1/2 & 0 \\ 0 & 1 \end{bmatrix} \begin{bmatrix} 1 & 0 \\ 0 & 1 \end{bmatrix} = \begin{bmatrix} -1 & -3 \\ 1/2 & 1 \end{bmatrix}.$$

EXERCISES

1. Find the inverses of the matrices

(a) $\begin{bmatrix} 0 & 1 & 2 \\ 3 & 2 & -1 \\ 2 & 4 & 0 \end{bmatrix}$ (b) $\begin{bmatrix} 0 & 0 & -1 \\ 0 & 1 & 0 \\ 1 & 0 & 0 \end{bmatrix}$ (c) $\begin{bmatrix} 2 & 0 & 4 & 0 \\ 1 & 1 & 2 & 0 \\ 0 & -1 & 0 & 3 \\ 0 & 0 & 1 & 2 \end{bmatrix}$

2. Find the inverse of the elementary matrix

$$\begin{bmatrix} 1 & 0 & 0 & 0 \\ 0 & 1 & 0 & -2 \\ 0 & 0 & 1 & 0 \\ 0 & 0 & 0 & 1 \end{bmatrix}$$

and check your result.

3. Prove that an $n \times n$ matrix A over F is nonsingular if and only if its column vectors form a basis for $V_n(F)$.

4. If A is an $n \times n$ matrix and if E is any matrix obtained from the $n \times n$ identity matrix I by a single elementary column operation, describe the matrix AE. (Call E an elementary column matrix.)

5. If A is a nonsingular $n \times n$ matrix, prove there exist elementary column matrices $E_1, E_2, \ldots, E_k$ such that $AE_1E_2\ldots E_k = I$.

6. If A and B are column-equivalent matrices prove there exists a nonsingular matrix P such that $B = AP$.

7-4 THE DETERMINANT

The determinant of a square matrix over a field is an element of the field. It is the most important and useful scalar associated with such a matrix. Being a scalar, a computation is therefore required to determine it. Two very useful purposes of determinants are to be found in Cramer's rule and in the calculation of the inverse of a nonsingular matrix. The determinant is also frequently useful in determining the rank of a matrix, something that is particularly important in the theory of quadratic forms. Unfortunately the more rows and columns there are in the square matrix, the more tedious becomes the computation of its determinant. However, sometimes such a computation is actually unavoidable.

Let F_n denote the set of all $n \times n$ matrices over a field F. (F_n is an algebra!)

Let $A = (a_{ij}) \in F_n$. Writing $a_i = (a_{i1}, a_{i2}, \ldots, a_{in})$ for the elements of the ith row of A, we shall also use the notation $A = (a_1, a_2, \ldots, a_n)$. We let I_n stand for the identity $n \times n$ matrix, that is the matrix whose elements on the principal diagonal are 1 and are 0 everywhere else.

We seek to define a function D from $F_n \to F$ which satisfies the following axioms:

1. $D(A)$ changes sign if two adjacent rows of A are interchanged; that is,

$$D(a_1, a_2, \ldots, a_{i+1}, a_i, \ldots, a_n) = -D(a_1, a_2, \ldots, a_i, a_{i+1}, \ldots, a_n).$$

2. $D(A) = 0$ if two adjacent rows of A are equal.
(This axiom is added to take care of fields of characteristic 2.)

3. D is a linear function of each row; that is, for each i from 1 to n,
    (a) $D(a_1, a_2, \ldots, a_i + b_i, \ldots, a_n)$
       $= D(a_1, \ldots, a_i, \ldots, a_n) + D(a_1, \ldots, b_i, \ldots, a_n)$, the $b_i \in F$.
    (b) $D(a_1, a_2, \ldots, ca_i, \ldots, a_n) = cD(a_1, a_2, \ldots, a_i, \ldots, a_n), c \in F$.

4. $D(I_n) = 1$.

Assuming temporarily the existence of such a function D, we first infer from these axioms some of its properties.

I. If any row of A consists entirely of zeros, then $D(A) = 0$.

Proof: Put $c = 0$ in axiom 3(b).

II. If any two rows of A are equal, then $D(A) = 0$

Proof: We can interchange adjacent rows of A until these two equal rows are made adjacent to each other. We end up with either $D(A)$ or $-D(A)$, since each interchange of adjacent rows changes the sign of $D(A)$, by Axiom 1. In either case $D(A) = 0$ or $-D(A) = 0$, by Axiom 2. Hence in either case, $D(A) = 0$.

III. Any scalar multiple of one row of A may be added to another distinct row of A, without changing the value $D(A)$ of D.

Proof: For $j \neq i$, multiply the jth row of A by a scalar c and add it to the ith row. We get

$D(a_1, \ldots, a_i + ca_j, a_{i+1}, \ldots, a_n)$
$\qquad = D(a_1, \ldots, a_i, \ldots, a_n) + cD(a_1, \ldots, a_j, a_{i+1}, \ldots, a_n).$

The second term on the right-hand side of this equation is zero, by II above, for the matrix $(a_1, \ldots, a_j, a_{i+1}, \ldots, a_n)$ has its ith row equal to its jth row, each is a_j. This proves III.

IV. If any two rows of A are interchanged to form a new matrix A' then $D(A') = -D(A)$.

Proof: Let $j > i$ and let us interchange a_i and a_j. Now the row a_i is moved down into the jth row by $j - i$ interchanges of adjacent rows causing $j - i$ changes of sign in $D(A)$. The row a_j is now the $(j - 1)$s row. Hence by $j - i - 1$ further interchanges of adjacent rows, the row a_j is moved up to become the ith row of the new matrix A'. This causes $j - i - 1$ further changes of sign in $D(A)$, for a total of $2(j - i) - 1$ changes of sign. Since $2(j - i) - 1$ is an odd integer,

$$D(A') = (-1)^{2(j-i)-1} D(A) = -D(A).$$

V. If A' is the matrix obtained from A by any permutation of the rows of A, then $D(A') = \pm D(A)$.

Proof: Let $A' = (a_{j_1}, a_{j_2}, \ldots, a_{j_n})$, where $\begin{pmatrix} 1 & 2 & \ldots & n \\ j_1 & j_2 & \ldots & j_n \end{pmatrix}$ is the given permutation. We now make interchanges of the rows of A' so as to turn the matrix A' into the matrix A. If $j_1 = 1$, then, naturally, we leave the row $a_{j_1} = a_1$ where it is. If $j_1 \neq 1$, then, interchange a_1 and a_{j_1}. This causes a change of sign of $D(A')$. If $j_2 = 2$, we leave $a_{j_2} = a_2$ where i is. If $j_2 \neq 2$, we interchange a_{j_2} and a_2, causing another change of sign of $D(A')$. Continuing in this same way, we finally reach $A = (a_1, a_2, \ldots, a_n)$. The total number of changes of sign is either an even integer (in which case the permutation is said to be even) and $D(A') = D(A)$, or it is an odd integer (and the permutation is then called odd) and $D(A') = -D(A)$.

Denote the rows of the identity matrix I_n by $e_1, e_2, \ldots, e_n$. Thus $e_i = (0, 0, \ldots, 0, 1, 0, \ldots, 0)$, where 1 occupies the ith position. For any scalar c, write

$$ce_i = c(0, \ldots, 0, 1, 0, \ldots, 0) = (0, 0, \ldots, 0, c, 0, \ldots, 0).$$

Thus we can express the ith row $a_i = (a_{i1}, \ldots, a_{in})$ of the matrix A in the form

$$a_i = a_{i1} e_1 + a_{i2} e_2 + \cdots + a_{in} e_n.$$

This holds for all i from 1 to n. Now

$$\begin{aligned} D(a_1, a_2, \ldots, a_n) &= D(a_{11} e_1 + \cdots + a_{1n} e_n, a_2, \ldots, a_n) \\ &= a_{11} D(e_1, a_2, \ldots, a_n) + a_{12} D(e_2, a_2, \ldots, a_n) \\ &\quad + \cdots + a_{1n} D(e_n, a_2, \ldots, a_n), \text{ by Axiom 3.} \\ &= \sum_{j=1}^{n} a_{1j_1} D(e_{j_1}, a_2, \ldots, a_n). \end{aligned}$$

Also
$$D(e_{j_1}, a_2, \ldots, a_n) = D(e_{j_1}, a_{21}e_1 + \cdots + a_{2n}e_n, \ldots, a_n)$$
$$= \sum_{j_2=1}^{n} a_{2j_2} D(e_{j_1}, e_{j_2}, a_3, \ldots, a_n).$$

Hence
$$D(a_1, \ldots, a_n) = \sum_{j_1=1}^{n} a_{1j_1} \sum_{j_2=1}^{n} a_{2j_2} D(e_{j_1}, e_{j_2}, a_3, \ldots, a_n)$$
$$= \sum_{j_1, j_2=1}^{n} a_{1j_1} a_{2j_2} D(e_{j_1}, e_{j_2}, a_3, \ldots, a_n).$$

Continuing in this same way, we eventually get

$$D(a_1, \ldots, a_n) = \sum_{j_1, \ldots, j_n=1}^{n} a_{1j_1} a_{2j_2}, \ldots, a_{nj_n} D(e_{j_1}, e_{j_2}, \ldots, e_{j_n}).$$

Now $\begin{pmatrix} 1 & 2 & \cdots & n \\ j_1 & j_2 & \cdots & j_n \end{pmatrix}$ is a permutation, and if this permutation is even, Then, by V above,

$$D(e_{j_1}, \ldots, e_{j_n}) = D(I_n) = I$$

by Axiom 4; while if it is odd,

$$D(e_{j_1}, \ldots, e_{j_n}) = -D(I_n) = -I$$

by Axiom 4. We therefore write

(3) $\displaystyle D(a_1, \ldots, a_n) = \sum \text{sgn}\begin{pmatrix} 1 & 2 & \cdots & n \\ j_1 & j_2 & \cdots & j_n \end{pmatrix} a_{1j_1} a_{2j_2} \cdots a_{nj_n},$

the summation extending over all $n!$ permutations of $1, 2, \ldots, n$. Half of these, $\dfrac{n!}{2}$, are even permutations and carry the plus sign; the other half are odd and carry the minus sign.

Thus any function D on $F_n \rightarrow F$ that satisfies the four axioms must have the values given by (3) above. It is therefore unique. That such a function exists is proved by showing the right-hand side of (3) defines a function that satisfies the four axioms.

An inspection of the right-hand side of (3) readily shows that each term in the sum contains, as a factor, one and only one element from each row of the matrix A. Hence if a row of A consists entirely of zeros, then every term in this sum is zero. This checks Axiom 1.

Now suppose two rows of A are equal, say $a_p = a_q$, $p < q$. Then $a_{pj_p} = a_{qj_p}$ and $a_{qj_q} = a_{pj_q}$. Hence
$$a_{1j_1} \cdots a_{pj_p} \cdots a_{qj_q} \cdots a_n = a_{1j_1} \cdots a_{pj_q} \cdots a_{qj_p} \cdots a_n.$$

Now each of these is a term in the sum (3) but they have opposite signs. For the permutation $\begin{pmatrix} 1 & 2 & \cdots & p & \cdots & q & \cdots & n \\ j_1 & j_2 & \cdots & j_q & \cdots & j_p & \cdots & n \end{pmatrix}$ is obtained by multiplying the permutation $\begin{pmatrix} 1 & 2 & \cdots & p & \cdots & q & \cdots \\ j_1 & j_2 & \cdots & j_p & \cdots & j_q & \cdots \end{pmatrix}$ by the transposition $\begin{pmatrix} j_p & j_q \\ j_q & j_p \end{pmatrix}$. Hence with their proper signs attached, according to the rule prescribed in (3), this pair of terms cancels. This true for the entire sum; that is, if two rows of A are equal, then the sum contains pairs of equal terms but with opposite signs. Hence the sum is This checks Axiom 2.

The verification of Axioms (3) and (4) is virtually trivial and the details are left to the reader.

The function D in (3) is called the **determinant function** and written **det**. Hence

(4)
$$\det A = \sum \operatorname{sgn} \begin{pmatrix} 1 & 2 & \cdots & n \\ j_1 & j_2 & \cdots & j_n \end{pmatrix} a_{1j_1} \cdots a_{nj_n}$$

and det satisfies the four axioms and has the five properties inferred from them. To say the least, the above formula for det A is not particularly suitable for actual computation.

The notation $|A|$ is frequently used for det A.

7-5 PROPERTIES OF THE DETERMINANT

In this section we derive the basic properties of a determinant and these properties can considerably lessen the labor of computing it.

THEOREM 7
$$\det(A') = \det A.$$

Proof: Using (1) we have

$$\det(A') = \Sigma \operatorname{sgn} \begin{pmatrix} 1\,2\,\cdots n \\ j_1 j_2 \cdots j_n \end{pmatrix} a_{j_1 1} a_{j_2 2} \cdots a_{j_n n}.$$

We can write each term in this sum as

$$a_{j_1 1} \cdots a_{j_n n} = a_{1 k_1} \cdots a_{n k_n}.$$

This corresponds to a permutation $\begin{pmatrix} j_1 \cdots j_n \\ 1 \cdots n \end{pmatrix}$ of the subscripts of $a_{j_1 1}, \ldots a_{j_n n}$.

Clearly
$$\begin{pmatrix} j_1 \ldots j_i \ldots j_n \\ 1 \ldots i \ldots n \end{pmatrix} = \begin{pmatrix} 1 \ldots i \ldots n \\ k_1 \ldots k_i \ldots k_n \end{pmatrix}.$$

Now $\begin{pmatrix} j_1 \ldots j_n \\ 1 \ldots n \end{pmatrix}$ is the inverse of the permutation $\begin{pmatrix} 1 \ldots n \\ j_1 \ldots j_n \end{pmatrix}$ and hence both are even or both are odd. Hence the permutations $\begin{pmatrix} 1 \ldots n \\ j_1 \ldots j_n \end{pmatrix}$ and $\begin{pmatrix} 1 \ldots n \\ k_1 \ldots k_n \end{pmatrix}$ are either both even or both odd, and therefore, in the formula for $\det(A')$, have the same sign attached to them. This proves

$$\det(A') = \Sigma \text{ sgn} \begin{pmatrix} 1 \ldots n \\ k_1 \ldots k_n \end{pmatrix} a_{1k_1} \ldots a_{nk_n}$$
$$= \det A.$$

Corollary. The axioms and the five properties deduced from them for determinants hold true if we replace "row" by "column" in each of them.

It is readily seen that the determinant of an elementary matrix E is -1, c, or 1, according as to whether the elementary row operation (1), (2), or (3) is used on 1.

Hence if A is any $n \times n$ matrix, then clearly $\det(EA)$ is $-\det A$, $c \det A$, or $\det A$ depending on whether E originates from elementary row operation (1) or (2) or (3). In all three cases, however, it follows that $\det(EA) = \det E \cdot \det A$.

Lemma 5. If A is an $n \times n$ matrix and E is an elementary matrix, then

$$\det(EA) = \det E \cdot \det A.$$

Proof: Form E and use the product rule to find EA. For elementary row operation (1) we have $\det E = -1$ and $\det(EA) = -\det A$. Hence $\det(EA) = \det E \cdot \det A$. Similarly for the other two operations.

Lemma 6. If P is a nonsingular $n \times n$ matrix and A is an arbitrary $n \times n$ matrix, then

$$\det(PA) = \det P \cdot \det A.$$

Proof: By Theorem 5, $P = E_1 E_2 \ldots E_n$, where the E_i are elementary matrices. Hence

$$\det(PA) = \det((E_1 E_2 \ldots E_n)A).$$

Since multiplication of matrices is associative

$$\begin{aligned}\det(PA) &= \det E_1(E_2E_3\cdots E_nA)\\ &= \det E_1 \cdot \det E_2 \cdot \:\cdots\: \cdot \det E_n \cdot \det A\\ &= \det P \cdot \det A, \text{ by Lemma 5.}\end{aligned}$$

If a matrix A has a row of zeros then it is singular, (Theorem 4), and from the formula for det A we see that det $A = 0$.

Let A be a singular $n \times n$ matrix. Then the row space of A has dimension less than n (Theorem 4). Hence we can find $(n-1)$ vectors of $V_n(F)$ that will span this row space of A. Using these vectors for $n-1$ of the rows, we can form a matrix B whose nth row is a row of zeros. Clearly the row space of B is the same as the row space of A. Hence A and B are row-equivalent (Theorem 2). Therefore $A = PB$, where P is a nonsingular matrix (Theorem 6). Hence by Lemma 6,

$$\det A = \det P \cdot \det B.$$

Since det $B = 0$ (B has a row of zeros), it follows that det $A = 0$. We have proved

THEOREM 8

If a matrix A is singular, then det $A = 0$.

We next prove the following important property about the determinant of a product.

THEOREM 9

If A and B are $n \times n$ matrices over a field F, then

$$\det(AB) = \det A \cdot \det B.$$

Proof: If A is nonsingular this theorem is merely a restatement of Lemma 6.

If A is singular, then by Lemma 2, AB is singular. Hence det $A = $ det $AB = 0$ (Theorem 8) and the statement of the theorem is still true.

We point out that it now follows from our results that the row vectors of a matrix A are linearly dependent if and only if det $A = 0$. This is equally true of the column vectors of A.

Exercise. Determine the dimensions of the row spaces of the following matrices:

(a) $\begin{bmatrix} 2 & 3 & -4 & 6 & 1 \\ 4 & -11 & 26 & -12 & 13 \\ 5 & -1 & 7 & 3 & 8 \end{bmatrix}$
(b) $\begin{bmatrix} 1 & 2 & 0 & 3 \\ -1 & 7 & -5 & 9 \\ 4 & -1 & 5 & 0 \\ 9 & 0 & 10 & 3 \end{bmatrix}$

Definitions. Let $A = (a_{ij})$ be an $n \times n$ matrix. The $(n-1) \times (n-1)$ matrix that is left after deleting the ith row and jth column of A (in other words, striking out the row and column containing the entry a_{ij}) is called the **minor** of the entry a_{ij}. The determinant of this minor multiplied by $(-1)^{i+j}$ is called the **cofactor** A_{ij} of a_{ij}.

Thus a minor is a matrix, whereas a cofactor is an element of the field F.

Since the cofactors $A_{ij} \in F$, we can form a matrix (A_{ij}). The transpose of this matrix $(A_{ij})' = (A_{ji})$ is called the **adjoint** of the matrix A. We write **adj** A. Thus to form the adjoint of $A = (a_{ij})$ we first replace every entry a_{ij} by its cofactor A_{ij} and then form the transpose of this matrix.

THEOREM 10

$$\det A = a_{i1}A_{i1} + \cdots + a_{in}A_{in}$$
$$= \sum_{j=1}^{n} a_{ij}A_{ij}, \text{ for each } i = 1, 2, \ldots, n.$$

If $i \neq k$, then

$$\sum_{j=1}^{n} a_{ij}A_{kj} = 0.$$

Proof: An examination of the terms in the sum

$$\det A = \Sigma \, \text{sgn} \begin{pmatrix} 1 & 2 & \ldots & n \\ j_1 & j_2 & \ldots & j_n \end{pmatrix} a_{1j_1} a_{2j_2} \ldots a_{nj_n}$$

for $\det A$ reveals that each term has exactly one factor from each row and exactly one factor from each column. Each row (column) occurs once and only once in each term of $\det A$. This means, for instance, that the entire sum can be regarded as a linear combination of the entries from, say, the ith row (column). Collecting successively terms that contain $a_{i1}, \ldots, a_{in}$ we find

(5) $\qquad \det A = a_{i1}A_{i1} + \cdots + a_{in}A_{in}, \quad i = 1, 2, \ldots, n.$

The corresponding formula for columns is

(6) $\qquad \det A = a_{1j}A_{1j} + \cdots + a_{nj}A_{nj}, \quad j = 1, 2, \ldots, n.$

Furthermore, if two rows are equal, say the ith and kth, then replace the ith row by the kth row in (1). Since A has two rows alike, $\det A = 0$, and now (5) becomes

$$\det A = a_{ki}A_{i1} + \cdots + a_{kn}A_{in} = 0, \quad k \neq i.$$

It is easy to see that a similar formula is true using columns in place of rows.

Formula (5) affords a very practical way of computing the determinant of a square matrix.

If
$$A = \begin{bmatrix} a_{11} & a_{12} & \cdots & a_{1n} \\ a_{21} & a_{22} & \cdots & a_{2n} \\ \multicolumn{4}{c}{\cdots\cdots\cdots} \\ a_{n1} & a_{n2} & & a_{nn} \end{bmatrix},$$

then the matrix adj A is the transpose of the matrix

$$\begin{bmatrix} A_{11} & A_{12} & \cdots & A_{1n} \\ A_{21} & A_{22} & & A_{2n} \\ \multicolumn{4}{c}{\cdots\cdots\cdots} \\ A_{n1} & A_{n2} & \cdots & A_{nn} \end{bmatrix}$$

where A_{ij} is the cofactor of a_{ij}. Hence

$$\text{adj } A = \begin{bmatrix} A_{11} & A_{21} & \cdots & A_{n1} \\ A_{12} & A_{22} & \cdots & A_{n2} \\ \multicolumn{4}{c}{\cdots\cdots\cdots} \\ A_{1n} & A_{2n} & \cdots & A_{nn} \end{bmatrix}.$$

It follows from Theorem 10 that

$$(\text{adj } A)A = A(\text{adj } A) = \begin{bmatrix} \det A & 0 & \cdots & 0 \\ 0 & \det A & & 0 \\ \multicolumn{4}{c}{\cdots\cdots\cdots\cdots} \\ 0 & 0 & \cdots & \det A \end{bmatrix},$$

a diagonal matrix.

We observe that a matrix A always commutes with its adjoint.

Example 3. Let

$$A = \begin{bmatrix} 3 & 2 & 0 \\ -1 & 0 & 1 \\ 2 & 4 & 5 \end{bmatrix}.$$

If we replace each entry in A by its cofactor we get the following matrix:

$$B = \begin{bmatrix} -4 & 7 & -4 \\ -10 & 15 & -8 \\ 2 & -3 & 2 \end{bmatrix}.$$

The matrix adj A is therefore the transpose of B:

$$\text{adj } A = \begin{bmatrix} -4 & -10 & 2 \\ 7 & 15 & -3 \\ -4 & -8 & 2 \end{bmatrix}.$$

We note that

$$A(\text{adj } A) = (\text{adj } A)A = \begin{bmatrix} 2 & 0 & 0 \\ 0 & 2 & 0 \\ 0 & 0 & 2 \end{bmatrix},$$

where det $A = 2$.

Example 4. Consider the matrix

$$A = \begin{bmatrix} 2 & -1 & 4 & 1 \\ -5 & 2 & -7 & 2 \\ -16 & 6 & -20 & -8 \\ 7 & -3 & 9 & 3 \end{bmatrix}.$$

If we perform each of the following operations on A:
  (a) multiply row one by 2 and add it to row two,
  (b) multiply row one by 6 and add it to row two,
  (c) multiply row one by -3 and add it to row four,
the resulting matrix is

$$B = \begin{bmatrix} 2 & -1 & 4 & 1 \\ -1 & 0 & 1 & 4 \\ -4 & 0 & 4 & -2 \\ 1 & 0 & -3 & 0 \end{bmatrix}.$$

By (III), Sec. 7-4, we know det $A =$ det B. Now use formula (6) for the second column; that is for $j = 2$, we find

$$\det A = +1 \begin{vmatrix} -1 & 1 & 4 \\ -4 & 4 & -2 \\ 1 & -3 & 0 \end{vmatrix},$$

the remaining terms being zero. If we multiply the first column by 3 and add it to the second column we get

$$\det A = \begin{vmatrix} -1 & -2 & 4 \\ -4 & -8 & -2 \\ 1 & 0 & 0 \end{vmatrix}.$$

Hence

$$\det A = \begin{vmatrix} -2 & 4 \\ -8 & -2 \end{vmatrix} = 4 - (-32) = 36.$$

EXERCISES

1. Show
$$\begin{vmatrix} 2 & 2 & -3 & -2 \\ 2 & 1 & -2 & 5 \\ 1 & 1 & -2 & 2 \\ -1 & -2 & 4 & 3 \end{vmatrix} = 4.$$

2. Prove by elementary-row operations that
$$\begin{vmatrix} x+y & 2v & x-2y & v+w \\ 3v & x-w & y-2v & v-w \\ 6v & 2v-2w & x+v & w \\ x+y & 2x & -5v & 3v-2w \end{vmatrix} = 0.$$

3. (a) Find the adjoint, adj A, of the matrix
$$A = \begin{bmatrix} 1 & -2 & 3 & 4 \\ 2 & 6 & 0 & -5 \\ 0 & -1 & 1 & 0 \\ -1 & -3 & 0 & 2 \end{bmatrix}$$

   (b) Compute $A \cdot (\text{adj } A)$.
   (c) Prove $\det(A \cdot (\text{adj } A)) = (\det A)^4$.

4. Prove
$$\begin{vmatrix} 2 & 6 & 6 & 5 \\ 1 & 1 & 11 & -9 \\ 0 & 0 & -3 & 5 \\ 0 & 0 & 6 & -12 \end{vmatrix} = \begin{vmatrix} 2 & 6 \\ 1 & 1 \end{vmatrix} \cdot \begin{vmatrix} -3 & 5 \\ 6 & -12 \end{vmatrix} = (-4)(6) = -24.$$

5. Prove

$$\begin{vmatrix} 1 & a^2 & a \\ 1 & b^2 & b \\ 1 & c^2 & c \end{vmatrix} = (a - b)(a - c)(b - c).$$

6. Prove the corollary to Theorem 7.

7. Prove that the column vectors of a square matrix A are dependent if and only if $\det A = 0$.

8. If A is an $n \times n$ matrix, prove

$$\det A = \sum_{i=1}^{n} a_{ij} A_{ij},$$

where A_{ij} is the cofactor of a_{ij}.

-6   INVERSE OF A MATRIX

We recall that a square matrix is called nonsingular or invertible if it has an inverse.

THEOREM 11

If A is an $n \times n$ matrix and $\det A \neq 0$ then A has an inverse matrix A^{-1} and

$$A^{-1} = \frac{1}{\det A} \operatorname{adj} A.$$

Moreover, if A has an inverse A^{-1} then $\det A \neq 0$.

Proof: First assume $\det A \neq 0$ and let A^{-1} be the matrix defined by $A^{-1} = \frac{1}{\det A} \operatorname{adj} A$. Let $A = (a_{ij})$. Then $\operatorname{adj} A = (A_{ji})$; that is, the element in the ith row and jth column of $\operatorname{adj} A$ is the cofactor A_{ji} of a_{ji}. Now

$$AA^{-1} = \frac{1}{\det A}(a_{ij})(A_{ji}) = \frac{1}{\det A}\left(\sum_{k=1}^{n} a_{ik} A_{jk}\right).$$

By Theorem 10,

$$\sum_{k=1}^{n} a_{ik} A_{jk} = 0, \quad \text{if} \quad i \neq j$$
$$= \det A, \quad \text{if} \quad i = j.$$

Hence

$$AA^{-1} = \frac{1}{\det A} \begin{bmatrix} \det A & 0 & 0 & \cdots & 0 \\ 0 & \det A & 0 & \cdots & 0 \\ \multicolumn{5}{c}{\dotfill} \\ 0 & 0 & 0 & \cdots & \det A \end{bmatrix} = I_n.$$

Similarly, we can show $A^{-1}A = I_n$. This proves A^{-1} is the inverse of A.

To prove the second part of the theorem, we assume A has an inverse A^{-1}. Then $AA^{-1} = I_n$. Hence by Theorem 9,

$$\det(AA^{-1}) = \det A \cdot \det(A^{-1}) = \det I_n = 1.$$

Therefore $\det A \neq 0$.

Thus A has an inverse A^{-1} if and only if $\det A \neq 0$.

This important result gives us another method of determining the inverse of a nonsingular matrix.

Let T be an endomorphism of a finite-dimensional vector space V over a field F; that is, T is a linear operator on V.

Let A and B be the matrices of T relative to two distinct bases for V. Then $B = PAP^{-1}$, where P is a nonsingular matrix; that is, $\det P \neq 0$. Then

$$\det B = \det(PAP^{-1}) = \det P \cdot \det A \cdot \det(P^{-1})$$
$$= \det P \cdot \det A \cdot (\det P)^{-1} = \det A.$$

In other words, similar matrices have equal determinants. It is customary therefore to call $\det A = \det B$, the **determinant of T, $\det T$**.

THEOREM 12

A linear operator T on a finite-dimensional vector space V is nonsingular (that is, T is an automorphism of V) if and only if $\det T \neq 0$.

Proof: If T is nonsingular, then it has an inverse T^{-1} which is also a linear transformation on V. If A is the matrix of T with respect to a basis for V, then A^{-1} must be the matrix of T^{-1} with respect to this basis since TT^{-1} is the identity linear transformation on V. The existence of the inverse matrix A^{-1} of A implies $\det A = \det T \neq 0$.

Conversely, if $\det T = \det A \neq 0$, then the matrix A has an inverse A^{-1}. The linear transformation that corresponds to the inverse matrix is the inverse T^{-1} of T. Since T has an inverse, it is nonsingular.

Example 5. For the matrix A in Example 3 we found that $\det A = 2$, and so A is an invertible matrix. From the results obtained there we find that

$$A^{-1} = \frac{1}{\det A}(\text{adj } A) = \frac{1}{2}\text{adj } A = \begin{bmatrix} -2 & -5 & 1 \\ \frac{7}{2} & \frac{15}{2} & \frac{-3}{2} \\ -2 & -4 & 1 \end{bmatrix}.$$

EXERCISES

1. A is an $n \times n$ matrix over a field F and $x \in F$. Is $\det(xA) = x \det A$? Find $\det(A + A + A)$. Find $\det(nA)$ where n is any positive integer. Check your answers.

2. A is a nonsingular $n \times n$ matrix over a field F. Prove $\det(\text{adj } A) = (\det A)^{n-1}$. Is this true if A is singular?

3. A^t is the transpose of the $n \times n$ matrix A. Is adj A^t the transpose of adj A? Prove your answer.

4. Prove $\det(\text{adj } A) = \det(\text{adj } A^t)$.

5. If A is a singular $n \times n$ matrix, show $A(\text{adj } A) = 0$.

6. Find the inverse of the matrix

$$\begin{bmatrix} 2 & 0 & 1 & 3 \\ 1 & -1 & 0 & 0 \\ 1 & 0 & 2 & 2 \\ 0 & 1 & -3 & 1 \end{bmatrix}$$

and check your result.

4-7 SYSTEMS OF LINEAR EQUATIONS

Consider a system of m linear equations in n "unknowns" $x_1, x_2, \ldots, x_n$ with coefficients a_{ij}, $i = 1, 2, \ldots, m$, $j = 1, 2, \ldots, n$ in a field F.

(5) $$\sum_{j=i}^{n} a_{ij}x_j = b_i, \quad i = 1, 2, \ldots, m, \quad b_i \in F.$$

The $m \times n$ matrix (a_{ij}) is called the **matrix of the system**. The number m of rows is the number of equations and the number n of columns is the number of "unknowns" x_j in the system.

The mn coefficients a_{ij} and $b_1, \ldots, b_m$ are to be regarded as given elements of F and to solve the system (5) means to determine the unknowns $x_1, \ldots, x_n$ as elements of F, such that the system (5) is satisfied.

Whether such a system has a solution in the field F, or not, can be elegantly characterized in terms of vector spaces and of a linear transformation between them.

Let T be the unique linear transformation of $V_n(F) \to V_m(F)$ that is represented, with respect to the standard bases in both these spaces, by the transpose of the matrix (a_{ij}). Then

$$(x_1, x_2, \ldots, x_n)T = (\Sigma a_{1j}x_j, \Sigma a_{2j}x_j, \ldots, \Sigma a_{mj}x_j),$$

and hence the system (5) is equivalent to the vector equation

(6) $\qquad (x_1, x_2, \ldots, x_n)T = (b_1, b_2, \ldots, b_m).$

Hence if the system (5) has a solution then there exists a vector $(x_1, x_2, \ldots, x_n) \in V_n(F)$ such that (6) is satisfied and conversely if there exists a vector $(x_1, \ldots, x_n) \in V_n(F)$ such that (6) is satisfied then the elements $x_1, x_2, \ldots, x_n$ of F form a solution of (5).

We can express this succinctly by saying that the system (5) has a solution in the field F if and only if the given vector $(b_1, \ldots, b_m)$ of $V_m(F)$ is in the codomain (range) im T of T. Moreover the vector $(x_1, \ldots, x_n)$ of $V_n(F)$ is unique if and only if T is injective. We have proved

THEOREM 13

The system (5) has a solution if and only if the vector $(b_1, \ldots, b_m)$ belongs to im T and the solution is unique if and only if T is injective.

If all the $b_i = 0$, $i = 1, 2, \ldots, m$, in (5) we get what is called a **homogeneous system of equations,**

(7) $\qquad \displaystyle\sum_{j=1}^{n} a_{ij}x_j = 0, \qquad i = 1, 2, \ldots, m.$

Obviously the homogeneous system (7) always has the "trivial" solution $x_1 = 0, x_2 = 0, \ldots, x_n = 0$. Our interest is therefore in whether or not this system has a nontrivial solution, one for which not all the x_i, $i = 1, \ldots, n$, are zero.

For the homogeneous case the equation (6) becomes

(6') $\qquad (x_1, x_2, \ldots, x_n)T = (0, 0, \ldots, 0),$

and a nontrivial solution of (7) therefore exists if and only if there exists a nonzero vector $(x_1, \ldots, x_n)$ of $V_n(F)$ satisfying (6'); that is, if and only if ker T contains a nonzero vector. We have proved

Corollary. The system (7) has a nontrivial solution if and only if T is not injective.

Now let r be the rank of the linear transformation T defined by (6). By definition, $r = \dim(\text{im } T)$. We proved in Chapter 3 that, if V is finite-dimensional vector space, $\dim V = \dim(\ker T) + \dim(\text{im } T)$.

Since dim $V = n$ and dim (im T) = r, it follows that dim (ker T) = $n - r$.

As explained near the beginning of this chapter, the rank r of T is equal to the rank of its matrix (a_{ji}). In fact the rank of (a_{ji}) is, by definition, the dimension of its row space; and the vectors of this row space are precisely the vectors on the right-hand side of (6). We therefore have the theorem.

THEOREM 14

A homogeneous system of m equations in n unknowns, whose coefficient matrix (a_{ij}) has the rank r, has exactly $n - r$ linearly independent solutions. If $m < n$, we have $n - r \geq n - m \geq 1$. Hence the homogeneous system always has nontrivial solutions if $m < n$.

We next use determinants to actually determine the solution of a system of n equations in n unknowns.

Let

(8) $$\sum_{j=1}^{n} a_{ij}x_j = b_i, \quad i = 1, 2, \ldots, n$$

be a system of n linear equations in n unknowns $x_1, x_2, \ldots, x_n$, where the a_{ij} and the b_i are elements of a field F. Let A be the matrix of the coefficients a_{ij}; that is, let $A = (a_{ij})$. Let X and B be the matrices $X = (x_1 x_2 \ldots x_n)$, $B = (b_1 b_2 \ldots b_n)$. Then as a matrix equation (8) has the form

$$\begin{bmatrix} a_{11} & a_{12} & \ldots & a_{1n} \\ a_{21} & a_{22} & \ldots & a_{2n} \\ \ldots & \ldots & \ldots & \ldots \\ \ldots & \ldots & \ldots & \ldots \\ a_{n1} & a_{n2} & \ldots & a_{nn} \end{bmatrix} \begin{bmatrix} x_1 \\ x_2 \\ \cdot \\ \cdot \\ x_n \end{bmatrix} = \begin{bmatrix} b_1 \\ b_2 \\ \cdot \\ \cdot \\ b_n \end{bmatrix};$$

that is,

(9) $$AX^t = B^t.$$

where X^t and B^t are the matrices that are the transposes of the matrices X and B respectively.

If the matrix A has an inverse A^{-1} then multiplying (9) on the left by the matrix A^{-1} we see that (9) has the unique matrix solution

$$X^t = A^{-1} B^t.$$

This matrix equation gives us n equations expressing each of the unknowns $x_1, x_2, \ldots, x_n$ explicitly in terms of the coefficients a_{ij} and $b_1, b_2, \ldots, b_n$.

THEOREM 15 (Cramer's Rule)

The system (8) has a unique solution if det $A \neq 0$ and this solution is given by

(10) $$x_i = \frac{\det B_i}{\det A}, \quad i = 1, 2, \ldots, n,$$

where B_i is the matrix obtained from A by replacing in A the ith column $a_{1i}, a_{2i}, \ldots, a_{ni}$ by the scalars $b_1, b_2, \ldots, b_n$; that is, put $a_{ki} = b_k$, $k = 1, 2, \ldots, n$.

Proof: That (10) is a solution is proved at once by substituting (10) into (8) and verifying that (10) satisfies (8).

Since we assume det $A \neq 0$, the matrix A has an inverse and the uniqueness of (10) follows from the uniqueness of the matrix solution of (9).

The system (8) above can be regarded as a linear transformation T of a vector $X = (x_1, x_2, \ldots, x_n)$ of $V_n(F)$ into the vector $B = (b_1, b_2, \ldots b_n)$ of $V_n(F)$, the matrix of T being A. We can write (8) as

(11) $$XT = B.$$

Here B and T are known and we are searching for X. If T is a nonsingular transformation, that is, det T = det $A \neq 0$, then (11) has a unique solution by Cramer's rule.

Consider now the homogeneous system of equations obtained by taking $b_1 = b_2 = \cdots = b_n = 0$ in (8).

(12) $$\sum_{j=1}^{n} a_{ij} x_j = 0, \quad i = 1, 2, \ldots, n.$$

In terms of a linear transformation, this is

(13) $$TX^t = \overline{0}_V.$$

If T is nonsingular the only solution is $X = \overline{0}_V$. On the other hand if det $A = 0$, that is T is singular, then T is neither injective nor surjective [This follows from the result: dim $V_n(F) = n$ = dim (ker T) + dim (im T).] Hence there exist nonzero vectors X in the kernel of T and these are solutions of (6). We conclude then that if the determinant A of the homogeneous system (12) is zero, then the system has nonzero solutions. We call them **nontrivial solutions**.

Let

$$\sum_{j=1}^{n} a_{ij} x_j = 0, \quad i = 1, 2, \ldots, n$$

be a homogeneous system of n equations in n unknowns. Its matrix is (a_{ij}). This system is equivalent to a linear combination of the column vectors of the matrix (a_{ij}) being equal to zero. If these vectors are linearly independent (form a basis of $V_n(F)$) then the only solution of the system is the trivial one, $x_1 = x_2 = \cdots = x_n = 0$, and conversely. This is Cramer's rule for the homogeneous case. On the other hand, if these vectors are dependent, there exists a nontrivial solution of the system. This means that if the matrix is singular (hence its determinant is zero) then the homogeneous system has, in addition to the trivial solution, a nontrivial solution, and conversely.

Hence the homogeneous system has a nontrivial solution if and only if det $(a_{ij}) = 0$; that is, if and only if the rank of the matrix (a_{ij}) is less than n.

EXERCISES

1. Find the inverses of the following matrices, being sure to check your results.

(a) $\begin{bmatrix} 1 & 0 & 0 & 0 \\ 0 & 0 & 2 & -1 \\ 0 & 0 & -1 & 1 \\ 0 & 1 & 0 & 0 \end{bmatrix}$ (b) $\begin{bmatrix} 2 & 0 & 1 & 3 \\ 1 & -1 & 0 & 0 \\ 1 & 0 & 2 & 2 \\ 0 & 1 & -3 & 1 \end{bmatrix}$

(c) $\begin{bmatrix} 2 & 1 & 0 & 0 & 0 \\ 0 & 0 & 0 & 1 & 0 \\ 0 & 0 & 3 & 0 & 1 \\ 0 & 0 & 2 & 0 & 1 \\ 0 & 1 & 0 & -1 & 0 \end{bmatrix}$.

2. T is a linear operator on $V_3(R)$ defined by

$$(x, y, z)T = (x + 2y + z, x - y, z + 2x).$$

Find (a) det (T), (b) $(x, y, z)T^{-1}$, (c) det (T^{-1}).

3. Are the following systems of equations solvable?
(a) $3x_1 - 2x_2 + 4x_3 = 1$
$4x_1 + 5x_2 - x_3 = 8$
$2x_1 - 9x_2 + 9x_3 = 3$

(b) $5x_1 - 6x_3 = 0$
$2x_1 + 9x_2 = -6$
$4x_1 + 2x_2 - 3x_3 = 5$

(c) $2x_1 + x_2 - 3x_3 = 2$
$5x_1 + x_3 = 1$.

4. Solve the following system of equations for x_4 by use of Cramer's rule.
$$x_1 + x_2 + x_3 + x_4 = 0$$
$$3x_1 + 2x_2 - x_3 + x_5 = -1$$
$$2x_1 + x_2 + 2x_3 + x_4 - 3x_5 = 1$$
$$3x_2 - x_3 + x_4 = 0$$
$$2x + x_3 - 2x_5 = 0$$

5. If A is a real orthogonal matrix prove that $\det A = \pm 1$.

6. Use induction on n to prove that

$$\begin{vmatrix} 1 & x_1 & x_1^2 & \ldots & n_1^{n-1} \\ 1 & x_2 & x_2^2 & \ldots & x_2^{n-1} \\ \vdots & & & & \vdots \\ 1 & x_n & x_n^2 & \ldots & x_n^{n-1} \end{vmatrix} = \begin{array}{l} (x_2 - x_1)(x_3 - x_1)(x_4 - x_1)\ldots(x_n - x_1) \\ (x_3 - x_2)(x_4 - x_2)\ldots(x_n - x_2) \\ (x_4 - x_3)\ldots(x_n - x_3) \\ \vdots \\ (x_n - x_{n-1}) \end{array}$$

(This is called the **Vandermonde determinant.**)

7. The **trace**, tr A, of an $n \times n$ matrix A is defined to be the sum of the entries on the main diagonal of A. If $A = (a_{ij})$, then

$$\text{tr } A = \sum_{i=1}^{n} a_{ii}.$$

Prove the following fundamental properties of the trace function.
If A and B are $n \times n$ matrices over F and if $x \in F$, then
(a) tr $(xA) = x$ tr A
(b) tr $(A + B) = $ tr $A + $ tr B
(c) tr $AB = $ tr BA.

8. If A is a nonsingular $n \times n$ matrix and B is an $n \times n$ matrix, prove

$$\text{tr } (ABA^{-1}) = \text{tr } B.$$

The trace function and the determinant function are the two most important scalar-valued functions on the algebra of $n \times n$ matrices over a field F.

Chapter 8

Eigenvalues and the Spectral Theorem

We are concerned in this chapter with properties of a single linear transformation and in particular with its **eigenvalues** and **eigenvectors**. (Some authors call them *characteristic* or *proper values* and *vectors*.) A knowledge of these enables us to decompose certain types of linear transformations into linear transformations of simpler types and the consummation of our study in this chapter is to be found in the two spectral theorems. In particular, if a basis consisting of eigenvectors of a linear operator on a vector space is possible for the space, then the linear operator is represented by a diagonal matrix with respect to this basis. Moreover the diagonal elements are the eigenvalues of the operator. However, we shall see in the next two chapters that eigenvalues rate top billing for their usefulness in the reduction of matrices and quadratic forms to certain canonical forms.

8-1 DEFINITIONS AND EXAMPLES

Definitions. Let T be a linear operator on a vector space V over a field F. A nonzero vector $\alpha \in V$ is called an **eigenvector** (proper vector, characteristic vector) of T if $\alpha T = c\alpha$ for some $c \in F$. A scalar c is called an **eigenvalue** of T if $\alpha T = c\alpha$ for some nonzero vector α. The set of all eigenvalues of T is known as the **spectrum** of T.

The eigenvalues and eigenvectors of T are defined by the equation $\alpha T = c\alpha$, $\alpha \in V$, $\alpha \neq \overline{0}_V$, $c \in F$. This equation states that the vector αT is dependent on the vector α. (In the language of "arrows" the equation means that the arrow for the vector αT must have either the same or opposite direction to the arrow for the vector α.) The equation $\alpha T = c\alpha$ is equivalent to $\alpha T = c(\alpha I)$; that is, to $\alpha(T - cI)$, where I is the identity operator on V. If α is an eigenvector of T and if c is the corresponding eigenvalue of T then $\alpha(T - cI) = \overline{0}_V$ and, since $\alpha \neq \overline{0}_V$, this means that $T - cI$ is not injective. It is a singular linear operator. Conversely, if for some $c \in F$, $T - cI$ is a singular operator (that is, $T - cI$ is not injective) then the kernel of $T - cI$ contains at least one nonzero vector $\alpha \in V$ such that $\alpha(T - cI) = \overline{0}_V$. This implies that $\alpha T = c\alpha$ and hence that the scalar c is an eigenvalue of T. We have proved

THEOREM 1

If T is a linear operator on a vector space V over F, then $c \in F$ is an eigenvalue of T if and only if the linear operator $T - cI$ is singular.

We now assume that V is an n-dimensional vector space over F and let T be a linear operator on V. By Theorem 1 we know that $c \in F$ is an eigenvalue of T if and only if the linear operator $T - cI$ is singular, and by Theorem 12, Chapter 7, $T - cI$ is singular if and only if the determinant $|T - cI|$ of $T - cI$ is 0. We have also seen in Sec. 7-6 that $|T - cI|$ is independent of the choice of basis for V. Choosing an arbitrary basis in V, let A be the matrix of T relative to this basis. Then clearly $A - cI$ is the matrix of $T - cI$ relative to this basis. By definition (Sec. 7-6), $|T - cI| = |A - cI|$ and hence $|A - cI| = 0$.

Let x be variable over the field F and form the matrix

$$A - xI = \begin{bmatrix} a_{11} - x & a_{12} & a_{13} & \cdots & a_{1n} \\ a_{21} & a_{22} - x & a_{23} & \cdots & a_{2n} \\ a_{31} & a_{32} & a_{33} - x & \cdots & a_{3n} \\ \cdots & \cdots & \cdots & \cdots & \cdots \\ \cdots & \cdots & \cdots & \cdots & \cdots \\ a_{n1} & a_{n2} & a_{n3} & \cdots & a_{nn} - x \end{bmatrix}.$$

The determinant $|A - xI|$ of the matrix $A - xI$ is evidently a polynomial in x of degree n with scalar coefficients, that is $|A - xI| \in F[x]$, the integral domain of polynomials in x over the field F (see Sec. 1-6). It is called the **characteristic polynomial of T**. We have proved

THEOREM 2

Let T be a linear operator on a finite-dimensional vector space V over a field F. A scalar $c \in F$ is an eigenvalue of T if and only if c is a root of the polynomial equation $|A - xI| = 0$, where A is the matrix representing T with respect to an arbitrary basis for V.

Thus the eigenvalues of T are those scalars that satisfy the equation $|A - xI| = 0$. This is called the **characteristic equation** of T.

We can also define the eigenvalues and eigenvectors for matrices. If A is a given $n \times n$ matrix over F then A determines the unique linear operator T_A on $V_n(F)$ whose matrix is A relative to the standard basis for $V_n(F)$.

Definition. The **eigenvalues and eigenvectors of a matrix** A are the eigenvalues and eigenvectors of the linear operator T_A.

The eigenvalues therefore of the matrix A are those roots of the polynomial equation $|T_A - xI| = |A - xI| = 0$ that are elements of the scalar field F. The polynomial $|A - xI|$ is called the **characteristic polynomial of the matrix** A and $|A - xI| = 0$ is called the **characteristic equation of** A.

Example 1. Let

$$A = \begin{bmatrix} 0 & 1 & 3 \\ 3 & 0 & 0 \\ 0 & 0 & 2 \end{bmatrix}$$

be regarded as a matrix over the field Q of rational numbers. Its eigenvalues are therefore the rational roots of $|A - xI| = (x^2 - 3)(2 - x) = 0$. Hence A has the one rational root 2 and this is the only eigenvalue of A. The eigenvectors corresponding to 2 are the vectors (x_1, x_2, x_3) of $V_3(Q)$ determined by the equation

$$(x_1, x_2, x_3)A = 2(x_1, x_2, x_3),$$

that is, by

$$(-3x_2, -x_1, -2x_1 - 3x_2 + 2x_3) = (2x_1, 2x_2, 2x_3).$$

The eigenvectors are therefore given by the solutions of the equations

$$-3x_2 = 2x_1, \qquad -x_1 = 2x_2, \qquad -2x_1 - 3x_2 + 2x_3 = 2x_3.$$

We easily find that all vectors of $V_3(Q)$ of the form $(0, 0, t)$, where $t \in Q$ and $t \neq 0$, are the eigenvectors of A.

If A is regarded as a matrix over the real field R—that is, the scalar field is taken to be R—then its eigenvalues are $2, \sqrt{3}, -\sqrt{3}$. We solve the matrix equations

$$(x_1, x_2, x_3)A = \sqrt{3}(x_1, x_2, x_3) \quad \text{and} \quad (x_1, x_2, x_3)A = -\sqrt{3}(x_1, x_2, x_3)$$

to get the corresponding eigenvectors. They are found to be respectively the vectors of the forms

$$(-\sqrt{3}t, t, -\sqrt{3}t) \quad \text{and} \quad (\sqrt{3}t, t, \sqrt{3}t),$$

where $t \in R$ and $t \neq 0$.

Since similar matrices represent the same linear operator on a finite-dimensional vector space with respect to different bases, we would expect similar matrices to have the same characteristic polynomial. However, we next give a direct proof of this.

THEOREM 3

Similar matrices have the same characteristic polynomial.

Proof: Let A and B be similar matrices; that is, let $B = PAP^{-1}$. Then

$$\begin{aligned} |B - xI| &= |PAP^{-1} - xPIP^{-1}| \\ &= |P(A - xI)P^{-1}| \\ &= |P| \cdot |A - xI| \cdot |P^{-1}| = |P| \cdot |A - xI| \cdot |P|^{-1} \\ &= |A - xI|. \end{aligned}$$

This proves the theorem.

The eigenvalues of a linear transformation are defined without use of a basis for the vector space, and this means therefore that the scalar roots of the characteristic polynomial are independent of a choice of a basis. Furthermore, in virtue of the last theorem, the characteristic polynomial itself is independent of such a choice. This further confirms our right to speak of the **characteristic polynomial of the linear transformation** rather than of the matrix of the linear transformation.

Corollary. The eigenvalues of a linear operator T on V are independent of the particular choice of basis for the vector space V.

Thinking geometrically we can regard a one-dimensional vector space as a line, and finding an eigenvector of an operator T on a space V, is then simply determining a line in V that is mapped into itself by T. If the dimension of V is n, and then if we can find n such distinct lines, we see at once that the matrix of T relative to these lines is diagonal. Moreover its diagonal entries are the eigenvalues of T, these eigenvalues being the scalars which indicate how much the lines are stretched or compressed.

It is possible that a matrix A over a field F may have no eigenvalues. For example, the matrix $\begin{pmatrix} 1 & -2 \\ 1 & -1 \end{pmatrix}$ over the real field has no eigenvalues. Its characteristic equation is $x^2 + 1 = 0$ and this equation has no real roots. On the other hand, over the complex field the matrix has the eigenvalues $i = \sqrt{-1}$ and $-i$.

Only those roots of the characteristic equation of a matrix over F that are elements of F are eigenvalues of the matrix.

As we have remarked, the eigenvectors of a linear transformation, in terms of vectors as "arrows" in the real plane, are those vectors whose directions are left unchanged, or are reversed, by the linear transformation.

Eigenvalues and the Spectral Theorem

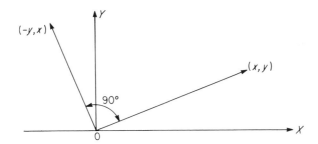

Example 2. The matrix $\begin{pmatrix} 0 & 1 \\ -1 & 0 \end{pmatrix}$ rotates nonzero vectors (x, y) of $V_2(R)$ through 90° into vectors $(-y, x)$ of $V_2(R)$. Here R is the real field. Hence this matrix has no eigenvectors. This is confirmed by its characteristic equation $x^2 + 1 = 0$, which has no real roots.

Example 3. The matrix $\begin{pmatrix} 1 & 1 \\ 0 & 1 \end{pmatrix}$ transforms the vector $(x, y) \in V_2(R)$ into the vector $(x, x + y)$. This is called a **shear parallel to the Y-axis**. The only vectors whose directions are left unchanged, or are reversed, by this linear transformation are the vectors parallel to the Y-axis. These are the eigenvectors. This is confirmed by the fact that the only eigenvalue is 1, and hence all eigenvectors are of the form $(0, y)$ where y is arbitrary.

Suppose, on the other hand, that there are two nonparallel vectors α_1 and α_2, in the real plane, whose directions are left unchanged, or are reversed, by some linear transformation T. They are then eigenvectors of T. Since two such vectors span $V_2(R)$ and $\alpha_1 T = x_1 \alpha_1$, $\alpha_2 T = x_2 \alpha_2$, we see that T has the diagonal matrix $\begin{pmatrix} x_1 & 0 \\ 0 & x_2 \end{pmatrix}$ with respect to this basis. Conversely, if T has a diagonal matrix with respect to some pair of nonparallel vectors α_1 and α_2, then $\alpha_1 T = x_1 \alpha_1$, $\alpha_2 T = x_2 \alpha_2$, and hence they are eigenvectors of T.

All this quite naturally suggests the next theorem.

THEOREM 4

An $n \times n$ matrix A over a field F is similar to a diagonal matrix D if and only if the eigenvectors of A span $V_n(F)$. Moreover the diagonal elements of D are the eigenvalues of A.

Proof: If the eigenvectors of A span $V_n(F)$ then there are n of them, $\alpha_1, \ldots, \alpha_n$, that form a basis for $V_n(F)$. If T is the linear operator on $V_n(F)$, whose matrix with respect to the standard basis is A, the $\alpha_i T = x_i \alpha_i$, $i = 1, 2, \ldots, n$, where the x_i are the eigenvalues of A. With respect to this eigenbasis T has a diagonal matrix whose diagonal elements are the eigenvalues of A. Hence A is similar to this diagonal matrix.

Conversely, suppose A is similar to a diagonal matrix D. Let $d_1, \ldots, d_n$ be the diagonal elements of D. For the standard basis $e_1 = (1, 0, \ldots, 0)$, $e_2 = (0, 1, 0, \ldots, 0), \ldots, e_n = (0, 0, \ldots, 0, 1)$, evidently $e_i D = d_i e_i$, $i = 1, 2, \ldots, n$. Hence the e_i are the eigenvectors of D and the d_i are the eigenvalues of D. Therefore since A is similar to D, the d_i are the eigenvalues of A.

Definition. A linear operator T on a finite-dimensional vector space V is said to be **diagonable** or a **diagonal operator** if it can be represented by a diagonal matrix relative to some basis of V.

The last theorem proves that a *linear operator T on a finite-dimensional vector space V is diagonable if and only if there exists a basis for V consisting of eigenvectors of T*.

Let A be an $n \times n$ diagonable matrix over a field F. Then there exists a basis $\alpha_1, \alpha_2, \ldots, \alpha_n$ of $V_n(F)$ consisting of eigenvectors of A. If $x_1, x_2, \ldots, x_n$ are the corresponding eigenvalues of A, then we have $\alpha_i A = x_i \alpha_i$, $i = 1, 2, \ldots, n$.

Let D be the diagonal matrix that is similar to A. Then there exists a nonsingular matrix $P = (p_{ij})$ over F for which $D = PAP^{-1}$, that is $DP = PA$, and of course A and D have the same eigenvalues. If $\epsilon_1, \epsilon_2, \ldots, \epsilon_n$ denote the standard basis of $V_n(F)$, then $\epsilon_i D = x_i \epsilon_i$, $i = 1, 2, \ldots, n$. Since $\epsilon_i DP = \epsilon_i PA$, we have $x_i(\epsilon_i P) = (\epsilon_i P)A$ and therefore, $\epsilon_i P = \alpha_i$, $i = 1, 2, \ldots, n$. These equations determine the matrix P as follows Since

$$\epsilon_1 = (1, 0, 0, \ldots, 0),$$
$$\epsilon_2 = (0, 1, 0, \ldots, 0),$$
$$\ldots\ldots\ldots\ldots\ldots,$$
$$\epsilon_n = (0, 0, \ldots, 0, 1),$$

we obtain for the rows of P,

$$(p_{11}, p_{12}, \ldots, p_{1n}) = \alpha_1,$$
$$(p_{21}, p_{22}, \ldots, p_{2n}) = \alpha_2,$$
$$\ldots\ldots\ldots\ldots\ldots\ldots,$$
$$(p_{n1}, p_{n2}, \ldots, p_{nn}) = \alpha_n,$$

Eigenvalues and the Spectral Theorem

and therefore

$$P = \begin{bmatrix} \alpha_1 \\ \alpha_2 \\ \vdots \\ \alpha_n \end{bmatrix}.$$

The rows of P are therefore the n independent eigenvectors of A.

Example 4. Consider the matrix

$$A = \begin{bmatrix} 2 & 0 & 1 \\ 0 & 0 & -1 \\ 0 & -1 & 0 \end{bmatrix}$$

over the real field. Its eigenvalues are $-1, 1, 2$ and a basis of eigenvectors is $(0,1,1)$, $(0,1,-1)$, and $(-3,1,-2)$. Thus if we take

$$P = \begin{bmatrix} 0 & 1 & 1 \\ 0 & 1 & -1 \\ -3 & 1 & -2 \end{bmatrix}$$

we find

$$PAP^{-1} = \begin{bmatrix} -1 & 0 & 0 \\ 0 & 1 & 0 \\ 0 & 0 & 2 \end{bmatrix},$$

the order $-1, 1, 2$ of arrangement of the eigenvalues on the main diagonal corresponding to the order $(0, 1, 1)$, $(0, 1, -1)$, $(-3, 1, -2)$ of arrangement of the eigenvectors as rows of P, for these vectors belong respectively to the eigenvalues $-1, 1, 2$.

Example 5. Let A be the real matrix

$$A = \begin{bmatrix} 2 & 3 & 1 \\ 1 & 4 & 5 \\ 2 & 6 & 1 \end{bmatrix}.$$

The eigenvalues of A are given by

$$|A - xI| = \begin{bmatrix} 2-x & 3 & 1 \\ 1 & 4-x & 5 \\ 2 & 6 & 1-x \end{bmatrix}$$

$$= -x^3 + 7x^2 + 21x - 27$$

$$= (-3 - x)(1 - x)(9 - x) = 0.$$

They are $-3, 1, 9$. For $x = -3$ the eigenvectors are found to be all vector of the form $(t, 3t, -4t)$, where t is any nonzero real number. For $x = 1$ they are the vectors $(-5t, t, 2t)$, $t \neq 0$; and for $x = 9$, they are $(t, 3t, 2t)$ $t \neq 0$. Since these vectors obviously span $V_3(R)$, A is similar over R to the diagonal matrix $\begin{bmatrix} -3 & 0 & 0 \\ 0 & 1 & 0 \\ 0 & 0 & 9 \end{bmatrix}$.

If T is the linear operator on $V_3(R)$ whose matrix is A with respect to the standard basis $(1, 0, 0)$, $(0, 1, 0)$, $(0, 0, 1)$, then the matrix of T with respect to the eigenvectors as a basis is the diagonal matrix $\begin{bmatrix} -3 & 0 & 0 \\ 0 & 1 & 0 \\ 0 & 0 & 9 \end{bmatrix}$.

Lemma 1. If x is an eigenvalue of T then the set of all vectors $\alpha \in V$ such that $\alpha T = x\alpha$ is a subspace S of V of dimension ≥ 1. (This subspace is called an **eigenspace** of T.)

Proof: If $\alpha T = x\alpha$, $\beta T = x\beta$, then $(\alpha + \beta)T = x(\alpha + \beta)$. Thus if $\alpha, \beta \in S$, then $\alpha + \beta \in S$. If $y \in F$, $(y\alpha)T = y(\alpha T) = yx\alpha = x(y\alpha)$. Hence if $\alpha \in S$, then $y\alpha \in S$. Hence S is a subspace. If x is an eigenvalue, then there exists $\alpha \neq 0_V$ such that $\alpha T = x\alpha$. Hence dim $S \geq 1$.

Lemma 2. If $\alpha_1, \ldots, \alpha_k$ are eigenvectors of T with distinct eigenvalues $x_1, \ldots, x_k$, then $\alpha_1, \ldots, \alpha_k$ is an independent set of vectors.

Proof: Assume they are dependent. Then we know there is a least index m such that α_m depends on the independent vectors

$$\alpha_1, \ldots, \alpha_{m-1}.$$

Then

$$\alpha_m = \sum_{1}^{m-1} y_i \alpha_i, \quad y_i \in F$$

Eigenvalues and the Spectral Theorem

and
$$\alpha_m T = x_m \alpha_m = \sum_{1}^{m-1} y_i x_i \alpha_i.$$

But
$$x \alpha_m = x_m \sum_{1}^{m-1} y_i \alpha_i = \sum_{1}^{m-1} y_i x_m \alpha_i.$$

Hence
$$\sum_{1}^{m-1} y_i (x_i - x_m) \alpha_i = \overline{0}_V.$$

Since $\alpha_1, \ldots, \alpha_{m-1}$ are independent vectors, this forces $y_i(x_i - x_m) = 0$, $i = 1, 2, \ldots, m - 1$. Since $x_i \neq x_m$ we have $y_i = 0$, $i = 1, 2, \ldots, m - 1$. Hence $\alpha_m = \overline{0}_V$. This is a contradiction since α_m is an eigenvector. Therefore the assumption that the α_i are dependent is impossible.

Corollary. If dim $V = n$, there can be at most n distinct eigenvalues of a linear operator T.

A field F is said to be **algebraically closed** if every polynomial equation over F has all its roots in F. The "fundamental theorem of algebra" proves that the complex field is algebraically closed. It can be proved that for every field F there exists an algebraically closed extension of F. The simple example $x^2 + 1 = 0$ shows that the real field is not algebraically closed. Hence if T is a linear operator on an n-dimensional vector space over the complex field, then its characteristic equation has all its roots in this field. Since there may be only one distinct root of this characteristic equation, we can assert that a linear operator on a finite-dimensional complex vector space has at least one eigenvector. On the other hand, a linear operator on a finite-dimensional real vector space may have no eigenvectors; that is, none of the roots of its characteristic equation may be real.

The roots of the characteristic equation of a matrix A over F are called the **characteristic roots** of A. A linear operator on a vector space over an algebraically closed field has at least one characteristic root. If F is an algebraically closed field then all characteristic roots are eigenvalues.

Note this distinction between characteristic roots and eigenvalues when the base field is not assumed to be algebraically closed.

We mention here two useful facts concerning the solution of certain polynomial equations:

I. The only possible rational roots of a monic polynomial equation
$$x^n + a_1 x^{n-1} + \cdots + a_n = 0,$$

where the a_i are integers, are the integral divisors (positive and negative) of a_n.

II. Let $f(x) = 0$ be a polynomial equation with real coefficients. If for two real numbers a and b, $f(a)$ and $f(b)$ have opposite signs, then the equation has at least one real root between a and b.

EXERCISES

1. Find the eigenvalues and the eigenvectors of the linear operator T on $V_2(R)$ defined by
$$(x, y) T = (2x + 3y, 4x - 2y).$$

2. Let
$$A = \begin{bmatrix} 2 & 0 & 0 \\ -3 & 1 & 0 \\ 0 & 1 & 0 \end{bmatrix}$$

be the matrix of a linear operator T on $V_3(R)$ with respect to the standard basis.
  (a) Find $(x, y, z) T$.
  (b) Find the eigenvalues and eigenvectors of T and prove that the eigenvectors span $V_3(R)$.
  (c) Prove that A is similar to the diagonal matrix
$$B = \begin{bmatrix} 1 & 0 & 0 \\ 0 & 2 & 0 \\ 0 & 0 & 0 \end{bmatrix}.$$

  (d) Find a matrix P for which $B = PAP^{-1}$.

3. Prove that zero is an eigenvalue of a linear operator T if and only if T is singular.

4. If x is an eigenvalue of a nonsingular linear operator T on a vector space V, prove that x^{-1} is an eigenvalue of T^{-1}. Prove that an eigenvector of T is an eigenvector of T^{-1} and conversely.

5. If A is a square matrix for which $A^k = 0$ for some positive integer k, prove that A is not similar to a diagonal matrix. (This is true for any field).

6. Use induction on k to prove that the eigenvectors $\alpha_1, \alpha_2, \ldots, \alpha_k$ of a linear operator T, that have distinct eigenvalues, are independent.

7. If the characteristic polynomial of a linear operator T on an n-dimensional vector space V over a field F has n distinct roots in F, prove that the eigenvectors, $\alpha_1, \alpha_2, \ldots, \alpha_n$ belonging to these eigenvalues, form a basis for V.

Eigenvalues and the Spectral Theorem

8. Find the eigenvalues of the matrices

(a) $\begin{bmatrix} 1 & 9 & -6 \\ 0 & 3 & 5 \\ 0 & 4 & 2 \end{bmatrix}$;   (b) $\begin{bmatrix} 2 & 0 & 0 \\ 9 & 5 & 6 \\ 8 & -3 & -1 \end{bmatrix}$.

9. S and T are linear operators on a finite-dimensional vector space and $ST = TS$. Prove that if α is an eigenvector of T with the eigenvalue x and if $\alpha S \neq \bar{0}_V$, then αS is an eigenvector of T with the eigenvalue x.

10. Prove that the eigenvectors of a linear operator on a vector space V, that belong to the same eigenvalue, form a subspace of V of dimension ≥ 1.

11. U and W are subspaces of a finite-dimensional vector space V and $V = U \oplus W$. S and T are linear operators on V and $S(U) \subset U$ and $T(W) \subset W$.

Prove that the characteristic polynomial of $S + T$ is the product of the characteristic polynomials of the restrictions S and T to U and W respectively.

12. U is a subspace of a finite-dimensional vector space V and T is a linear operator on V. If $T(U) \subset U$, let T_1 be the restriction of T to U.

(a) Prove that T induces a linear operator T_2 on V/U given by $(\alpha + U)T_2 = \alpha T + U$.

(b) Prove that the characteristic polynomial of T is the product of the characteristic polynomials of T_1 and T_2.

13. S and T are linear operators on an n-dimensional vector space V and $ST = TS$. Prove that the characteristic roots of ST are $\alpha_1\beta_1, \alpha_2\beta_2, \ldots, \alpha_n\beta_n$ where $\alpha_1, \alpha_2, \ldots, \alpha_n$ and $\beta_1, \beta_2, \ldots, \beta_n$ are the characteristic roots of S and T respectively.

14. Prove that the matrix $\begin{bmatrix} 1 & 1 & 0 \\ 0 & 0 & 1 \\ 0 & 1 & 0 \end{bmatrix}$ is not similar to a diagonal matrix.

15. S and T are linear operators on a vector space V. Prove that ST and TS have the same eigenvalues. Do they have the same eigenvectors?

8-2 THE MINIMAL POLYNOMIAL

Let V be an n-dimensional vector space. We have seen that the set Hom (V, V) of all linear operators on V is an n^2-dimensional vector space over the same scalar field as that of V. If $T \in$ Hom (V, V) then the set I, T, $T^2, \ldots, T^{n^2}$ of $n^2 + 1$, vectors of Hom (V, V) must be linearly dependent. Hence there must be a vector T^k of this set, $1 \leq k \leq n^2$ for which

$$T^k = a_0 I + a_1 T + \cdots + a_{k-1} T^{k-1},$$

where the coefficients a_i are scalars. If we write

$$f(x) = x^k - a_{k-1} x^{k-1} - \cdots - a_1 x - a_0,$$

then we have a nonzero monic polynomial $f(x)$ such that $f(T) = 0$. Here 0 stands for the linear operator on V which maps every vector of V into the zero vector.

We have therefore proved that for any linear operator T on a finite-dimensional vector space, there exists a monic polynomial $f(x)$ such that $f(T) = 0$. In the set of positive integers that are the degrees of such polynomials, there must be a least positive integer (the well-ordering property of the positive integers). Let $m(x)$ be a monic polynomial of this least degree. Then $m(T) = 0$. Moreover $m(x)$ is unique. For if $m'(x)$ is another polynomial, monic and of the same degree as $m(x)$, and for which $m'(T) = 0$, then $[m(T) - m'(T)] = 0$. However, $m(x) - m'(x)$ is a polynomial of degree less than the degree of $m(x)$, and if we divide it by its leading coefficient, we get a monic polynomial $h(x)$ for which $h(T) = 0$. Since the degree of $h(x)$ is less than that of $m(x)$, this contradicts the definition of $m(x)$. Hence $m(x)$ is unique.

Definition. The **minimal polynomial** of the linear operator T on a finite-dimensional vector space is the unique monic polynomial $m(x)$ of least degree, such that $m(T) = 0$.

THEOREM 5

If $f(x)$ is any polynomial over F such that $f(T) = 0$, then the minimal polynomial $m(x)$ is a divisor of $f(x)$.

Proof: Let $f(x) = m(x)q(x) + r(x)$, where either $r(x)$ is a polynomial of degree less than the degree of $m(x)$ or $r(x) = 0$. Then

$$f(T) = m(T)q(T) + r(T)$$

and hence $r(T) = 0$. Unless $r(x) = 0$, this is a contradiction of the fact that $m(x)$ is the nonzero polynomial of least degree for which $m(T) = 0$. Hence $r(x) = 0$ and therefore $m(x)$ is a divisor of $f(x)$.

Definition. The **minimal polynomial of an n × n matrix** A, over a field F, is the minimal polynomial of the linear operator on $V_n(F)$ whose matrix is A with respect to some basis of $V_n(F)$.

This latter definition is justified by the next theorem.

THEOREM 6

Similar matrices over a field F have the same minimal polynomial.

Proof: Similar matrices are simply different representations of the same linear operator with respect to different bases in $V_n(F)$, hence they must have the same minimal polynomial.

We come now to one of the most important theorems of linear algebra, for it implies that the minimal polynomial is a divisor of the characteristic polynomial.

Eigenvalues and the Spectral Theorem

THEOREM 7 (Cayley-Hamilton Theorem)

Let A be an $n \times n$ matrix over a field F and let
$$c(x) = \det(A - xI) = c_0 + c_1 x + \cdots + c_{n-1} x^{n-1} + x^n$$
be the characteristic polynomial of A. Then
$$c(A) = c_0 I + c_1 A + \cdots + c_{n-1} A^{n-1} + A^n = 0_n,$$
where I is the identity $n \times n$ matrix and 0_n is the $n \times n$ zero matrix.

(This theorem is often imprecisely expressed by the statement that every square matrix satisfies its characteristic equation.)

Proof: Let $B(x)$ denote the adjoint matrix of the matrix $A - xI$. The elements of $B(x)$ are therefore the cofactors of the elements of $A - xI$ and hence they are polynomials in x of degree $\leq n - 1$. Hence
$$B(x) = B_0 + B_1 x + \cdots + B_{n-1} x^{n-1}$$
where the B_i are constant $n \times n$ matrices; that is, the B_i are independent of x. Now
$$B(x) \cdot (A - xI) = \det(A - xI) \cdot I.$$
That is
$$(B_0 + B_1 x + \cdots + B_{n-1} x^{n-1})(A - xI) = c(x) \cdot I.$$

This is an identity in x, and hence the matrix coefficients of corresponding powers of x, on the two sides of the identity, must be equal. Hence we get the following equations:
$$B_0 A = c_0 I$$
$$B_1 A - B_0 = c_1 I$$
$$\cdots\cdots\cdots\cdots\cdots\cdots$$
$$\cdots\cdots\cdots\cdots\cdots\cdots$$
$$B_{n-2} A - B_{n-3} = c_{n-2} I$$
$$B_{n-1} A - B_{n-2} = c_{n-1} I$$
$$-B_{n-1} = I.$$

Multiplying these matrix equations respectively by $1, A, A^2, \ldots, a^{n-1}, A^n$, we obtain
$$B_0 A = c_0 I$$
$$B_1 A^2 - B_0 A = c_1 A$$
$$\cdots\cdots\cdots\cdots\cdots\cdots$$
$$\cdots\cdots\cdots\cdots\cdots\cdots$$
$$B_{n-2} A^{n-1} - B_{n-3} A^{n-2} = c_{n-2} A^{n-2}$$
$$B_{n-1} A^n - B_{n-2} A^{n-1} = c^{n-1} A^{n-1}$$
$$-B_{n-1} A^n = A^n.$$

Adding these $n + 1$ matrix equations we get
$$0_n = c_0 I + c_1 A + \cdots + A^n = c(A),$$
which is the theorem.

If T is the linear operator on $V_n(F)$ whose matrix is A with respect to some basis in $V_n(F)$, this last equation can be written as $c(T) = 0$, where 0 is the zero linear operator on $V_n(F)$. This follows, since by the isomorphism theorem of linear transformations with matrices, $c(A)$ is the matrix corresponding to the linear transformation $c(T)$.

Corollary. The minimal polynomial $m(x)$ is a divisor of the characteristic polynomial $c(x)$.

Proof: This follows at once from Theorem 5, since $c(T) = 0$.

If $c(x)$ is the characteristic polynomial of an $n \times n$ matrix A, then the Cayley-Hamilton theorem states that $c(A) = 0$, where 0 stands for the $n \times n$ zero matrix. Thus the minimal polynomial of A is the monic polynomial $m(x)$ of least degree for which the matrix equation $m(A) = 0$ is true. Sometimes $m(x) = c(x)$ and always $m(x)$ is a divisor of $c(x)$. We give an example where $m(x) \neq c(x)$.

Let A be the matrix $\begin{bmatrix} -1 & 7 & 0 \\ 0 & 2 & 0 \\ 0 & 3 & -1 \end{bmatrix}$. Its characteristic polynomial is

$(1 + x)^2(2 - x)$. However, its minimal polynomial is
$$-(1 + x)(2 - x) = x^2 - x - 2.$$
To see this simply find the 3×3 matrix determined by $A^2 - A - 2I$. It turns out to be the 3×3 zero matrix; that is, $A^2 - A - 2I = 0$.

Lemma 3. Let $c(x)$ be the characteristic polynomial and $m(x)$ the minimal polynomial of an $n \times n$ matrix A over a field F. Then $c(x)$ is a divisor of the polynomial $[m(x)]^n$.

Proof: Let $m(x) = x^r + c_1 x^{r-1} + \cdots + c_{r-1} x + c_r$. Form the set of r matrices

$$B_0 = I, \quad B_j = A^j + c_1 A^{j-1} + \cdots + c_{j-1} A + c_j I, \quad j = 1, 2, \ldots, r-1.$$

Then
$$B_0 = I, \quad B_j - AB_{j-1} = c_j I, \quad j = 1, 2, \ldots, r-1.$$
Now
$$AB_{r-1} = A^r + c_1 A^{r-1} + \cdots + c_{r-2} A^2 + c_{r-1} A = m(A) - c_r I = -c_r I.$$
Put
$$B(x) = x^{r-1} B_0 + x^{r-2} B_1 + \cdots + x B_{r-2} + B_{r-1}.$$

Then
$$(xI - A)B(x) = x^r B_0 + x^{r-1}(B_1 - AB_0) + \cdots$$
$$+ x(B_{r-1} - AB_{r-2}) - AB_{r-1}$$
$$= x^r I + x^{r-1} c_1 I + \cdots + x c_{r-1} I + c_r I$$
$$= m(x) I$$

Now take the determinant of both sides of this last equation. Since the determinant of $B(x)$ is a polynomial in x, we get

$$c(x) \cdot \det B(x) = [m(x)]^n$$

and we see that $c(x)$ is a divisor of $[m(x)]^n$.

An immediate consequence of this lemma is the following

Corollary. The minimal polynomial and the characteristic polynomial have the same irreducible factors.

To emphasize the importance of this last corollary, we express it another way:

The eigenvalues of an $n \times n$ matrix A are roots of the minimal polynomial of A.

EXERCISES

1. T is a linear operator on $V_3(R)$ whose matrix with respect to the standard basis is
$$A = \begin{bmatrix} -1 & 0 & 2 \\ -3 & -2 & 6 \\ -1 & 0 & 2 \end{bmatrix}.$$

(a) Find the characteristic and minimal polynomials of T.
(b) Find the eigenvalues and eigenvectors of T.
(c) Prove that A is similar to the diagonal matrix
$$\begin{bmatrix} 0 & 0 & 0 \\ 0 & 1 & 0 \\ 0 & 0 & -2 \end{bmatrix}.$$

2. If the degree of the minimal polynomial of an $n \times n$ matrix A is n, prove that A is similar to a diagonal matrix.

3. Find the characteristic polynomial and the eigenvalues of the lower triangular matrix
$$A = \begin{bmatrix} a_{11} & 0 & 0 & \cdots & 0 \\ a_{21} & a_{22} & 0 & \cdots & 0 \\ \cdots & \cdots & \cdots & \cdots & \cdots \\ \cdots & \cdots & \cdots & \cdots & \cdots \\ a_{n1} & a_{n2} & a_{n3} & \cdots & a_{nn} \end{bmatrix}.$$

4. Prove that the characteristic polynomial and the minimal polynomial of a linear operator T on a finite-dimensional vector space have the same irreducible factors.

5. Prove that a linear operator on a finite-dimensional vector space is nonsingular if and only if the constant term of its minimal polynomial is not zero.

8-3 THE ADJOINT OPERATOR

We next use Theorem 13, Chapter 6, to define a very important linear operator T^* associated with a given **linear operator T on a real finite-dimensional inner-product vector space**.

For each fixed vector $\beta \in V$, the function f of $V \to F$ defined by $\alpha f = \alpha T \cdot \beta$ is evidently a linear form on V; that is, $f \in V^*$. Hence by Theorem 12, Chapter 6, there exists a unique vector $\beta' \in V$ such that $\alpha T \cdot \beta = \alpha \cdot \beta'$, for all $\alpha \in V$.

Let us now define a mapping T^* of $V \to V$ by

$$\beta T^* = \beta'$$

for all $\beta \in V$, where β' is uniquely given by $\alpha T \cdot \beta = \alpha \cdot \beta'$, for all $\alpha \in V$.

It is a straightforward exercise to verify that T^* is a linear operator on V. It is called the **adjoint operator** of T. The all important relationship between a linear operator T and its adjoint T^* is given by

(1) $$\alpha T \cdot \beta = \alpha \cdot (\beta T^*),$$

for all $\alpha, \beta \in V$.

Thus, if T is a linear operator on a finite-dimensional inner-product vector space V, there exists a unique linear operator T^* on V such that for all $\alpha, \beta \in V$, $\alpha T \cdot \beta = \alpha \cdot \beta T^*$. Clearly T^* depends on both T and on the inner product of V.

If $T^* = T$ then T is said to be a **real self-adjoint** or a **real symmetric linear operator**.

If a basis is chosen in V it is to be expected that there would exist a relation between the matrices of T and T^* with respect to this basis. For an orthonormal basis this relation turns out to be a very simple one as the next theorem proves.

THEOREM 8

If V is a real n-dimensional inner-product vector space, then with respect to an orthonormal basis the matrix of a linear operator T is the transpose of the matrix of its adjoint operator T^*.

Eigenvalues and the Spectral Theorem

Proof: Let $\alpha_1, \alpha_2, \ldots, \alpha_n$ be an orthonormal basis for V. Let (a_{ij}) be the matrix of T relative to this basis. Then

$$\alpha_i T = \sum_{j=1}^{n} a_{ij}\alpha_j = \sum_{1}^{n} ((\alpha_i T) \cdot \alpha_j)\alpha_j.$$

Hence $a_{ij} = (\alpha_i T) \cdot \alpha_j$. Let (b_{ij}) be the matrix of T^* relative to this same basis. Then

$$\alpha_i T^* = \sum_{j=1}^{n} b_{ij}\alpha_j = \sum_{1}^{n} ((\alpha_i T^*) \cdot \alpha_j)\alpha_j.$$

Hence $b_{ij} = (\alpha_i T^*) \cdot \alpha_j = (\alpha_j T) \cdot \alpha_i = a_{ji}$, using (1) above, and thus $(b_{ij}) = (a_{ji}) = (a_{ij})^t$.

Corollary. T is a self-adjoint operator if and only if its matrix (a_{ij}) relative to an orthonormal basis is **symmetric**; that is $a_{ij} = a_{ji}$ for all i and j.

For example, any linear operator T on $V_n(R)$ that has a symmetric basis with respect to the standard basis of $V_n(R)$ is self-adjoint.

THEOREM 9

Let T be a self-adjoint linear operator on a real n-dimensional inner-product vector space V. Then there exists an orthonormal basis of V consisting of the eigenvectors of T.

Proof: We shall prove this theorem by use of induction on n. Clearly the theorem is trivially true for $n = 1$. Assume dim $V = n > 1$ and let α_1 be an eigenvector of T.[†] Let U be the subspace of V spanned by the unit vector $\beta_1 = \alpha_1 / |\alpha_1|$. Let $U^\perp$ be the orthogonal complement of U, so that $V = U \oplus U^\perp$, and therefore dim $U^\perp = n - 1$ (dim $V =$ dim U + dim $U^\perp$).

For $\alpha \in U$, $\alpha = y_1\alpha_1$ where y_1 is a real number. Hence $\alpha T = y_1(\alpha_1 T)$. Since α_1 is an eigenvector of T, $\alpha_1 T = x_1\alpha_1 \in W$. Hence $\alpha T \in W$ for all $\alpha \in W$.

Now let $\gamma \in U^\perp$. Then $\gamma \cdot \alpha = 0$, for all $\alpha \in U$. Now $\gamma T^* \cdot \alpha = \gamma \cdot (\alpha T)$ and, since $\alpha T \in U$, $\gamma T^* \cdot \alpha = 0$ for all $a \in U$. Hence $\gamma T = \gamma T^*$, $\gamma \in U^\perp$. Thus the restriction of T to $U^\perp$ is a linear operator on $U^\perp$ and it is, of course, self-adjoint.

Under the induction hypothesis, the theorem is true for a vector space of dimension $n - 1$. Hence $U^\perp$ has an orthonormal basis consisting of eigenvectors $\beta_2, \ldots, \beta_n$ of the restriction of T to $U^\perp$ and these are eigenvectors of T itself. Since $\beta_1 \cdot \beta_i = 0$, $i = 2, \ldots, n$, the

[†] We are assuming that a real self-adjoint operator has an eigenvector. This is proved in the next section when we study hermitian operators.

set $\beta_1, \ldots, \beta_n$ is orthonormal and consists of eigenvectors of T. This completes the proof.

In the last theorem, since the orthonormal basis consists of eigenvectors of T, we have equations of the form

$$\beta_i T = x_i \beta_i, \quad x_i \in R, \quad i = 1, 2, \ldots, n.$$

Hence with respect to this basis the linear operator T has a diagonal matrix and the elements of R on the principal diagonal are the eigenvalues of T. This proves the following.

Corollary 1. *All the eigenvalues of a real self-adjoint linear operator are real.*

In terms of matrices, all this can be expressed by saying that a real symmetric matrix A is similar over the real field to a diagonal matrix, whose elements on the principal diagonal are the eigenvalues of A. (Corollary to Theorem 8.) These eigenvalues are therefore real (the product of real matrices is a real matrix) and we have

Corollary 2. *All the eigenvalues of a real symmetric matrix are real.*

A real symmetric $n \times n$ matrix A is similar to a diagonal matrix and hence its eigenvectors span $V_n(R)$, the n-dimensional vector space over the real field R equipped with the usual inner product. Let $e_1, e_2, \ldots, e_n$ be the standard basis of $V_n(R)$. According to the last theorem, there exists an orthonormal basis $\beta_1, \beta_2, \ldots, \beta_n$ for $V_n(R)$ consisting of eigenvectors of A.

Let P be the linear operator on $V_n(R)$ defined by $e_i P = \beta_i$, $i = 1, 2, \ldots, n$. If (p_{ij}) is the matrix of P with respect to these bases, then

$$\beta_i = \sum_{j=1}^{n} p_{ij} e_j, \quad i = 1, 2, \ldots, n.$$

Since the β_i are orthonormal vectors,

$$\sum_{j=1}^{n} p_{ij}^2 = 1, \quad i = 1, 2, \ldots, n,$$

and if $i \neq k$ then

$$\sum_{j=1}^{n} p_{ij} p_{kj} = 0.$$

Thus (p_{ij}) is an orthogonal matrix and hence P is an orthogonal operator. If T is the linear operator whose matrix with respect to the two bases is A, then we have $PTP^{-1} = D$, where D is the diagonal linear operator whose matrix is diagonal with the eigenvalues $x_1, \ldots, x_n$ of T on the main

Eigenvalues and the Spectral Theorem

diagonal. Since P is orthogonal, $P^{-1} = P^t$, and hence $PTP^t = D$ or, in terms of matrices, $PAP^t = D$. We have proved

THEOREM 10

If T is a real symmetric linear operator on $V_n(R)$, then there exists an orthogonal operator P on $V_n(R)$ such that $PTP^t = D$, where D is a diagonal operator.

EXERCISES

1. $V_2(R)$ is the real vector space with the standard inner product. A linear operator T on $V_2(R)$ is defined by $(x, y)T = (ax + by, cx + dy)$, where a, b, c, d are fixed real numbers. Find $(x, y)T^*$ where T^* is the adjoint of T.

2. $V_3(R)$ is the real vector space with the standard inner product. Prove that the linear operator T defined on $V_3(R)$ by $(x, y, z)T = (2x + y + 3z, x - z, 3x - y + 4z)$ is self-adjoint.

3. Is the sum of two self-adjoint linear operators a self-adjoint linear operator? What about the product? Prove your answers.

4. Find an orthogonal operator P on $V_3(R)$ for which PTP^t is a diagonal operator, where T is the self-adjoint operator defined in Exercise 2.

5. $\alpha_1, \alpha_2, \ldots, \alpha_n$ is an orthonormal basis of a real vector space V. ϕ is the isomorphism of $V \to V^*$ that carries this basis into its dual $f_1, \ldots, f_n$ for V^*. An inner product is defined in V^* by

$$\left(\sum_{i=1}^n x_i f_i\right) \cdot \left(\sum_{i=1}^n y_i f_i\right) = \sum_{i=1}^n x_i y_i, \quad x_i, y_i \in R.$$

(a) Prove that ϕ is an isometry.

(b) If T is a given linear operator on V, prove that T determines a unique linear operator S on V^* for which the accompanying diagram is commutative; that is, $T\phi = \phi S$.

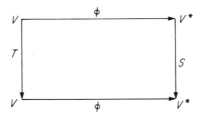

(c) If T is self-adjoint, prove that S is self-adjoint.

6. T is a linear operator on V. If $\alpha T = x\alpha$ and $\beta T^* = y\beta$ with $x \neq y$, prove that α and β are orthogonal vectors.

8-4 HERMITIAN OPERATORS

Let C be the **complex field**; that is, the set of all ordered pairs (x, y) of real numbers under the following definitions:

$$(x_1, y_1) = (x_2, y_2) \text{ if and only if } x_1 = x_2, \quad y_1 = y_2,$$
$$(x_1, y_1) + (x_2 y_2) = (x_1 + x_2, \quad y_1 + y_2),$$
$$(x_1, y_1) \cdot (x_2, y_2) = (x_1 x_2 - y_1 y_2, \quad x_1 y_2 + x_2 y_1).$$

It is an easy exercise for the reader to show that C is actually a field and that it contains a subfield (the set of all pairs of the form $(x, 0)$) isomorphic to the real field. One often writes the pair (x, y) in the form $x + yi$, where $i^2 = -1$. The notation $\bar{z}$ will be used for the conjugate of a complex number z; that is, if $z = x + yi$ then $\bar{z}$ is defined as $\bar{z} = x - yi$.

NOTE: We shall assume C is an algebraically closed field. We make use of this assumption to prove

THEOREM 11

A linear operator T on a finite dimensional vector space V over the complex field has a (nonzero) eigenvector.

Proof: Let $m(x)$ be the minimal polynomial of T. Since the complex field is algebraically closed, all the roots of $m(x)$ are in this field and we can write

$$m(x) = (x - x_1)(x - x_2) \ldots (x - x_r).$$

Hence

$$0 = m(T) = (T - x_1 I)(T - x_2 I) \ldots (T - x_r I).$$

At least one of the factors $T - x_k I$ must be a singular linear operator on V, for otherwise we have the contradiction that $m(T) = 0$ is nonsingular. (See Corollary to Theorem 10, chapter 3.) Hence there exists $\alpha \neq \bar{0}_V$ such that $\alpha(T - x_k I) = \bar{0}_V$; that is, $\alpha T = x_k \alpha$. This proves that $\alpha \neq \bar{0}_V$ is an eigenvector of V.

Let $A = (a_{ij})$ be an $n \times n$ matrix over C. The matrix $\bar{A} = (\bar{a}_{ij})$ is called the **conjugate** of A and we use the notation A^* for the **conjugate transpose** of A. Hence

$$A^* = \bar{A}^t.$$

A square matrix A is said to be **hermitian** if $A^* = A$.

Definition. An **inner product** on a vector space V over the complex field C is a function on $V \times V \to C$, whose values in C we shall denote by $\alpha \cdot \beta$ for $\alpha, \beta \in V$, and which has the following properties:

(a) $\beta \cdot \alpha = \overline{\alpha \cdot \beta}$ (the conjugate of $\alpha \cdot \beta$),

(b) $(\alpha + \beta) \cdot \gamma = \alpha \cdot \gamma + \beta \cdot \gamma$, $\alpha, \beta, \gamma \in V$,
(c) $(z\alpha \cdot \beta) = z(\alpha \cdot \beta)$, $z \in C$,
(d) $\alpha \cdot \alpha \geq 0$ and $\alpha \cdot \alpha = 0$ if and only if $\alpha = \bar{0}_V$.

The inner product is therefore a complex number. It is easy to show that these four properties imply the following two results:

(1) $(\alpha \cdot z\beta) = \bar{z}(\alpha \cdot \beta)$, $z \in C$, $\alpha, \beta \in V$,
(2) $\alpha \cdot (\beta + \gamma) = \alpha \cdot \beta + \alpha \cdot \gamma$.

Example 6. Let $\alpha_1, \alpha_2, \ldots, \alpha_n$ be a basis of an n-dimensional vector space V over the complex field C. If

$$\beta = x_1\alpha_1 + \cdots + x_n\alpha_n \quad \text{and} \quad \gamma = y_1\alpha_1 + \cdots + y_n\alpha_n$$

are two vectors of V, let us define an inner product in V by

(2) $\beta \cdot \gamma = x_1\bar{y}_1 + x_2\bar{y}_2 + \cdots + x_n\bar{y}_n$.

It can be easily verified that this definition fulfills the requirements (a), (b), (c), (d) for an inner product in V.

We again note that the basis $\alpha_1, \alpha_2, \ldots, \alpha_n$ of V, that is used to define the inner product (2), is an orthonormal basis with respect to this choice of inner product.

In particular if we choose the standard basis $(1, 0, 0, \ldots, 0)$, $(0, 1, 0, \ldots, 0), \ldots,$ $(0, 0, \ldots, 0, 1)$ for $V_n(C)$, then if $\alpha = (x_1, x_2, \ldots, x_n)$ and $\beta = (y_1, y_2, \ldots, y_n)$ are vectors in $V_n(C)$, we get the inner product

$$\alpha \cdot \beta = x_1\bar{y}_1 + \cdots + x_n\bar{y}_n$$

for $V_n(C)$.

This is called the **standard inner product for** $V_n(C)$.

Definition. Since for $\alpha \in V$, $\alpha \cdot \alpha \geq 0$, we define the **length** or **norm** $|\alpha|$ of the vector α by $|\alpha| = (\alpha \cdot \alpha)^{1/2}$.

Definition. Two vectors α and β are said to be **orthogonal** if $\alpha \cdot \beta = 0$.

Since $\beta \cdot \alpha = \overline{\alpha \cdot \beta}$, it follows that $\alpha \cdot \beta = 0$ implies $\beta \cdot \alpha = 0$.

Definition. A finite-dimensional inner-product vector space over the complex field is called a **unitary space**.

As in the real case we can show that to a linear-operator T on a unitary space $V_n(C)$, there corresponds a unique linear operator T^* on $V_n(C)$ for which

$$\alpha T \cdot \beta = \alpha \cdot (\beta T^*), \quad \text{for all } \alpha, \beta \in V_n(C).$$

As in the real case, it can be shown that if $A = (a_{ij})$ is the matrix of T with respect to an orthonormal basis of $V_n(C)$, then the matrix of T^* is the **conjugate transpose** $A^* = (\bar{a}_{ij})^t$ of the matrix A of T (with respect to this same basis).

Definition. The operator T^* is called the **adjoint** of the operator T.

If $T = T^*$, the linear operator T is said to be **hermitian** or **self-adjoint**. In terms of matrices, if $A = A^*$, then the matrix A is said to be **hermitian**.

Observe that if T is hermitian, then $\alpha T \cdot \alpha$ is real for all $\alpha \in V_n(C)$. This follows at once from

$$\alpha T \cdot \alpha = \alpha(\alpha T^*) = \alpha \cdot (\alpha T) = \overline{\alpha T \cdot \alpha}.$$

THEOREM 12

The eigenvalues of a hermitian linear operator T are real.

Proof: By Theorem 11 we know T has at least one eigenvector α. Let $\alpha T = x\alpha$, $\alpha \neq 0_V$. Then $\alpha T \cdot \alpha = (x\alpha) \cdot \alpha = x(\alpha \cdot \alpha)$. Since $\alpha T \cdot \alpha$ and $\alpha \cdot \alpha$ are both real and $\alpha \cdot \alpha \neq 0$, it follows that x is real.

Lemma 4 A real self-adjoint linear operator T on $V_n(R)$ has at least one eigenvector.

Proof: Let A be the matrix of T relative to an orthonormal basis of $V_n(R)$. Then by Theorem 8, Section 3, we have $A = A^t$. Since A can be regarded as a matrix over the complex field C, we can regard A as the matrix of a hermitian operator T' on $V_n(C)$. The eigenvalues of T' are real and are the roots of the characteristic polynomial $|A - xI| = O$. These are therefore also the eigenvalues of T and hence T has at least one real eigenvalue. Thus T has at least one eigenvector.

Corollary. All the characteristic roots of a real self-adjoint linear operator on $V_n(R)$ are real.

Definition. A nonsingular linear operator T on a unitary space is called a **unitary operator** if $T^*T = I$, alternatively, if $T^{-1} = T^*$.

Geometrically, a unitary operator is a linear transformation that preserves lengths of vectors.

Definition. A complex square matrix A is called **unitary** if $A^*A = I$, that is, if $A^* = A^{-1}$.

The unitary matrix in the complex field corresponds to the orthogonal matrix in the real field. However, here A^* is the conjugate transpose of A.

It is very easy to make obvious modifications in the proof of Theorem 9 to obtain the following result:

Eigenvalues and the Spectral Theorem

THEOREM 13

Let T be a hermitian linear operator on $V_n(C)$, where C is the complex field. Then there exists an orthonormal basis of $V_n(C)$ consisting of the eigenvectors of T.

Again with the same modifications we can prove

THEOREM 14

If T is a hermitian linear operator on $V_n(C)$, then there exists a unitary operator P on $V_n(C)$ such that $PTP^* = D$, where D is a diagonal operator.

With respect to the two chosen bases, D has the diagonal matrix whose entries on the main diagonal are the eigenvalues (they are all real) of T.

The obvious modifications for hermitian linear operators (mentioned above) are made necessary by the changed form for the inner product, and the replacement of orthogonal operators by unitary operators.

We point out that if $A = (a_{ij})$ is a unitary $n \times n$ matrix and if I is the $n \times n$ identity matrix, then $AA^* = I$ implies

$$\sum_{k=1}^{n} a_{jk} \overline{a_{jk}} = 1, \quad j = 1, 2, \ldots, n,$$

and

$$\sum_{k=1}^{n} a_{jk} \overline{a_{rk}} = 0, \quad r \neq j.$$

In words, this says that each row vector of A is of length 1, and any two row vectors are orthogonal.

The relation $A^*A = I$ implies the same thing is true of columns.

EXERCISES

1. Derive the properties (1) and (2) of the complex inner product.
2. Prove that the Schwarz inequality is still true for the complex inner product.
3. Prove that the vectors $(1, i)$ and $(1 - i, 0)$ of $V_2(C)$ are independent.
   Assuming the standard inner product in $V_2(C)$ find an orthonormal basis for the subspace of $V_2(C)$ spanned by the vectors $(1, i)$ and $(1 - i, 0)$.
4. With the standard inner product in $V_3(C)$, find an orthonormal basis for the subspace spanned by the vectors $(2, 3i, 6)$ and $(3 - 4i, 0, 0)$.
5. Prove that the formula (3) defines an inner product in $V_n(C)$.
6. Let V be the complex vector space of all complex-valued functions of a

real variable x, defined and integrable on some closed interval $[a,b]$. Prove that an inner product is defined in V by

$$f \cdot g = \int_a^b f(x) \cdot \overline{g(x)}\, dx$$

where $f, g \in V$ and $\bar{g}$ is the complex conjugate of g.

7. Define the length of a vector for the vector space V of Exercise 6 and verify the Schwarz inequality for this space.

8. Prove for the complex field C, that for each linear operator T on $V_n(C)$ there exists a unique linear operator T^* on $V_n(C)$, for which

$$\alpha T \cdot \beta = \alpha \cdot \beta T^*, \quad \alpha, \beta \in V_n(C).$$

9. If $A = (a_{ij})$ is the matrix of T with respect to any orthonormal basis for $V_n(C)$, prove that $\bar{A}^t = (\bar{a}_{ij})^t$ is the matrix of T^* with respect to this same basis. (The matrix of T^* is the conjugate transpose of the matrix of T.)

10. Prove Theorem 13.

11. Prove Theorem 14.

12. Prove $T = T^*$ if and only if $\alpha T \cdot \alpha$ is real for every $\alpha \in V_n(C)$.

13. Prove that a unitary operator on a finite-dimensional vector space preserves inner products.

14. If S and T are operators on a vector space V prove $(ST)^* = T^*S^*$.

15. Prove the product of two unitary operators on a vector space V is a unitary operator.

16. If T is a unitary operator and S is a self-adjoint operator prove that TST^{-1} is a self-adjoint operator.

17. Show by an example that the product of two self-adjoint operators is not self-adjoint.

18. If S and T are self-adjoint operators and if $ST = TS$, prove that the product ST is a self-adjoint operator.

19. A self-adjoint linear operator P on V is called **positive definite** if $\alpha P \cdot \alpha \geq 0$ for all $\alpha \in V$. If S is a nonsingular linear operator on V, prove that $S = TP$ where P is a nonsingular positive definite operator and T is a unitary operator. Show also that $S = P'T'$ where P' is nonsingular and positive definite and T' is unitary.

20. A linear operator T on the complex vector space $V_3(C)$ with the standard inner product is defined by

$$(x, y, z)T = (2x - (1+i)y + iz, \quad x + (3-4i)z, \quad (2+3i)x - iy).$$

Find $(x, y, z)T^*$.

21. If A is a positive definite real matrix, prove that A has an inverse that is positive definite.

8-5 THE SPECTRAL THEOREM

We have shown that an $n \times n$ matrix A over a field F is similar to a diagonal matrix, if and only if the eigenvectors of A span $V_n(F)$. Moreover, the diagonal elements are the eigenvalues of A. We have also shown that a real symmetric matrix is similar to a diagonal matrix; hence its eigenvectors span $V_n(R)$, where R is the real field. In other words, there exists a basis for $V_n(R)$ consisting of eigenvectors of the real symmetric matrix. In terms of linear operators, this says that a real symmetric linear operator T on $V_n(R)$ must have real eigenvalues and its eigenvectors span $V_n(R)$.

Let T be a real symmetric linear operator on the inner-product vector space $V_n(R)$ and let $x_1, x_2, \ldots, x_r$, $r \leq n$, denote its distinct (real) eigenvalues. Let S_1 denote the eigenspace [it is a subspace of $V_n(R)$] spanned by the eigenvectors of T that belong to the eigenvalue x_1. Now $V_n(R) = S_1 \oplus S_1^\perp$ (direct sum). Here $S_1^\perp$ stands, as usual, for the orthogonal complement of S_1.

Let S_2 be the eigenspace spanned by the eigenvectors of T that belong to the eigenvalue x_2. For $\alpha_1 \in S_1$, $\alpha_2 \in S_2$, we have $\alpha_1 T = x_1 \alpha_1$ and $\alpha_2 T = x_2 \alpha_2$. Since T is symmetric, $\alpha_1 \cdot \alpha_2 T = \alpha_1 T \cdot \alpha_2$. Hence $\alpha_1 \cdot x_2 \alpha_2 = x_1 \alpha_1 \cdot \alpha_2$; that is, $x_2(\alpha_1 \cdot \alpha_2) = x_1(\alpha_1 \cdot \alpha_2)$. Since $x_2 \neq x_1$, $\alpha_1 \cdot \alpha_2 = 0$. Thus α_1 and α_2 are orthogonal. Hence S_2 is a subspace of $S_1^\perp$, and so

$$S_1^\perp = S_2 \oplus S_2^\perp$$

where $S_2^\perp$ is the orthogonal complement of S_2 in $S_1^\perp$. Continuing from $x_1, x_2, \ldots$ to x_r we find that

$$V_n(R) = S_1 \oplus S_2 \oplus \ldots \oplus S_r \quad \text{(direct sum)},$$

for the vectors of $S_1, S_2, \ldots, S_r$ span $V_n(R)$. If $i \neq j$, S_i is orthogonal to S_j.

Every vector $\alpha \in V_n(R)$ thus has a unique expression of the form

$$\alpha = \alpha_1 + \alpha_2 + \cdots + \alpha_r,$$

where $\alpha_i \in S_i$, $i = 1, 2, \ldots, r$.

Example 7. Let

$$A = \begin{bmatrix} 0 & 1 & 1 \\ 1 & 0 & -1 \\ 1 & -1 & 0 \end{bmatrix}$$

be the matrix of a linear operator T on $V_3(R)$ with respect to the standard basis, where R is the real field. The characteristic polynomial of A is found to be $-x^3 + 3x - 2$ and its eigenvalues are therefore 1, 1, and -2. Call $x_1 = 1$, $x_2 = -2$. The eigenvectors corresponding to $x_1 = 1$ are found to be of the form $(t, u, t - u)$, where t and u are any real numbers, not both of which are 0. The eigenvectors for $x_2 = -2$ are found to have the form $(t, -t, -t)$, where t is any nonzero real number.

Thus the eigenspace S_1 is spanned by the vectors $(1,0,1)$ and $(0,1,-1)$ and the eigenspace S_2 is spanned by the vector $(1,-1,-1)$.

For each $i = 1, 2, \ldots, r$, the mapping P_i of $V_n(R) \to V_n(R)$ defined by

$$\alpha P_i = \alpha_i$$

is readily seen to be a linear operator on $V_n(R)$. As we know, it is called the *projection* of $V_n(R)$ on S_i. Clearly $P_i P_j = 0$, $i \neq j$, where 0 is the zero linear operator on $V_n(R)$. Thus the P_i form orthogonal projections. It also follows at once that $P_1 + P_2 + \cdots + P_r = I$, where I stands for the identity operator on $V_n(R)$. Let $\alpha \in V_n(R)$ and let

$$\alpha = \alpha_1 + \alpha_2 + \cdots + \alpha_r,$$

where $\alpha_i \in S_i$, $i = 1, 2, \ldots, r$. Now

$$\alpha T = \sum_{i=1}^r \alpha_i T = \sum_{i=1}^r x_i \alpha_i.$$

Also,

$$\alpha \left(\sum_{i=1}^r x_i P_i \right) = x_1 \alpha P_1 + \cdots + x_r \alpha P_r = \sum_{i=1}^r x_i \alpha_i.$$

Hence $T = \sum_{i=1}^r x_i P_i$. Moreover, $TP_i = P_i T$, $i = 1, 2, \ldots, r$. For $\alpha = \alpha_1 + \cdots + \alpha_r \in V_n(R)$, we have

$$\alpha T P_i = (\alpha_1 T + \cdots + \alpha_r T) P_i = (x_1 \alpha_1 + \cdots + x_r \alpha_r) P_i = x_i \alpha_i.$$

Also, $\alpha P_i T = (\alpha P_i) T = \alpha_i T = x_i \alpha_i$.

This combination of results

(a) $P_i P_j = 0$, $i \neq j$

(b) $\sum_{i=1}^r P_i = 1$

(c) $T = \sum_{i=1}^r x_i P_i$,

is often called the **spectral theorem for real symmetric linear operators** on a real finite-dimensional inner-product vector space.

Normal Operators

We shall assume throughout the rest of this section that V is a complex finite-dimensional inner-product vector space.

Definition. A linear operator T on V is called a **normal operator** if and only if $TT^* = T^*T$; that is, if and only if T commutes with its adjoint T^*.

Example. Hermitian and unitary operators are both normal operators.

Let S and T be linear operators on V. The proofs of the following formulas are straightforward and are left as exercises for the reader.

Exercise

(a) $(S + T)^* = S^* + T^*$
(b) $(xT)^* = \bar{x}T^*$, x a complex number
(c) $(ST)^* = T^*S^*$
(d) $(T^*)^* = T$
(e) $I^* = I$

Definition. If T is a linear operator on V, then a subspace W of V is said to be **T-invariant** if $\alpha T \in W$ for every $\alpha \in W$.

Lemma 5. Let T be a linear operator on V. If W is a T-invariant subspace of V, then its orthogonal complement $W^\perp$ is a T^*-invariant subspace of V.

Proof: Let $\alpha \in W$. If $\beta \in W$, then $\beta T \in W$. Hence $\beta \cdot \alpha T^* = \beta T \cdot \alpha = 0$. Therefore $\alpha T^* \in W^\perp$.

Lemma 6. If T is a normal operator on V, then $\alpha T = \bar{0}_V$ if and only if $\alpha T^* = \bar{0}_V$.

Proof: Now $\alpha T \cdot \alpha T = \alpha \cdot (\alpha T)T^* = \alpha \cdot (\alpha T^*)T$. Since $(T^*)^* = T$ (see the exercise above), we have $\alpha \cdot (\alpha T^*)T = \alpha T^* \cdot \alpha T^*$. Thus $|\alpha T| = |\alpha T^*|$. This proves the lemma.

Lemma 7. If T is a normal operator then $T - xI$, where x is a scalar, is a normal operator.

Proof: $(T - xI)(T - xI)^* = (T - xI)(T^* - \bar{x}I) = TT^* - \bar{x}T - xT^* + x\bar{x}I$. (See the above exercise.) Also $(T - xI)^*(T - xI) = (T^* - \bar{x}I)(T - xI) = T^*T - \bar{x}T - xT^* + x\bar{x}I$. Since $TT^* = T^*T$, we see that $(T - xI)(T - xI)^* = (T - xI)^*(T - xI)$, and therefore $T - xI$ is a normal operator.

Lemma 8. Let T be a normal operator on V. Then $\alpha \in V$ is an eigenvector of T if and only if α is an eigenvector of T^*. In fact if $\alpha T = x\alpha$, then $\alpha T^* = \bar{x}\alpha$, x a scalar.

Proof: $T - xI$ is a normal operator (Lemma 6). Hence $\alpha(T - xI) = \bar{0}_V$ if and only if $\alpha(T^* - \bar{x}I) = \bar{0}_V$ (Lemma 5).

We are ready to prove our main theorem.

THEOREM 15

Let T be a normal operator on a unitary space V. There exists an orthonormal basis of V consisting of eigenvectors of T; that is, the matrix of T with respect to this basis is a diagonal matrix whose diagonal entries are the eigenvalues of T, each eigenvalue appearing as many times as its multiplicity.

Proof: We use induction on the dimension of V. The theorem is trivially true if dim $V = 1$. Assume it true for vector spaces of dimension $n - 1, n > 2$. Let dim $V = n$.

Since the field is complex, the operator T has at least one eigenvalue and hence an eigenvector α. Since $\alpha \neq 0_V$, we can assume it to be a unit vector. Let W be the subspace of V spanned by the vector α. Since α is an eigenvector of T, W is a T-invariant subspace of V; and since α is an eigenvector of T^* (Lemma 7), W is also a T^*-invariant subspace. Hence $W^\perp$ is invariant under $(T^*)^* = T$ (Lemma 1).

Now $V = W \oplus W^\perp$. Hence dim V = dim W + dim $W^\perp$. Since dim $W = 1$, dim $W^\perp = n - 1$. Hence by our induction hypothesis, there exists an orthonormal basis $\alpha_2, \ldots, \alpha_n$ of $W^\perp$ consisting of eigenvectors of the restriction of T to $W^\perp$, and therefore eigenvectors of T. Since $\alpha_1 \cdot \alpha_i = 0, i = 2, 3, \ldots, n$, then $\alpha_1, \alpha_2, \ldots, \alpha_n$ is an orthonormal basis of V and the α_i are eigenvectors of T.

As we have shown before, in the case of real symmetric (self-adjoint) operators on a real finite-dimensional inner-product vector space, this last theorem is equivalent to the spectral theorem for normal operators.

THEOREM 16 (SPECTRAL THEOREM)

Let T be a normal operator on a unitary space V. There exist orthogonal projections $P_1, P_2, \ldots, P_r$ on V and scalars $x_1, x_2, \ldots, x_r$, such that

(1) $$T = \sum_{i=1}^{r} x_i P_i$$

(2) $$\sum_{i=1}^{r} P_i = I$$

(3) $$P_i P_j = 0 \quad \text{(the zero operator)}, i \neq j.$$

Since hermitian and unitary operators are normal operators the spectral theorem holds for them. In particular, it specializes into the previous spectral theorem for real symmetric operators on a real finite dimensional vector space.

Example 9. Let T be the operator on the inner product $V_2(F)$ defined by $(x_1, x_2)T = (ax_1 + bx_2, bx_1 + cx_2)$, $a,b,c \in F$. The matrix of T relative to the standard basis $(1,0)$, $(0,1)$ is $\begin{pmatrix} a & b \\ b & c \end{pmatrix}$, hence T is self-adjoint.

On the other hand, the operator $(x_1, x_2)T = (ax_1 - bx_2, bx_1 + ax_2)$ has the matrix $\begin{pmatrix} a & b \\ -b & a \end{pmatrix}$ and is therefore not self-adjoint. The transpose of $\begin{pmatrix} a & b \\ -b & a \end{pmatrix}$ is $\begin{pmatrix} a & -b \\ b & a \end{pmatrix}$ and this yields the adjoint operator T^* where $(x_1, x_2)T^* = (ax_1 + bx_2, -bx_1 + ax_2)$. It is easy to show however that T is normal.

Example 10. Let T be an operator on $V_2(F)$ defined by $(y_1, y_2)T = (4y_1 + 2y_2, 2y_1 + y_2)$. It is easily verified that $\alpha T \cdot \beta = \alpha \cdot \beta T$, for all vectors α and β of $V_2(F)$, so that T is self-adjoint. Alternatively, the matrix $\begin{pmatrix} 4 & 2 \\ 2 & 1 \end{pmatrix}$ of T is symmetric and so $T = T^*$. The eigenvalues of T are found to be $x_1 = 0$ and $x_2 = 5$ and $\alpha_1 = (1, -2)$, $\alpha_2 = (2, 1)$ are the corresponding eigenvectors. Thus the eigenspace S_1 is spanned by $(1, -2)$ and S_2 is spanned by $(2, 1)$. Clearly $(1, -2) \cdot (2, 1) = 0$ and $S_2 = S_1^\perp$. The eigenvectors span $V_2(F)$ and hence $V_2(F) = S_1 \oplus S_2$. If P_1 and P_2 are the projections of $V_2(F)$ on S_1 and S_2 respectively, then $P_1 + P_2 = I_V$ and $P_1 P_2 = 0$, where now 0 stands for the operator mapping every element of $V_2(F)$ into $\bar{0}_V$. Thus P_1 and P_2 are orthogonal. Now $\alpha_1 T = \bar{0}_V$ and $\alpha_2 T = 5\alpha_2$. $S_1 = \ker T$ and hence $S_2 = \operatorname{im} T$.

If $\alpha \in V$, then $\alpha = \beta_1 + \beta_2$, $\beta_1 \in S_1$ and $\beta_2 \in S_2$. Hence $\alpha T = \beta_1 T + \beta_2 T = 0 + 5\beta_2 = 5(\alpha P_2) = \alpha(5 P_2)$. Hence $T = 5P_2$. We can write $T = x_1 P_1 + x_2 P_2$ where $x_1 = 0$, $x_2 = 5$ are the eigenvalues. This is a decomposition of T into orthogonal projections for which $P_1 + P_2 = I_V$.

Example 11. Let T be the operator on the inner-product vector space $V_3(F)$, F the real field, defined by
$$(y_1, y_2, y_3)T = (2y_1 - y_2, -y_1 + 2y_2 - y_3, -y_2 + 2y_3).$$
Relative to the basis $(1,0,0)$, $(0,1,0)$, $(0,0,1)$ the matrix of T is
$\begin{bmatrix} 2 & -1 & 0 \\ -1 & 2 & -1 \\ 0 & -1 & 2 \end{bmatrix}$ and hence T is self-adjoint. The characteristic equation is $(2 - x)^3 - 2(2 - x) = 0$ and the eigenvalues are found to be $x_1 = 2$, $x_2 = 2 + \sqrt{2}$, $x_3 = 2 - \sqrt{2}$. The eigenspaces S_1, S_2, S_3 corresponding to these eigenvalues are spanned by the eigenvectors $(1, 0, -1)$, $(1, -\sqrt{2}, 1)$ and $(1, \sqrt{2}, 1)$ and $V_3(F) = S_1 \oplus S_2 \oplus S_3$. Let P_1, P_2, P_3 be the projections of $V_3(F)$ on S_1, S_2, S_3 respectively. This means that if

$\alpha \in V_3(F)$ and $\alpha = \alpha_1 + \alpha_2 + \alpha_3$ where $\alpha_i \in S_i$, $i = 1, 2, 3$, then $\alpha P_i = \alpha_i$, $i = 1, 2, 3$.

It can easily be verified that $P_i P_j$, $i \neq j$, is the zero operator and that $P_1 + P_2 + P_3$ is the identity operator. Moreover, it follows readily that $T = x_1 P_1 + x_2 P_2 + x_3 P_3$.

EXERCISES

1. If S is a linear operator that commutes with the normal linear operator T, prove that S commutes with T^*.

2. S and T are normal operators and $ST = TS$. Prove that $S + T$ and ST are normal operators.

3. Prove that a linear operator T on a finite-dimensional complex inner-product vector space is normal if and only if $|\alpha T| = |\alpha T^*|$ for every $\alpha \subset V$.

4. A linear operator T on the complex vector space $V_2(C)$ with the standard inner product is defined by $(x, y)T = (x - iy, -ix + y)$, $i = \sqrt{-1}$. Prove that T is a normal operator, and prove there exists an orthonormal basis for $V_2(C)$ made up of eigenvectors of T.

5. Prove that the eigenvectors, belonging to distinct eigenvalues of a normal operator T, are orthogonal.

6. Let x be an eigenvalue of a normal operator T, and let S be the set of all solutions of $\alpha T = x\alpha$. Prove that S and $S^\perp$ are invariant under T; that is, $T(S) \subset S$ and $T(S^\perp) \subset S^\perp$.

7. Find an example of a normal operator that is neither unitary nor hermitian.

8. A normal operator on a finite-dimensional complex vector space $V_n(C)$ is unitary if and only if all its eigenvalues have absolute value 1.

9. Prove that if a normal operator T on an inner-product vector space V has the property that U is a T-invariant subspace of V (that is $T(U) \subset U$) then $U^\perp$ is a T-invariant subspace of V.

10. If the characteristic polynomial of a linear operator T on an n-dimensional vector space over a field F is irreducible over F, find the minimal polynomial and the eigenvalues of T.

11. T is a linear operator on a vector space V over F. If $f(x) \in F[x]$, prove that the set of all vectors $\alpha \in V$ such that $\alpha f(T) = \overline{0}_V$ is a T-invariant subspace of V; that is, show these vectors form a subspace U of V and that $T(U) \subset U$.

12. T is a linear operator on a two-dimensional vector space over the complex field and is defined by $\alpha T = 3\alpha + 5\beta$, $\beta T = -2\alpha + \beta$, where α and β form a basis for V.
    (a) Find the characteristic and minimal polynomials of T.
    (b) Find the eigenvalues of T.
    (c) Is T diagonable?
    (d) Find the eigenspaces of T and show they are orthogonal complements.

Chapter 9

Bilinear and Quadratic Forms

9-1 DEFINITIONS AND NOTATION

We have already studied linear forms on a vector space V over a field F and have shown that they form a vector space V^* over F, called the dual space to V. Also we left as an exercise the proof that the bilinear forms on V (the bilinear mappings of $V \times V \to F$) form a vector space over F.

We shall be concerned here, however, with the bilinear forms themselves, particularly the symmetric forms. These will lead us to a definition of a quadratic form on a finite-dimensional vector space, and our principal goal in this chapter is the reduction and simplification of quadratic forms over the real and complex fields.

We repeat in detail the definition of a bilinear function.

Definition. A **bilinear function** on a vector space V over a field F is a mapping f of $V \times V \to F$ that satisfies the following conditions:
For all $\alpha, \beta, \in V$ and all $x \in F$,

(1)
$$(\alpha + \beta, \gamma)f = (\alpha, \gamma)f + (\beta, \gamma)f,$$
$$(\alpha, \beta + \gamma)f = (\alpha, \beta)f + (\alpha, \gamma)f,$$
$$(x\alpha, \beta)f = x((\alpha, \beta)f) = (\alpha, x\beta)f.$$

A bilinear function f is called **symmetric** if for all $\alpha, \beta \in V$, $(\alpha, \beta)f = (\beta, \alpha)f$. It is called **skew-symmetric** if $(\alpha, \beta)f = -(\beta, \alpha)f$.

We shall now assume that V is an n-dimensional vector space and take $\alpha_1, \alpha_2, \ldots, \alpha_n$ to be a basis for V. If f is a bilinear function on V, then the $n \times n$ matrix $A = (a_{ij})$ defined by

(2) $\qquad a_{ij} = (\alpha_i, \alpha_j)f, \qquad i, j = 1, 2, \ldots, n,$

is called the **matrix of f** with respect to this basis.

If f is symmetric, then $a_{ij} = a_{ji}$ for all $i, j = 1, 2, \ldots, n$. This means its matrix A has the property $A^t = A$, and A is called a **symmetric matrix**. Conversely, if the matrix A of f is symmetric then f is a symmetric bilinear function. It is evident that if f has a symmetric matrix with respect to one basis of V, then f will have a symmetric matrix with respect to any basis of V.

If $X = x_1\alpha_1 + \cdots + x_n\alpha_n$ and $Y = y_1\alpha_1 + \cdots + y_n\alpha_n$ are any vectors in V, then $(X,Y)f$ is linear in each of the vectors X and Y, and so we have

$$(X,Y)f = (x_1\alpha_1 + \cdots + x_n\alpha_n,\ y_1\alpha_1 + \cdots + y_n\alpha_n)f$$

$$= \sum_{i,j=1}^{n} (x_i\alpha_i, y_j\alpha_j)f$$

$$= \sum_{i,j=1}^{n} x_i y_j (\alpha_i, \alpha_j)f$$

$$= \sum_{i,j=1}^{n} a_{ij} x_i y_j, \quad \text{by (2)}.$$

We call

(3) $$(X,Y)f = \sum_{i,j=1}^{n} a_{ij} x_i y_j$$

a **bilinear form on** V in the $2n$ variables x_i, y_j, $i, j = 1, 2, \ldots, n$. Since the a_{ij}, x_i, y_j are scalars we have $(X, Y)f \in F$, as it should.

We are departing here from our usual practice of using Greek letters for vectors, in anticipation of the fact that it will often be found convenient to regard a vector in an n-dimensional vector space as a matrix. Thus if $\alpha_1, \alpha_2, \ldots, \alpha_n$ is a basis of the vector space and if $X = x_1\alpha_1 + \cdots + x_n\alpha_n$, then we can write X in the matrix notation $X = (x_1 x_2 \cdots x_n)$. The use of a capital X for the vector is then consistent with our notation of capital letters for matrices. We also remark that sometimes we write V for an n-dimensional vector space over a field F, and sometimes we write $V_n(F)$ for such a space. The two spaces are isomorphic and actually their only distinction is notational. However, the notation $V_n(F)$ does serve to emphasize the matrix aspect of its vectors X.

Example 1. If V is a finite-dimensional inner-product vector space and if X and Y are any vectors in V, then $(X,Y)f = X \cdot Y$ is a symmetric bilinear form on V.

Example 2. If $2x_1 y_1 + 3x_1 y_2 - 4x_2 y_3 + 5x_3 y_1$ represents a bilinear form on a real three-dimensional vector space V, then the bilinear function f of $V \times V \to R$ is determined by

$$(X,Y)f = 2x_1 y_1 + 3x_1 y_2 - 4x_2 y_3 + 5x_3 y_1,$$

where $X = x_1\alpha_1 + x_2\alpha_2 + x_3\alpha_3$ and $Y = y_1\alpha_1 + y_2\alpha_2 + y_3\alpha_3$ are arbitrary vectors of V expressed in terms of some basis $\alpha_1, \alpha_2, \alpha_3$ of V. The matrix of this bilinear form is $\begin{bmatrix} 2 & 3 & 0 \\ 0 & 0 & -4 \\ 5 & 0 & 0 \end{bmatrix}$. It is not symmetric, and so the bilinear form is not symmetric.

If we write the vectors X and Y of the n-dimensional vector space V as row vectors (matrices)

$$X = (x_1 x_2 \cdots x_n) \quad \text{and} \quad Y = (y_1 y_2 \cdots y_n)$$

then we can write (3) in the matrix form

$$(X, Y)f = XAY^t,$$

where $A = (a_{ij})$ and Y^t is the column vector $\begin{bmatrix} y_1 \\ y_2 \\ \vdots \\ y_n \end{bmatrix}$, the transpose of the row vector Y.

Definition. Let f be a given symmetric bilinear function on a vector space V over F. The mapping Q of $V \to F$ defined by

$$Q(\alpha) = (\alpha, \alpha)f$$

is called a **quadratic function on** V (the quadratic function associated with the symmetric bilinear function f.)

Assume now that V is again an n-dimensional vector space over F with a basis $\alpha_1, \alpha_2, \ldots, \alpha_n$. The matrix $A = (a_{ij})$, where $a_{ij} = (\alpha_i, \alpha_j)f$, $i,j = 1, 2, \ldots, n$, is called the **matrix of** Q with respect to this basis. Since f is symmetric, *the matrix of Q is symmetric.*

If $X = x_1\alpha_1 + \cdots + x_n\alpha_n$ is any vector of V, then $Q(X) = (X, X)f$ and, as in (3), we obtain

$$Q(X) = \sum_{i,j=1}^{n} a_{ij} x_i x_j.$$

As before we can write this in matrix form as follows:

Let $X = (x_1 x_2 \cdots x_n)$ and let A be the symmetric matrix $A = (a_{ij})$. Then

$$Q(X) = XAX^t$$

where

$$X^t = \begin{bmatrix} x_1 \\ x_2 \\ \vdots \\ x_n \end{bmatrix}$$

is the transpose of X.

Definition. $Q(X)$ is called a **quadratic form** (**q-form**).

We see that the matrix of Q with respect to the basis $\alpha_1, \alpha_2, \ldots, \alpha$ is the matrix of the coefficients of the quadratic form $Q(X)$ for $X = x_1\alpha_1 + \cdots + x_n\alpha_n$.

Definition. Two $n \times n$ matrices A and B over a field F are said to be **congruent** if there exists a nonsingular matrix P such that $A = PBP^t$.

Lemma 1. If A and B are $n \times n$ matrices over a field F, then $(AB)^t = B^t A^t$.

Proof: It is easy to verify that the element in the ith row and jth column of each of the matrices $(AB)^t$ and $B^t A^t$ is $\sum_{i=1}^{n} a_{ji} b_{ij}$. Hence the two matrices are equal.

Lemma 2. If A is a nonsingular $n \times n$ matrix, then $(A^t)^{-1} = (A^{-1})^t$.

Proof: If I denotes the identity matrix, we have

$$I = (AA^{-1})^t = (A^{-1})^t A^t, \text{ by Lemma 1.}$$

Also, $I = (A^{-1}A)^t = A^t(A^{-1})^t$. Hence $(A^{-1})^t$ is the inverse of A^t. (Since $\det A^t = \det A$, A^t is also a nonsingular matrix.)

What is the effect on the quadratic form $Q(X) = XAX^t$ if we change the basis of V? Suppose $\alpha_1, \ldots, \alpha_n$ is the basis of V for which $Q(X) = XAX^t$. Let P be the automorphism of V that changes the basis from $\alpha_1, \ldots, \alpha_n$ to a new basis $\beta_1, \ldots, \beta_n$. Then

$$\alpha_i P = \beta_i = \sum_{j=1}^{n} p_{ij} \alpha_j, \quad i = 1, 2, \ldots, n.$$

A change of basis is equivalent to a change in the coordinates x_i of the vector $X = (x_1, x_2, \ldots, x_n)$. Suppose this vector becomes the vector $Y = (y_1, y_2, \ldots, y_n)$ with respect to the new basis $\beta_1, \ldots, \beta_n$. As a matrix equation we can write this transformation as

$$X = YP.$$

Bilinear and Quadratic Forms

Then
$$Q(Y) = (YP)A(YP)^t$$
$$= Y(PAP^t)Y^t.$$

Hence the effect of the change of basis is that the matrix A of Q becomes the matrix PAP^t under the transformation $X = YP$.

Note that the new matrix PAP^t is symmetric just as A is. For $(PAP^t)^t = P^{tt}A^tP^t = PAP^t$.

If we write a basis $\alpha_1, \ldots, \alpha_n$ formally as a matrix in the form $\alpha = (\alpha_1, \alpha_2, \ldots, \alpha_n)$, then we can write a change of basis in the symbolic form

$$\beta = \alpha P^t$$

where $\beta = (\beta_1, \ldots, \beta_n)$ is a new basis.

Two real quadratic forms are said to be **equivalent** if there exists a nonsingular linear transformation P that transforms one into the other. We have seen that if they are equivalent then their matrices are congruent. Conversely, if their matrices are congruent $A = PBP^t$, then P is nonsingular and P transforms one into the other. *Thus two real quadratic forms are equivalent if and only if their matrices are congruent.*

EXERCISES

1. If A and B are congruent matrices and if B and C are congruent matrices, show that A and C are congruent matrices.

2. Find a nonsingular matrix P for which $P^t = P^{-1}$.

3. Similar matrices have the same eigenvalues. Show by a counterexample that this is not true of congruent matrices.

4. If f and g are linear functions on $V \to F$ prove that the function ϕ defined on $V \times V \to F$ by $(\alpha, \beta)\phi = (\alpha f) \cdot (\beta g)$, $\alpha, \beta \in V$ is bilinear.

5. Prove that the quadratic forms
$$3x_1^2 + 2x_1x_3 + 2x_2^2 \quad \text{and} \quad 5y_1^2 - 8y_1y_2 + 2y_1y_3 + 5y_2^2 - 2y_2y_3$$
are equivalent.

6. Show that being congruent is an equivalence relation on the set of all $n \times n$ symmetric matrices over a field.

7. Describe a skew-symmetric matrix A and prove that if P is an orthogonal matrix, then $P^{-1}AP$ is a skew-symmetric matrix.

8. If A and B are symmetric matrices, prove that the product AB is symmetric if and only if $AB = BA$.

9. Show that the function f of $V_3(R) \to R$ defined by
$$(X, Y)f = 3x_1y_1 + 2x_1y_3 - x_2y_3 - 2x_3y_1 + x_3y_2 + 4x_3y_3$$
where $X = (x_1, x_2, x_3)$, $Y = (y_1, y_2, y_3)$ is a skew-symmetric bilinear form.

10. f is a bilinear form on an n-dimensional vector space V over F. The matrix of f is A with respect to some basis of V. Prove that the matrix of f is PAP^t with respect to a new basis of V, where P is the automorphism changing the basis of V.

9-2 THE REDUCTION OF QUADRATIC FORMS UNDER GROUPS

Let S and T be two automorphisms of a vector space V over a field F. Their product (map composition) is given by $\alpha(ST) = (\alpha S)T$, $\alpha \in V$. Now ST is also an automorphism. For if $\alpha(ST) = \beta(ST)$, $\alpha, \beta \in V$, then $\alpha S = \beta S$, since T is an automorphism. Hence $\alpha = \beta$, since S is an automorphism. This proves ST is injective. It is also surjective, for let α be any vector of V. Then there exists $\beta \in V$ such that $\beta T = \alpha$ and $\gamma \in V$ such that $\gamma S = \beta$. This is true since both S and T are surjective. Hence $\alpha = \gamma(ST)$, and so ST is surjective. We have already seen (Chapter 3) that ST is linear. Hence ST is an automorphism. The identity map 1_V on V is clearly an automorphism and the inverse T^{-1} of an automorphism T is an automorphism. Since map composition is an associative binary operation, it follows that the set of automorphisms of V forms a group. We repeat a definition given earlier in Chapter 3.

Definition. The group of automorphisms of the n-dimensional vector space $V_n(F)$ is called the **full linear group** $L_n(F)$. The full linear group is often called the *general linear group*.

We are going to study the effects of applying automorphisms of $V_n(F)$ to quadratic forms (q-forms, as we shall often call them), hoping thereby to reduce them to q-forms of a simpler type, which usually means to forms with more of the coefficients $a_{ij} = 0$. Since these automorphisms form the full linear group $L_n(F)$, we can apply successively any finite number of automorphisms to a given Q-form and their total effect is that of a single automorphism. We speak of this process as the reduction of q-forms *"under the full linear group."*

We shall also consider the reduction of a q-form on $V_n(F)$ under the orthogonal group. In this case we shall assume $V_n(F)$ is an inner-product vector space, usually with the standard inner product. The orthogonal group (the set of all orthogonal transformations on $V_n(F)$) is a subgroup of the full linear group $L_n(F)$. Naturally we would expect to achieve more by using the full linear group but it is both of interest and importance to see how much can be done with this restriction to its subgroup, the orthogonal group.

An automorphism T of an n-dimensional vector space V over F can be represented by a nonsingular $n \times n$ matrix P over F. If $Q(X)$ is a q-form on V with $n \times n$ symmetric matrix A ($A = A^t$) then T transforms $Q(X)$ into a q-form on V whose matrix is PAP^t and this matrix is also symmetric. Thus our problem can be restated in terms of matrices as follows: how much can we simplify—perhaps even reduce to diagonal form—the matrix PAP^t by nonsingular matrices P of different types?

We shall see that under the full linear group a symmetric matrix A can be reduced to diagonal form, and that the number of nonzero terms on the main diagonal of the diagonal form is equal to the rank of A.

First we make a few remarks about the rank of a matrix A and show that when A is multiplied by a nonsingular matrix, the product matrix has the same rank as A.

If U is a subspace of a finite-dimensional vector space V and if P is an automorphism of V (nonsingular linear transformation of $V \to V$), then it is easy to see that P maps U into a subspace of the same dimension as U. For dim U = dim (ker P) + dim ($P(U)$), where $P(U)$ is the image of U under P. Since ker $P = O$, the result follows at once. Thus P preserves dimension.

Now let T be a linear transformation of $V \to V$ (an endomorphism of V) of rank r. Then, by definition, dim (im T) = r. Hence if P and Q are automorphisms of V, then it follows that dim (im PTQ) = r. For im $P = V$ and hence im PT = im T, and, since dim (im T) = r, Q maps im T into a subspace of dimension r.

In terms of matrices this translates at once into:

If A is an $n \times n$ matrix of rank r and if P and Q are nonsingular $n \times n$ matrices, then PAQ is a matrix of rank r. In particular then if $Q = P^t$ (transpose of P), the matrices A and PAP^t have the same rank; that is, *congruent matrices have the same rank*.

If then a quadratic form with a matrix of rank r is reduced by nonsingular transformations to a diagonal form, there are exactly r nonzero entries on the main diagonal of the diagonal form. Any more or less than r would mean the rank of the diagonal form is $> r$ or $< r$.

Since the rank of the matrix of a real q-form is invariant under nonsingular linear transformations, we make the following

Definition. The *rank* of the matrix of a real quadratic form is called the **rank of the form.**

Definition. The **determinantal rank** of an $m \times n$ matrix A over a field is the order (number of rows) of the largest square submatrix of A (obtained by deleting rows and columns of A) whose determinant is not 0.

Lemma 3. The determinantal rank of a matrix is equal to its rank.

Proof: Let A be an $m \times n$ matrix over a field. Let rank of $A = r$ and determinantal rank of $A = s$.

Now at least one $s \times s$ submatrix of A has a nonzero determinant. We can permute the rows of A and permute the columns of A so that this submatrix B is in the upper left hand corner of the resulting new matrix A'. Such permutations clearly do not change the values of r and s, that is, rank of $A' = r$ and determinantal rank of A' is s. Now a linear combination with nonzero coefficients of the first s rows of A' that is zero would imply the existence of a similar relation among the rows of the submatrix B. But this is impossible, since det $B \neq 0$ and so these s rows of A must be linearly independent. Thus $s \leq r$.

On the other hand, there exist r linearly independent rows in A. We permute the rows of A so that these linearly independent rows appear as the first r rows of the resulting new matrix A'. Let C be the $r \times n$ matrix formed with these rows. Now the rank of $C = r$. Since the row rank of $C =$ the column rank of C (= rank of C), we can determine r linearly independent columns of C. Now permute the columns of A', and hence also of C, to obtain an $r \times r$ matrix C'. Again we have rank of $C' = r$. Hence det $C' \neq 0$, and therefore $r \leq s$.

Since $s \leq r$, this proves $r = s$.

The fact stated in this last lemma frequently provides a very quick way of determining the rank of a matrix.

We begin the consideration of reducing q-forms under the full linear group with two illustrative examples. These will help to explain the general method. In both examples the method of "completion of squares" is used to diagonalize the q-form.

The reader must bear in mind, as we emphasized in Sec. 5-1, that the matrix of the linear equations defining a linear transformation is the transpose of the matrix of the linear transformation.

Example 2. Let

$$Q(X) = 3x_1^2 - 12x_1x_2 + 5x_2^2$$
$$= 3(x_1^2 - 4x_1x_2 + 4x_2^2) - 7x_2^2 = 3(x_1 - 2x_2)^2 - 7x_2^2$$

The matrix of Q is $A = \begin{pmatrix} 3 & -6 \\ -6 & 5 \end{pmatrix}$ and det $A \neq 0$. Hence the rank of A is 2.

Apply to $Q(X)$ the linear transformation T defined by $y_1 = x_1 - 2x_2$, $y_2 = x_2$. We have $(x_1, x_2) = (y_1, y_2)T$. The matrix of T is $P = \begin{pmatrix} 1 & 0 \\ 2 & 1 \end{pmatrix}$. Since det $P \neq 0$, T is nonsingular. Hence $T \in L_2(R)$, the full linear group, where R is the real field. $Q(X)$ is transformed by T into the diagonal form

$$Q(Y) = 3y_1^2 - 7y_2^2.$$

A check by use of matrices reveals very easily that $PAP' = \begin{pmatrix} 3 & 0 \\ 0 & -7 \end{pmatrix}$, as of course it should.

Bilinear and Quadratic Forms

Note that the matrix P of T is obtained by solving for x_1 and x_2 in terms of y_1 and y_2. We get $x_1 = y_1 + 2y_2$, $x_2 = y_2$, and now P is the transpose of $\begin{pmatrix} 1 & 2 \\ 0 & 1 \end{pmatrix}$.

Example 3. Let
$$Q(X) = 6x_1x_2 + 12x_1x_3 + x_2^2 + 3x_3^2 = (3x_1 + x_2)^2 + 3(2x_1 + x_3)^2 - 21x_1^2.$$
The matrix of Q is
$$A = \begin{bmatrix} 0 & 3 & 6 \\ 3 & 1 & 0 \\ 6 & 0 & 3 \end{bmatrix},$$
and det $A \neq 0$. Hence the rank of Q is 3.

Now define a linear transformation T by $y_1 = 3x_1 + x_2$, $y_2 = 2x_1 + x_3$, $y_3 = x_1$. We have $(x_1, x_2, x_3) = (y_1, y_2, y_3)T$, where $x_1 = y_3$, $x_2 = y_1 - 3y_3$, $x_3 = y_2 - 2y_3$. Hence the matrix P of T is
$$P = \begin{bmatrix} 0 & 1 & 0 \\ 0 & 0 & 1 \\ 1 & -3 & -2 \end{bmatrix}.$$

Since det $P \neq 0$, T is nonsingular, and hence $T \in L_3(R)$, the full linear group. If we apply T to $Q(X)$ we get
$$Q(Y) = y_1^2 + 3y_2^2 - 21y_3^2.$$
Again, a check by matrices shows that
$$PAP^t = \begin{bmatrix} 1 & 0 & 0 \\ 0 & 3 & 0 \\ 0 & 0 & -21 \end{bmatrix}.$$

Note also that there are three diagonal terms in the reduced diagonal form $Q(Y)$, corresponding to the rank three of A.

EXERCISES

1. Find the rank of the linear operator T on $V_3(R)$ defined by $(x, y, z)T = (x - 2y + z, -x + 2y, z)$.

2. Find the rank of the quadratic form
$$Q(X) = 2x_1^2 + 2x_2x_4 - 4x_3x_4.$$

3. Find a nonsingular matrix P such that PAP^t is a diagonal matrix where $A = \begin{pmatrix} 1 & -2 \\ -2 & 3 \end{pmatrix}$.

4. Prove that the set of orthogonal transformations on V is a group.

9-3 THE REDUCTION OF A QUADRATIC FORM UNDER THE FULL LINEAR GROUP

Let
$$Q(X) = \sum_{i,j=1}^{n} a_{ij} x_i x_j$$

be a quadratic form in n variables over the field F; that is, the coefficients $a_{ij} \in F$. $X = (x_1, x_2, \ldots, x_n)$ is then a vector in $V_n(F)$.

We shall assume $Q(X)$ is a nonzero q-form; that is, there is at least one coefficient of $Q(X)$ that is not zero.

If this nonzero term is a diagonal term, $a_{kk} \neq 0$, then the nonsingular linear transformation $X = X'P$ defined by $x_1 = x'_k$, $x_k = x'_1$, $x_j = x'_j$ for all $j \neq 1$ and $j \neq k$, transforms $Q(X)$ into a q-form whose new coefficient $a'_{11} \neq 0$.

If all the diagonal terms of $Q(X)$ are zero then there is a coefficient $a_{ij} \neq 0$, $i \neq j$. In this case the nonsingular transformation $X = X'P$ defined by $x_i = x'_i$, $x_j = a_{ij} x'_i + a_{ji} x'_j$, $x_k = x'_k$ for all $k \neq i$ and $k \neq j$, transforms $Q(X)$ into a q-form whose new coefficient $a'_{ii} \neq 0$.

The upshot of these remarks is that there is no loss of generality in assuming at the outset that if $Q(X) \neq 0$, then $a_{11} \neq 0$; for if $a_{11} = 0$, nonsingular linear transformations will transform $Q(X)$ into a q-form in which this first diagonal term is not zero.

It is easy to see that we can write $Q(X)$ in the form

$$Q(X) = \frac{1}{a_{11}} \left[a_{11}x_1 + \sum_{j=2}^{n} a_{1j}x_j \right] \left[a_{11}x_1 + \sum_{i=2}^{n} a_{i1}x_i \right] + Q'(x_2, x_3, \ldots, x_n)$$

where $Q'(x_2, \ldots, x_n)$ is a quadratic form in the $n - 1$ variables $x_2, x_3, \ldots, x_n$.

If we now transform $Q(X)$ by the nonsingular linear transformation

$$x'_1 = a_{11}x_1 + \sum_{j=2}^{n} a_{1j}x_j, \quad x'_k = x_k \text{ for } k = 2, 3, \ldots, n$$

(the determinant of this linear transformation is easily verified to be $a_{11} \neq 0$) we obtain the new q-form†

$$\frac{1}{a_{11}} x'^2_1 + Q'(x'_2, \ldots, x'_n).$$

†We are assuming that the characteristic of the field is not two.

We can now continue by using this same argument on the q-form $Q'(x_2, \ldots, x_n')$ in $n - 1$ variables. After a finite number of steps we can ultimately reduce the original q-form to the diagonal form

(4) $$g_1 y_1^2 + g_2 y_2^2 + \cdots + g_r^2 y_r^2,$$

where $r \leq n$ and the $g_i \in F$. The positive integer r is the **rank of Q(X)**.

If F is the real field R, put $z_i = \sqrt{g_i}\, y_i$ for all $g_i > 0$, and put $z_i = \sqrt{-g_i}\, y_i$ for all $g_i < 0$ in (4) and we see that a real q-form can be reduced to the form

(5) $$z_1^2 + \cdots + z_p^2 - z_{p+1}^2 - \cdots - z_r^2.$$

On the other hand, if F is the complex field the transformation $z_i = \sqrt{g_i}\, y_i$ will reduce (4) to the form

(6) $$z_1^2 + z_2^2 + \cdots + z_r^2.$$

We can conclude then that under the full linear group, a real q-form can be reduced to the diagonal form (5), while a complex q-form can be reduced to the diagonal form (6).

We know that the rank of a real quadratic form is invariant under the full linear group. We shall now prove that the number of positive terms, when it is reduced to diagonal form, is also an invariant under this group. This is called **Sylvester's law of inertia**. For this reason the diagonal form (5) is termed **canonical** under the full linear group.

THEOREM 1

The number of positive terms is the same for every diagonal form of a given real quadratic function.

Proof: Assume there are two diagonal forms

(7) $$z_1^2 + \cdots + z_p^2 - z_{p+1}^2 - \cdots - z_r^2$$
(8) $$y_1^2 + \cdots + y_q^2 - y_{q+1}^2 - \cdots - y_r^2$$

for the same quadratic function Q. Then there exists a nonsingular linear transformation P carrying (7) into (8). We can therefore regard $(z_1, z_2, \ldots, z_n)$ as the coordinates of a vector α with respect to some basis of V, and $(y_1, y_2, \ldots, y_n)$ as the coordinates of the same vector with respect to a second basis.

Now $Q(\alpha) \geq 0$, if $z_{p+1} = \cdots = z_r = 0$. The vectors α that satisfy these $r - p$ conditions clearly form a subspace S_1 of V of dimension $n - (r - p)$, where n is the dimension of V.

Also $Q(\alpha) < 0$, if $y_1 = \cdots = y_q = y_{r+1} = \cdots = y_n = 0$ and the vectors α satisfying these $n - r - q$ conditions form a subspace S_2 of dimension $n - (n - r + q) = r - q$.

Now $\dim S_1 + \dim S_2 = n + (p - q)$. Hence if $q < p$, then $\dim S_1 + \dim S_2 > n$. Since $\dim (S_1 + S_2) \leq n$, it follows by Lemma 10, Chapter 2, that $\dim (S_1 \cap S_2) \neq 0$. Hence there would exist $\alpha \in S_1 \cap S_2$, $\alpha \neq \overline{0}_V$. This gives the contradiction $Q(\alpha) \geq 0$ and $Q(\alpha) < 0$. Therefore $q < p$ is false. Similarly, by interchanging the roles of p and q in this reasoning, we see that $p < q$ is also false. Hence $p = q$.

The **signature of a real quadratic form** is defined as the number of positive terms minus the number of negative terms in its diagonal (canonical) form. Thus the signature is $2p - r$. The signature therefore uniquely determines the canonical form.

Example 4. As an example illustrating the theorem, consider the q-form $Q(X) = 4x_1x_2 + 6x_1x_3$. This is a case where all the diagonal terms are 0. The matrix of Q is

$$A = \begin{bmatrix} 0 & 2 & 3 \\ 2 & 0 & 0 \\ 3 & 0 & 0 \end{bmatrix}.$$

It is easy to see that its rank is 2.

Following faithfully the method explained in this section we first transform $Q(X)$ into a q-form in which the diagonal coefficient $a_{11} \neq 0$. For this purpose apply the linear transformation T, given by $x_1 = y_1$, $x_2 = 2y_1 + 2y_2$, $x_3 = y_3$. Since $(x_1, x_2, x_3) = (y_1, y_2, y_3)T$ we find the matrix P_1 of T is

$$P_1 = \begin{bmatrix} 1 & 2 & 0 \\ 0 & 2 & 0 \\ 0 & 0 & 1 \end{bmatrix}, \det P_1 \neq 0.$$

We obtain

$$Q(Y) = 4y_1(2y_1 + 2y_2) + 6y_1y_3$$
$$= 8y_1^2 + 8y_1y_2 + 6y_1y_3.$$

The matrix of $Q(Y)$ is

$$B = \begin{bmatrix} 8 & 4 & 3 \\ 4 & 0 & 0 \\ 3 & 0 & 0 \end{bmatrix}.$$

As a check we find $P_1 A P_1^t = B$.

Now to continue with the reduction to diagonal form. Again following the method in the proof of the theorem, we find $\frac{1}{8}(8y_1 + 4y_2 +$

$3y_3)^2 = Q(Y) + 2y_2^2 + 3y_2y_3 + \tfrac{9}{8} y_3^2$. Hence, in the notation used in the proof, we get

$$Q(Y) = \tfrac{1}{8} (8y_1 + 4y_2 + 3y_3)^2 + Q'(Y)$$

where

$$Q'(Y) = -2(y_2^2 + \tfrac{3}{2} y_2y_3 + \tfrac{9}{16} y_3^2) = -2(y_2 + \tfrac{3}{4} y_3)^2.$$

Now we apply a second linear transformation S defined by $z_1 = 8y_1 + 4y_2 + 3y_3$, $z_2 = y_2 + \tfrac{3}{4} y_3$, $z_3 = y_3$. We find $y_1 = \tfrac{1}{8} z_1 - \tfrac{1}{2} z_2$, $y_2 = z_2 - \tfrac{3}{4} z_3$, $y_3 = z_3$, and so the matrix P_2 of S is given by

$$P_2 = \begin{bmatrix} \tfrac{1}{8} & 0 & 0 \\ -\tfrac{1}{2} & 1 & 0 \\ 0 & -\tfrac{3}{4} & 1 \end{bmatrix}.$$

Again we see det $P_2 \neq 0$. Hence both T and S are nonsingular and hence belong to $L_3(R)$, the full linear group on $V_3(R)$.

We obtain

(a) $$Q(Z) = \tfrac{1}{8} z_1^2 - 2z_2^2.$$

(b) $$P_2 B P_2^t = \begin{bmatrix} \tfrac{1}{8} & 0 & 0 \\ 0 & -2 & 0 \\ 0 & 0 & 0 \end{bmatrix}.$$

The rank of A is 2 and there are 2 nonzero terms in the diagonal form of Q.

EXERCISES

1. Find the rank and signature of the real quadratic form
$$18x_3^2 - x_1^2 - 4x_1x_2 - x_2^2.$$
2. Use the method illustrated in Example 4 to reduce the quadratic form
$$Q(X) = 9x_1^2 + 12x_1x_2 + 4x_2^2 - 2x_3^2$$
to the diagonal form. Check your result.

9-4 REDUCTION OF A QUADRATIC FORM UNDER THE ORTHOGONAL GROUP

Applications of the reduction of quadratic forms in mathematical physics, mechanics, engineering, occur in the context of a metric space. In this section we shall assume our vector space is a real inner-product finite-dimensional vector space (usually known as euclidean space) and consider the effect of an orthogonal transformation on a quadratic form in this space. The orthogonal transformations form a group and so we are again studying the reduction of a quadratic form under a group. Now, however, we seek to reduce the form using a subgroup of the full linear group, a more restrictive requirement.

In Chapter 8 it was proved that if T is a real symmetric operator then there exists an orthogonal operator P such that $PTP^t = D$ where D is a diagonal operator.

If we translate this into terms of matrices we have the following theorem:

THEOREM 2

If A is a real symmetric $n \times n$ matrix over a field F, then there exists an orthogonal matrix P such that $PAP^t = D$, where D is a diagonal matrix.

Since $P^t = P^{-1}$, A and D are similar matrices and hence have the same eigenvalues. Hence the entries on the main diagonal of D will be the eigenvalues of A.

Thus a real q-form can always be reduced to diagonal form by an orthogonal transformation.

We have shown in the section on orthogonal matrices in Chapter 6 that, starting with a given nonzero vector $\alpha = (\alpha_1, \alpha_2, \ldots, \alpha_n) \in V_n(R)$, an orthogonal matrix can be constructed (in fact in many ways), whose first row (or column) consists of the components of the unit vector $\alpha/|\alpha|$. Here $V_n(R)$ is the n-dimensional inner-product vector space over the real field R. We use this fact to give another proof for Theorem 2.

The method of proof is by induction. The theorem is trivially true for a 1×1 matrix, for it is already in the diagonal form. Assume it is true for an $(n - 1) \times (n - 1)$ matrix and use induction on the order n of the matrix A.

Since all the eigenvalues x_i of A are real, there exists a nonzero vector α_1 such that $\alpha_1 A = x_1 \alpha_1$. Normalize α_1 and use it as the first column of a real orthogonal matrix B. The first column of AB is $x_1 \alpha_1$ and the first column of $B^t AB$ consists of the inner products of the successive row vectors of B^t (that is, column vectors of B) with the vector $x_1 \alpha_1$. Hence the first column of $B^t AB$ is $(x_1, 0, 0, \ldots, 0)$. Since $B^t AB$ is a symmetric matrix, we must have, in block notation,

$$B^t AB = \left[\begin{array}{c|c} x_1 & 0 \\ \hline 0 & A_1 \end{array}\right]$$

where A_1 is the minor of x_1 in $B^t AB$. A_1 is a real $(n - 1) \times (n - 1)$ symmetric matrix. $B^t AB$ and A are similar matrices and so have the same eigenvalues. Hence the eigenvalues of A_1 must be $x_2, x_3, \ldots, x_n$. By our induction hypothesis there exists a real orthogonal matrix C such that

$$C^t A_1 C = \begin{bmatrix} x_2 & 0 & 0 & \cdots & 0 \\ 0 & x_3 & 0 & \cdots & 0 \\ \cdots & \cdots & \cdots & \cdots & \cdots \\ 0 & 0 & 0 & \cdots & x_n \end{bmatrix}.$$

Putting

$$H = \left[\begin{array}{c|c} 1 & 0 \\ \hline 0 & C \end{array}\right]$$

we see that H is a real orthogonal $n \times n$ matrix. Moreover,

$$H^t(B^t A B)H = \left[\begin{array}{c|c} x_1 & 0 \\ \hline 0 & C^t A C \end{array}\right] = \begin{bmatrix} x_1 & 0 & \cdots & 0 \\ 0 & x_2 & \cdots & 0 \\ \cdots & \cdots & \cdots & \cdots \\ 0 & \cdots & \cdots & x_n \end{bmatrix}$$

Since B and H are orthogonal matrices, then BH is orthogonal (the orthogonal matrices form a group). Putting $P = BH$, we have

$$P^t A P = \begin{bmatrix} x_1 & 0 & \cdots & 0 \\ 0 & x_2 & \cdots & 0 \\ \cdots & \cdots & \cdots & \cdots \\ 0 & 0 & \cdots & x_n \end{bmatrix}.$$

This completes the proof.

We know that P is the matrix formed from the eigenvectors of A, using them as the rows for P.

Example 5. The matrix

$$A = \begin{bmatrix} 1 & 2 & 0 \\ 2 & 0 & \sqrt{2} \\ 0 & \sqrt{2} & 1 \end{bmatrix}$$

has the eigenvalues $1, -2, 3$. They are the roots of the characteristic equation $x^3 - 2x^2 - 5x + 6 = 0$. The eigenvectors $(1, 0, -\sqrt{2})$, $(2, -3, \sqrt{2})$, $(\sqrt{2}, \sqrt{2}, 1)$ are linearly independent. Normalize them and use them for the rows of the matrix

$$P = \begin{bmatrix} 1/\sqrt{3} & 0 & -\sqrt{2}/\sqrt{3} \\ 2/\sqrt{15} & -3/\sqrt{15} & \sqrt{2}/\sqrt{15} \\ \sqrt{2}/\sqrt{5} & \sqrt{2}/\sqrt{5} & 1/\sqrt{5} \end{bmatrix}.$$

P is an orthogonal matrix and

$$PAP^t = \begin{bmatrix} 1 & 0 & 0 \\ 0 & -2 & 0 \\ 0 & 0 & 3 \end{bmatrix}.$$

Notice that we have written the rows of P as the normalized eigen vectors corresponding respectively to the eigenvalues $1, -2, 3$ of A, and hence the arrangement of these eigenvalues on the principal diagonal of PAP^t is precisely in this same order.

Let $Q(\alpha)$ be a real q-form on the real euclidean vector space $V_n(R)$ with the standard basis $\epsilon_1, \epsilon_2, \ldots, \epsilon_n$, where $\alpha = x_1\epsilon_1 + x_2\epsilon_2 + \cdots + x_n\epsilon_n$ is any vector in $V_n(R)$. Then we have proved there exists an orthogonal operator P such that, with respect to a new basis $\beta_1, \beta_2, \ldots, \beta_n$ of $V_n(R)$ given by (Section 1) $\beta_i = \epsilon_i P^t = \epsilon_i P^{-1}$, $i = 1, 2, \ldots, n$, the q-form is diagonalized. Moreover, since P is orthogonal, the new basis $\beta_1, \beta_2, \ldots, \beta_n$ is orthonormal. The vectors $\beta_1, \beta_2, \ldots, \beta_n$, are called the **principal axes** of Q.

Example 6. The equation of an ellipse with center at the origin with respect to the axes OX and OY is

$$ax^2 + bxy + cy^2 = d, \quad b \neq 0.$$

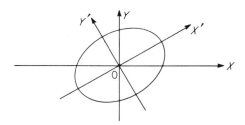

Its principal axes are OX' and OY' with respect to which its equation simplifies into $a'x'^2 + c'y'^2 = d'$.

Bilinear and Quadratic Forms

EXERCISES

1. Find the eigenvalues and eigenvectors of the matrix
$$A = \begin{bmatrix} 1 & 0 & -1 & 0 \\ 0 & 0 & -1 & 1 \\ -1 & -1 & -1 & -1 \\ 0 & 1 & -1 & 0 \end{bmatrix}.$$
Now find the orthogonal matrix P for which PAP^t is a diagonal matrix.

2. For each of the following real quadratic forms find an orthogonal transformation that reduces it to diagonal form
   (a) $2x_1 x_2$
   (b) $2x_1 x_3 - x_2^2$

3. Find an orthogonal transformation that reduces the quadratic form
$$2x_1^2 - 72x_1 x_2 + 25x_2^2 + 41x_3^2$$
to the diagonal form $50y_1^2 - 25y_2^2 + 25y_3^2$. How many such transformations are there?

9-5 POSITIVE DEFINITENESS

Definition. A real quadratic form $Q(X)$ on a real n-dimensional vector space V is said to be **positive definite** if $Q(X) > 0$ for all nonzero vectors X of V.

Definition. A real symmetric $n \times n$ matrix A is said to be **positive definite** if $XAX^t > 0$ for every nonzero vector X of $V_n(R)$.

Definition. A linear operator P on $V_n(R)$ is called **positive definite** if $XP \cdot X > 0$ for all nonzero vectors X of $V_n(R)$.

THEOREM 3

If a real symmetric matrix A is positive definite then any congruent matrix PAP^t is also positive definite.

Proof: Let X be a vector in $V_n(R)$ and let $X = YP$, where P is the matrix of a nonsingular linear operator on $V_n(R)$. Then $XAX^t = Y(PAP^t)Y^t$. Since $X = \bar{0}_V$ if and only if $Y = \bar{0}_V$, it follows that if $XAX^t > 0$ for all nonzero vectors X, then $Y(PAP^t)Y^t > 0$ for all nonzero vectors Y.

A simple and natural consequence of this theorem is

Corollary. A real quadratic form is positive definite if and only if its matrix is positive definite.

We have proved earlier that if A is a real symmetric matrix, there exists an orthogonal matrix P such that $PAP^t = D$, where D is a diagonal matrix whose diagonal entries are the eigenvalues $\lambda_1, \lambda_2, \ldots, \lambda_n$ of A. By the previous theorem A is positive definite if and only if D is positive definite.

Now
$$XDX^t = \lambda_1 x_1^2 + \cdots + \lambda_n x_n^2 > 0,$$
for every $X \neq \bar{0}_V$, if and only if all λ_i, $i = 1, 2, \ldots, n$, are positive. For if one of the eigenvalues, λ_1 say, is negative or 0, then for $X = (1, 0, 0, \ldots, 0)$ either $XDX^t = \lambda_1 < 0$ or $XDX^t = 0$. We have proved

THEOREM 4

A real symmetric matrix A is positive definite if and only if all its eigenvalues are positive.

Thus a real q-form is positive definite if and only if it can be reduced to the diagonal form $\lambda_1 x_1^2 + \cdots + \lambda_n x_n^2$, where all $\lambda_i > 0$. We can push this further and, by the nonsingular transformation
$$x_i' = \sqrt{\lambda_i} x_i, \qquad i = 1, 2, \ldots, n,$$
reduce a positive definite real q-form to the diagonal form $x_1'^2 + \cdots + x_n'^2$. Conversely, any real q-form that is reducible under the full linear group to this form is positive definite. For if A is the matrix of the q-form then the new matrix $PAP^t = I$, where I is the $n \times n$ identity matrix. Since I is positive definite, A must be; (Theorem 3). Hence

THEOREM 5

A real symmetric $n \times n$ matrix is positive definite if and only if it is congruent to the $n \times n$ identity matrix.

THEOREM 6

A real symmetric matrix A is positive definite if and only if $A = BB^t$, where B is a real nonsingular matrix.

Proof: A is positive definite if and only if $PAP^t = I$, where P is nonsingular; that is, A is positive definite if and only if
$$A = P^{-1} I (P^t)^{-1} = P^{-1}(P^t)^{-1} = P^{-1}(P^{-1})^t.$$
Take B to be the nonsingular matrix P^{-1}.

Example 6. The symmetric matrix
$$A = \begin{bmatrix} 2 & -1 & 0 \\ -1 & 2 & -1 \\ 0 & -1 & 2 \end{bmatrix}$$

is the matrix of the quadratic form
$$2x_1^2 - 2x_1x_2 + 2x_2^2 - 2x_2x_3 + 2x_3^2.$$
Its characteristic polynomial is $(2 - x)(x^2 - 4x + 2)$ and the eigenvalues are $2, 2 \pm \sqrt{3}$. Since they are all positive, the quadratic form is positive definite.

EXERCISES

1. Find the quadratic form whose matrix is
$$\begin{bmatrix} 1 & 1 & 0 & -1 \\ 1 & 3 & 0 & -1 \\ 0 & 0 & 1 & 0 \\ -1 & -1 & 0 & 3 \end{bmatrix}$$
and prove that it is positive definite.

2. If B is a real $n \times n$ matrix, prove that the matrix BB^t is symmetric.

7-6 THE SIMULTANEOUS REDUCTION OF QUADRATIC FORMS

Let B be a positive definite real symmetric $n \times n$ matrix. Hence all the eigenvalues $\mu_1, \ldots, \mu_n$ of B are positive. We know there exists an orthogonal matrix such that
$$PBP^t = D[\mu_1, \mu_2, \ldots, \mu_n]$$
where $D[\mu_1, \ldots, \mu_n]$ is a diagonal matrix whose diagonal entries are $\mu_1, \mu_2, \ldots, \mu_n$.

Let S be the matrix
$$S = D\left[\frac{1}{\sqrt{\mu_1}}, \frac{1}{\sqrt{\mu_2}}, \ldots, \frac{1}{\sqrt{\mu_n}}\right].$$
Then
$$SPBP^tS^t = SD[\mu_1, \ldots, \mu_n]S^t = I.$$
If we write $SP = C$, then $CBC^t = I$.

Now let A be a real symmetric matrix and consider the polynomial equation in x,
$$|A - xB| = 0.$$
Since C is a nonsingular matrix, this equation is equivalent to the equation
$$|C||A - xB||C^t| = |CAC^t - xCBC^t|$$
$$= |CAC^t - xI| = 0,$$
that is, the two equations have the same roots.

Since A is a real symmetric matrix, CAC^t is a real symmetric matrix and therefore all the roots of the equation $|CAC^t - xI| = 0$ are real. Hence all the roots of the equation $|A - xB| = 0$ are real. Denote them by $x_1, x_2, \ldots, x_n$.

There exists an orthogonal matrix K such that

$$K(CAC^t)K^t = D[x_1, x_2, \ldots, x_n],$$

that is, $(KC)A(KC)^t = D[x_1, x_2, \ldots, x_n]$. Moreover,

$$(KC)B(KC)^t = KIK^t = KK^t = I$$
$$= D[1, 1, \ldots, 1].$$

(NOTE: $KK^t = I$, because K is an orthogonal matrix.)

Setting $KC = S$, we have proved

THEOREM 7

If A and B are real symmetric matrices and if B is positive definite, then there exists a nonsingular matrix S such that
   (i) SAS^t is a diagonal matrix whose diagonal entries are real and are the roots of the equation $|A - xB| = 0$.
   (ii) $SBS^t = I$, the identity matrix.

We describe this theorem by saying that two real symmetric matrices, at least one of which is positive definite, can be simultaneously reduced to diagonal form.

Translating this in terms of quadratic forms it means that we can simultaneously reduce two real quadratic forms, one at least of which is positive definite, to the sum of square terms only. This is a particularly important and useful fact in mathematical physics. For instance, in certain problems in mechanics the kinetic energy and the potential energy are quadratic forms and the kinetic energy is a positive definite form. Hence there exists a change of variables in terms of which the kinetic energy is the sum of square terms all with the coefficient 1, while the potential energy is the sum of square terms with real coefficients. In terms of the new variables the differential equations of the system are greatly simplified and often very easily solved.

Note that while there exists an orthogonal matrix that reduces a single symmetric matrix to diagonal form, the matrix that simultaneously reduces two symmetric matrices, one of which is positive definite, to diagonal forms is not an orthogonal matrix, although it is nonsingular. This is to say that a pair of such symmetric matrices are simultaneously reducible under the full linear group, but not under the orthogonal group.

For a practical method of effecting the actual simultaneous reduction of two quadratic forms and a method that is highly suitable for

Bilinear and Quadratic Forms

applications, the reader is referred to Chapter 1 of R. Courant and D. Hilbert, *Methods of Mathematical Physics* (New York: Wiley, 1962).

9-7 HERMITIAN FORMS

Let V be an n-dimensional vector space over the complex field C with the basis $\alpha_1, \alpha_2, \ldots, \alpha_n$ and let $X = x_1\alpha_1 + \cdots + x_n\alpha_n$ be any vector of V.
$H(X)$ is called a **hermitian form on** V if

$$H(X) = \sum_{j,k=1}^{n} a_{jk}\bar{x}_j x_k, \qquad a_{jk} \in C$$

where the *matrix* $A = (a_{jk})$ *is a hermitian matrix*. This means that if A^* is the transpose of the conjugate of the matrix A, then $A = A^*$. Here $X = (x_1, x_2, \ldots, x_n)$ where the x_i are complex numbers and $\bar{x}_j$ denotes the complex conjugate of x_j.

The rank of the matrix (a_{jk}) is again called the **rank of the hermitian form** $H(X)$.

Exercise. Let A and B be complex matrices and let A^* designate the transpose of the conjugate of A. Prove that $(AB)^* = B^*A^*$.

So many of the results for hermitian forms are so similar to the corresponding ones for real quadratic forms, that we can leave the details (with obvious modifications of the real case) to the reader.

We can write a hermitian form with matrix A as $H(X) = XAX^*$. Hence if P is a nonsingular linear operator on the n-dimensional vector space V and if $X = YP$, then we find, as before, that in matrix notation

$$H(Y) = (YP)A(YP)^* = (YP)A(P^*Y^*)$$
$$= Y(PAP^*)Y^*.$$

Hence $H(Y)$ has the matrix PAP^*. This matrix is also hermitian, for $(PAP^*)^* = P^{**}A^*P^* = PAP^*$.

THEOREM 8

A hermitian quadratic form can be reduced to real diagonal form by a unitary transformation.

Proof: This theorem is equivalent to the following. If A is a hermitian matrix, then there exists a unitary matrix P such that $PAP^* = D$, where D is a diagonal matrix whose entries on the main diagonal are the eigenvalues (all real) of A. Since $P^* = P^{-1}$, A and D are similar matrices. (See Theorem 14 of Chapter 8.)

Definition. A hermitian quadratic form $H(X)$ is said to be **positive definite** if $H(X) > 0$ for all nonzero vectors of V.

As in Theorem 4, we can state that $H(X)$ is positive definite if and only if all the eigenvalues of its matrix A are positive or, alternatively, if and only if $H(X)$ can be reduced by a unitary transformation to the diagonal form

$$\lambda_1 \bar{z}_1 z_1 + \lambda_2 \bar{z}_2 z_2 + \cdots + \lambda_n \bar{z}_n z_n,$$

where all the $\lambda_i > 0$, $i = 1, 2, \ldots, n$.

Let $H(X)$ be a hermitian quadratic form on the complex inner product vector space $V_n(C)$, where $X = x_1 \epsilon_1 + \cdots + x_n \epsilon_n$ and $\epsilon_1, \epsilon_2, \ldots, \epsilon_n$ is the standard basis. Then, as in the real case (Sec 9-4) there exists a linear operator P on $V_n(C)$ such that with respect to a new orthonormal basis $\beta_1, \beta_2, \ldots, \beta_n$ of $V_n(C)$ with $\beta_i = \epsilon_i P^{-1}$, $i = 1, 2, \ldots, n$, $H(X)$ is diagonalized. The vectors $\beta_1, \beta_2, \ldots, \beta_n$ are called the **principal axes** of H and the unitary operator P is called the **principal axis transformation**.

EXERCISES

1. Reduce the following real quadratic forms to diagonal form and in each case find the nonsingular linear transformations that effect the reduction. Find the rank and signature of each form.
   (a) $6x_1 x_3 - 7x_2^2 - 8x_3 x_4 (5y_1^2 - 5y_2^2 - 7y_3^2)$
   (b) $6x_1 x_4 + 6x_2 x_3 + 16 x_2 x_4 (9y_1^2 - 9y_2^2 + y_3^2 - y_4^2)$.

2. For each of the previous quadratic forms find an orthogonal transformation that reduces it to diagonal form.

3. Reduce the following hermitian matrices to diagonal forms $(i^2 = -1)$:

   (a) $\begin{bmatrix} 1 & 1+i \\ 1-i & 1 \end{bmatrix}$ (b) $\begin{bmatrix} i & -1 \\ -1 & i \end{bmatrix}$

   (c) $\begin{bmatrix} 2 & 1 & 0 \\ -i & 1 & 1-i \\ 0 & 1+i & 0 \end{bmatrix}$

4. If A is a nonsingular matrix over the complex field and if A^* is the transpose of the conjugate of A, prove that the matrix $A^* A$ is positive definite.

5. Find two distinct diagonal forms for the quadratic form

$$x_1^2 + x_2^2 + 2x_3^2 + 2x_1 x_3 + 2x_1 x_2.$$

6. Reduce the following quadratic forms to the diagonal form:
   (a) $2x_1 x_2 + 2x_1 x_3 - 2x_2 x_3$
   (b) $2x_1^2 - 2x_2^2 + 3x_3^2 + 8x_1 x_2 + 12 x_1 x_3 + 4x_2 x_3$

7. Reduce each of the following quadratic forms to diagonal form by an orthogonal transformation:
   (a) $2x_1^2 + x_3^2 - 4x_1 x_3 - 4x_2 x_3$

(b) $2x_1^2 + 2x_2^2 + 2x_3^2 + 2x_4^2 + 2x_1x_2 - 4x_1x_4 - 4x_2x_3 + 2x_3x_4$

8. Show how the bilinear functions of $V \times V \to F$ can form a vector space over F and prove that the symmetric bilinear functions form a subspace of this space.

9. Show how the quadratic forms of $V \to F$ form a vector space over F.

10. If (a_{ij}) is an $n \times n$ symmetric matrix such that

$$a_{11} > 0, \quad \begin{vmatrix} a_{11} & a_{12} \\ a_{21} & a_{22} \end{vmatrix} > 0, \ldots, \det(a_{ij}) > 0,$$

prove that $a_{nn} > 0$.

11. T is a linear operator on a vector space V and Q is a quadratic form on V. Define a quadratic form Q' on V by $Q'(\alpha) = Q(\alpha T)$, $\alpha \in V$. Prove that the mapping $Q \to Q'$ is a linear operator on the vector space of all quadratic forms on V. Prove that this operator is bijective if and only if T is bijective.

Chapter **10**

Canonical Forms for Linear Transformations

10-1 INVARIANT SUBSPACES

It is important to find those subspaces of a vector space V over a field A that are invariant relative to some given linear operator T on V. In the case of a finite-dimensional vector space, these subspaces will determine both a direct sum decomposition of V, as well as a basis with respect to which the matrix of T has an important canonical form. (In the context of matrices the term "canonical" or "normal" means vaguely that the form possesses certain distinguished qualities that are, for instance, unique under equivalence or invariant under similarity.) We shall restrict ourselves here to the two canonical forms called the **rational normal form** and the **normal form of Jordan**. A canonical form of a matrix usually has advantages over the original in that it possesses a certain uniformity in its construction and, moreover, it usually has more zero entries! If it is simply related to the original matrix, this uniformity can often permit us to systematize our study of linear operators. We shall be exclusively concerned in this chapter with the reduction of a linear operator to a canonical form by means of a suitable choice of basis for its vector space. In terms of either operators or matrices, this means that we are concerned here only with their reduction to canonical forms under the relation of similarity.

We shall assume throughout this chapter that V is a finite-dimensional vector space.

Definition: A subspace U of a vector space V is said to be **invariant** under a linear operator T on V, if U is mapped into itself by T. This means that $\alpha T \in U$ for every $\alpha \in U$. We call U a **T-invariant subspace** of V.

If U is a T-invariant subspace of V, then the restriction of T to U is clearly a linear operator on U.

Exercise. Prove this last statement.

Definition: A **polynomial function** f of a linear operator T on a vector space V is a linear operator on V of the form

$$f(T) = a_0 + a_1 T + \cdots + a_m T^m, \quad m \geq 0,$$

where the a_i are scalars. We call this a **polynomial in T** over the scalar field F of V.

It is quite evident that if U is a T-invariant subspace of V, then U is invariant under the linear operator $f(T)$, where f is a polynomial function.

Exercise. Prove this last statement.

Let T be a linear operator on a vector space V over F and let $f(T)$ be a polynomial over F. It is easy to verify that the set of all vectors $\alpha \in V$ such that $\alpha f(T) = \bar{0}_V$ forms a subspace of V. It is called the **null space** of $f(T)$. Since $f(T)$ is itself a linear operator on V, its null space is merely the kernel of $f(T)$.

Exercise. Prove that the set of all vectors $\alpha \in V$ such that $\alpha f(T) = \bar{0}_V$ is a subspace of V.

Definition: A T-invariant subspace U of V is called **T-cyclic** if there exists a vector $\alpha \in U$ such that every vector of U can be written in the form $\alpha f(T)$, where $f(T)$ is a polynomial in T over the scalar field. The T-cyclic subspace U is said to be **generated by the vector α**.

Let U be a T-cyclic subspace of V generated by the vector α. Since V is finite-dimensional there must be a largest integer k such that $\alpha, \alpha T, \alpha T^2, \ldots, \alpha T^{k-1}$ are independent vectors in U. This means that all vectors $\alpha T^n, n \geq k$, are linear combinations of these k independent vectors, and hence that any vector $\alpha f(T) \in U$ is a linear combination of these k vectors. The vectors $\alpha, \alpha T, \ldots, \alpha T^{k-1}$ therefore form a basis for U and $\dim U = k$.

Every vector $\alpha \in V$ generates a T-cyclic subspace U of V, and it consists of all vectors of the form $\alpha f(T)$ for all polynomials $f(T)$ in T. For $\alpha f(T) + \alpha g(T) = \alpha[f(T) + g(T)]$ and the sum of two polynomials in T is a polynomial in T; also for any scalar r, $r(\alpha f(T)) = \alpha[rf(T)]$ and $rf(T)$ is a polynomial in T.

A T-cyclic subspace is the simplest type of T-invariant subspace. A T-cyclic subspace U generated by the vector α can also be defined as the intersection of the family of T-invariant subspaces of V that contain the generating vector α. (This family is not empty, since V itself belongs to the family.) U is therefore the smallest T-invariant subspace containing α.

Example 1. Let T be the linear operator on $V_3(R)$ defined by

$$(x, y, z)T = (x - y, z, x + y).$$

Then $V_3(R)$ is a T-cyclic space generated by the vector $\alpha = (1,0,0)$. For $\alpha T = (1,0,1)$ and $\alpha T^2 = (1,1,2)$ and the three vectors α, αT, αT^2 are linearly independent.

Assume now that $\alpha \neq \overline{0}_V$ and let U be the T-cyclic subspace of V generated by the vector α. Since V is finite-dimensional, U is finite-dimensional. Let dim $U = k$. Then the vectors $\alpha, \alpha T, \ldots, \alpha T^{k-1}$ form a basis for U and we have

$$\alpha T = b_0 \alpha + b_1(\alpha T) + \cdots + b_{k-1}(\alpha T^{k-1}),$$

where the b_i are scalars. Write $m_\alpha(x) = x^k - b_{k-1}x^{k-1} - \cdots - b_0$. Clearly $\alpha(m_\alpha(T)) = \overline{0}_V$, and $m_\alpha(x)$ is the polynomial of least degree for which $\alpha m_\alpha(T) = \overline{0}_V$. The polynomial $m_\alpha(x)$ is called the **minimal polynomial** of the restriction T_U of T to U.

The vectors $\alpha, \alpha T, \ldots, \alpha T^{k-1}$ form a basis for the T-cyclic subgroup U generated by α. Relative to this basis the matrix of T_U is clearly determined by

$$\alpha T_U = \alpha T$$
$$(\alpha T) T_U = \alpha T^2$$
$$\cdots\cdots\cdots\cdots\cdots\cdots$$
$$(\alpha T^{k-1}) T_U = \alpha T^k = b_0 \alpha + b_1(\alpha T) + \cdots + b_{k-1}(\alpha T^{k-1}).$$

That is, the matrix is

(1) $\begin{bmatrix} 0 & 1 & 0 & 0 & \cdots & 0 & 0 \\ 0 & 0 & 1 & 0 & \cdots & 0 & 0 \\ \cdots\cdots\cdots\cdots\cdots\cdots\cdots\cdots \\ \cdots\cdots\cdots\cdots\cdots\cdots\cdots\cdots \\ 0 & 0 & 0 & 0 & \cdots & 1 & 0 \\ 0 & 0 & 0 & 0 & \cdots & 0 & 1 \\ b_0 & b_1 & b_2 & b_3 & \cdots & b_{k-2} & b_{k-1} \end{bmatrix}$.

This type of matrix is often called a **companion matrix**. It is more precisely the companion matrix of the minimal polynomial $m_\alpha(x)$. If $m(x)$ is the minimal polynomial of T, then $m(T) = 0$. Hence $\alpha m(T) = \overline{0}_V$. Since $\alpha m_\alpha(T) = \overline{0}_V$, it follows that $m_\alpha(x)$ is a divisor of $m(x)$. We have therefore proved

THEOREM 1

Let T be a linear operator on a finite-dimensional vector space V and let $m(x)$ be the minimal polynomial of T on V.

Let U be a T-cyclic subspace of V, generated by the vector α and let $m_\alpha(x)$ be the minimal polynomial of T as an operator on U. Then

(a) With respect to the basis $\alpha, \alpha T, \ldots, \alpha T^{d-1}$ of U, where d is the degree of $m_\alpha(x)$, the matrix of T is the companion matrix of the polynomial $m_\alpha(x)$.

(b) $m_\alpha(x)$ is a divisor of $m(x)$.

In general for the minimal polynomial we have the factoring

$$m(x) = p_1(x)^{e_1} p_2(x)^{e_2} \cdots p_r(x)^{e_r},$$

where the $p_i(x)$ are monic irreducible polynomials over the field F and $e_i > 0, i = 1, 2, \ldots, r$.

In particular, however, if F is an algebraically closed field (the complex field, for example) then $m(x)$ has the form

$$m(x) = (x - \lambda_1)^{d_1}(x - \lambda_2)^{d_2} \cdots (x - \lambda_t)^{d_t},$$

and we shall prove (Sec. 9-3) in this case that with respect to a certain basis, the matrix of T has a particularly simple form, called the Jordan normal form.

THEOREM 2

Let T be a linear operator on a finite-dimensional vector space V over F. If the minimal polynomial $m(x)$ of T can be factored into the product of two monic relatively prime polynomials over F, that is, $m(x) = f(x) \cdot g(x)$, then $V = W_1 \oplus W_2$, where W_1 is the null space of $f(T)$ and W_2 is the null space of $g(T)$. W_1 and W_2 are T-invariant subspaces.

Proof: Since $f(x), g(x)$ are relatively prime, there exist polynomials $r(x)$ and $s(x)$ such that

$$1 = r(x)f(x) + s(x)g(x).$$

Thus

(2) $\qquad I = r(T)f(T) + s(T)g(T),$

where I is the identity operator on V. For $\alpha \in V$,

$$\alpha I = \alpha = (\alpha r(T))f(T) + (\alpha s(T))g(T).$$

Put $\beta = \alpha r(T)f(T), \gamma = (\alpha s(T))g(T)$. Then

$$\beta g(T) = \alpha r(T)m(T) = \overline{0}_V$$
$$\gamma f(T) = \alpha s(T)m(T) = \overline{0}_V.$$

Hence $\beta \in W_2$, the null space of $g(T)$, and $\gamma \in W_1$, the null space of $f(T)$. Hence $V = W_1 + W_2$; that is, V is spanned by the subspaces W_1 and W_2.

Suppose $\eta \in W_1 \cap W_2$. Then we have, using (2),

$$\eta = \eta r(T)f(T) + \eta s(T)g(T)$$
$$= \eta f(T)r(t) + \eta g(T)s(T)$$
$$= \overline{0}_V + \overline{0}_V = \overline{0}_V.$$

Hence $W_1 \cap W_2 = \bar{0}_V$ and therefore
$$V = W_1 \oplus W_2.$$

Each subspace W_1 and W_2 is invariant under T. For if $\alpha \in W_1$, then $(\alpha T)f(T) = (\alpha f(T))T = \bar{0}_V T = \bar{0}_V$. Hence $\alpha T \in W_1$. In the same way we prove W_2 is T-invariant.

Let T_1, T_2 be the restrictions of T to the subspaces W_1, W_2, respectively. Then T_1 is a linear operator on W_1 and T_2 is a linear operator on W_2. Hence the minimal polynomial $m_1(x)$ of T_1 is a divisor of $f(x)$, and the minimal polynomial $m_2(x)$ of T_2 is a divisor of $g(x)$.

Let $\alpha \in V$. Then $\alpha = \alpha_1 + \alpha_2, \alpha_1 \in W_1, \alpha_2 \in W_2$. Hence
$$\alpha m_1(T)m_2(T) = (\alpha_1 m_1(T))m_2(T) + (\alpha_2 m_2(T))m_1(T)$$
$$= \bar{0}_V + \bar{0}_V = \bar{0}_V.$$

Therefore by Theorem 4, Chapter 8, $m(x) = f(x)g(x)$ is a divisor of $m_1(x)m_2(x)$. Since $f(x)$ and $g(x)$ are relatively prime, $f(x)$ is a divisor of $m_1(x)$ and $g(x)$ is a divisor of $m_2(x)$. Hence $m_1(x) = f(x)$ and $m_2(x) = g(x)$. We have proved

Corollary. The restrictions, T_1 and T_2, of T to W_1 and W_2, respectively, have the minimal polynomials $f(x)$ and $g(x)$, respectively.

Example 2. Let T be the linear operator on $V_3(R)$ defined by $(x, y, z)T = (x + y + z, z - x, y)$. With respect to the standard basis, the matrix A of T is

$$A = \begin{bmatrix} 1 & -1 & 0 \\ 1 & 0 & 1 \\ 1 & 1 & 0 \end{bmatrix}.$$

The minimal polynomial of T is $(x + 1)(-x^2 + 2x - 2)$. It is also the characteristic polynomial.

For $\alpha = (1, 1, 1)$, we find $\alpha T = (3, 0, 1)$ and $\alpha T^2 = (4, -2, 0) = -2\alpha + 2\alpha T$. Let U be the T-cyclic subspace generated by $\alpha = (1, 1, 1)$. U is the subspace spanned by $(1, 1, 1)$ and $(3, 0, 1)$. It is the null space of $T^2 - 2T + 2$.

Note that the matrix of T, as a linear operator on U, is the companion matrix $\begin{pmatrix} 0 & 1 \\ -2 & 2 \end{pmatrix}$.

If W is the null space of $T + 1$, we find W is spanned by the vector $(0, 1, -1)$. For

$(x, y, z)(T + 1) = (2x + y + z, -x + y + z, y + z) = (0, 0, 0)$, if and only if $x = 0$ and $y + z = 0$. Also we see that $V = U \oplus W$, in confirmation of Theorem 2.

Example 3. Let T be the linear operator on $V_4(R)$ defined by $(x_1, x_2, x_3, x_4)T = (x_1 - x_4, x_1, -2x_2 - x_3 - 4x_4, 4x_2 + x_3)$. The matrix of T is

$$\begin{bmatrix} 1 & 1 & 0 & 0 \\ 0 & 0 & -2 & 4 \\ 0 & 0 & -1 & 1 \\ -1 & 0 & -4 & 0 \end{bmatrix}.$$

The minimal polynomial of T is found to be $(x^2 + 2)(x^2 + 1)$.

It is found that the null space S_1 of $T^2 + 2$ is the set of vectors of the form $(x_1, x_2, 2x_1 - 6x_2, x_1 + 2x_2)$ and the null space S_2 of $T^2 + 1$ is the set of vectors of the form $(x_1, x_2, x_1 - 5x_2, x_1 + x_2)$, where x_1 and x_2 are arbitrary real numbers. Hence, by Theorem 2, $V_4(R) = S_1 \oplus S_2$.

EXERCISES

1. T is a linear operator on the vector space V. Prove that a T-cyclic subspace of V is one-dimensional if and only if it is generated by a nonzero eigenvector of T.

2. The vectors $\alpha_1, \alpha_2, \alpha_3$ form a basis of a vector space V and T is a linear operator on V defined by

$$(x_1\alpha_1 + x_2\alpha_2 + x_3\alpha_3)T = (x_1 + x_2)\alpha_1 - x_3(\alpha_2 + \alpha_3).$$

   (i) Find the T-cyclic subspaces generated by each of the vectors $\alpha_1, \alpha_2, \alpha_3$ and their dimensions.
   (ii) Prove V is a T-cyclic space.
   (iii) Find each of the minimal polynomials $m_{\alpha_i}, i = 1, 2, 3$.
   (iv) Find the minimal polynomial of T and verify that it is divisible by each $m_{\alpha_i}, i = 1, 2, 3$.
   (v) Find the kernel of T and its dimension.
   (vi) Prove that the null space of $T^2 - 2T$ is spanned by the vector $\alpha_1 - \alpha_2$.

3. T is a linear operator on V and $m_\alpha(x)$ is the minimal polynomial of the restriction of T to the T-cyclic subspace generated by the vector α. If α is in the null space of $f(T)$, where f is a polynomial function, prove
   (i) U is a subspace of W
   (ii) $m_\alpha(x)$ is a divisor of $f(x)$.

4. T is a linear operator on an n-dimensional vector space V. Prove that V is a T-cyclic space if and only if the degree of its minimal polynomial is n.

5. Prove that the eigenspaces of a linear operator T on a finite-dimensional vector space are T-invariant subspaces.

6. T is a linear operator on V and $V = U \oplus W$. If P is the projection of V on U, prove that $PT = TP$ if and only if both U and W are T-invariant subspaces.

7. If W is T-invariant subspace of V show how T induces a linear operator T' on the quotient space V/W and define T'.

8. Prove that the minimal polynomial of the induced operator T' is a divisor of the minimal polynomial of T.

9. Find the minimal polynomial of the linear operator T on $V_3(R)$ (with the standard basis) defined by

$$(x, y, z)T = (x - 2z, 2x + z, y + 2z)$$

and use Theorem 2 to determine explicitly a direct sum decomposition of $V_3(R)$. Find the dimensions of the subspaces in this direct sum.

10-2. DECOMPOSITION THEOREMS

Given a linear operator T on a finite dimensional vector space V over a field F, we now begin the study of decomposition theorems of V with respect to this operator, and these culminate in the final decomposition of V into T-cyclic subspaces. With respect to a basis, determined by this final decomposition of V, we can express T in the first of our canonical forms, called the **rational canonical form** by reason of the fact that only the rational operations of the field F are used in the process. A specialization of this rational canonical form will lead, in Sec. 10-3, to the second and last of our canonical forms, the so-called Jordan canonical form.

THEOREM 3 (Primary Decomposition Theorem)

Let T be a linear operator on a finite-dimensional vector space V over F. Let

$$m(x) = p_1(x)^{e_1} p_2(x)^{e_2} \ldots p_k(x)^{e_k}, e_i > 0,$$

where the $p_i(x)$ are distinct monic irreducible polynomials over F, be the minimal polynomial of T. Then

$$V = W_1 \oplus \ldots \oplus W_k$$

where W_i is the null space of $p_i(T)^{e_i}$, $i = 1, 2, \ldots, k$. Each W_i is a T-invariant subspace and the restriction T_i of T to W_i is a linear operator on W_i, whose minimal polynomial is $p_i(x)^{e_i}$.

Proof: The $p_i(x)^{e_i}$ are relatively prime and the theorem follows by repeated applications of the previous theorem and its corollary.

Exercise. Prove that the null spaces W_i of $p_i(T)^{e_i}$, $i = 1, 2, \ldots, k$, have the dimensions n_i, where n_i is the power of $p_i(x)$ appearing in the characteristic polynomial of T. (Of course, $e_i \leq n_i$.)

Lemma 1. A finite-dimensional vector space V is a direct sum of two subspaces U and W, if and only if any basis $\alpha_1, \ldots, \alpha_j$ of U and any basis $\beta_1, \ldots, \beta_k$ of W combine to form a basis $\alpha_1, \ldots, \alpha_j, \beta_1, \ldots, \beta_k$ of V. (Thus dim $V = j + k$.)

Proof: Suppose $V = U \oplus W$. Then the vectors $\alpha_1, \ldots, \alpha_j, \beta_1, \ldots, \beta_k$ certainly span V. Moreover, they are independent. For, suppose

(3) $\quad x_1\alpha_1 + \cdots + x_j\alpha_j + x_{j+1}\beta_1 + \cdots + x_{j+k}\beta_k = \bar{0}_V,$

where the x_i are scalars, then

(4) $\quad \displaystyle\sum_{i=1}^{j} x_i\alpha_i = -\sum_{i=1}^{k} x_{j+i}\beta_i.$

The vector on the left of (4) belongs to U and the vector on the right belongs to W. Since $U \cap W = \bar{0}_V$, it follows that all the $x_i = 0$. Thus (3) implies that all the $x_i = 0$. Hence the α_i and β_i are independent and therefore form a basis for V.

Conversely, if they form a basis for V, then $V = U + W$. Since these vectors are independent, it follows from (4) that the only vector in $U \cap W$ is $\bar{0}_V$. Hence $V = U \oplus W$.

For instance, in Example 3 of Sec. 10-1, a basis of S_1 is seen to be $(1, -1, 0, -1)$, $(2, -1, 6, 0)$ and a basis of S_2 is $(1, -1, 6, 0)$, $(5, 1, 0, 6)$. Since these combine to form a basis of $V_4(R)$, we again see that $V_4(R) = S_1 \oplus S_2$.

It is easy to generalize this theorem to any finite number of subspaces, and we obtain

THEOREM 4

Let $W_1, \ldots, W_r$ be subspaces of a finite dimensional vector space V. Then

$$V = W_1 \oplus W_2 \oplus \ldots \oplus W_r$$

if and only if for any choices of bases for the W_i, these r bases combine to form a basis for V.

In fact, if V is spanned by the subspaces $W_1, \ldots, W_r$, that is $V = W_1 + \cdots + W_r$, and if the dimension of V is the sum of the dimensions of the W_i, then $V = W_1 \oplus \ldots \oplus W_r$.

We mention next a useful method of writing a matrix in terms of submatrices or **blocks,** as they are called. If we draw horizontal and vertical lines between the rows and columns of a matrix, we can subdivide it into blocks or submatrices and rewrite the matrix in terms of these blocks. For instance,

$$\begin{bmatrix} a_{11} & a_{12} & a_{13} & a_{14} & a_{15} & a_{16} \\ a_{21} & a_{22} & a_{23} & a_{24} & a_{25} & a_{26} \\ a_{31} & a_{32} & a_{33} & a_{34} & a_{35} & a_{36} \\ a_{41} & a_{42} & a_{43} & a_{44} & a_{45} & a_{46} \\ a_{51} & a_{52} & a_{53} & a_{54} & a_{55} & a_{56} \\ a_{61} & a_{62} & a_{63} & a_{64} & a_{65} & a_{66} \end{bmatrix} = \begin{bmatrix} B_1 & B_2 & B_3 \\ B_4 & B_5 & B_6 \end{bmatrix}$$

where

$$B_1 = \begin{bmatrix} a_{11} & a_{12} \\ a_{21} & a_{22} \end{bmatrix}, \quad B_2 = \begin{bmatrix} a_{13} & a_{14} & a_{15} \\ a_{23} & a_{24} & a_{25} \end{bmatrix}, \quad B_3 = \begin{bmatrix} a_{16} \\ a_{26} \end{bmatrix}$$

and so on.

One can readily devise a means of multiplying suitable block matrices. We shall not need it.

THEOREM 5

Let T be a linear operator on a finite-dimensional vector space V, and suppose $V = W_1 \oplus \cdots \oplus W_r$, where the W_i are T-invariant subspaces of V. If B_i is the matrix of the restriction of T to W_i with respect to some basis in W_i, then there exists a basis for V, with respect to which the matrix of T has the block form

$$B = \begin{bmatrix} B_1 & 0 & 0 & \cdots & 0 \\ 0 & B_2 & 0 & \cdots & 0 \\ \vdots & & & & \vdots \\ 0 & 0 & 0 & \cdots & B_r \end{bmatrix}.$$

Each B_i is a block and the B_i are placed so that their diagonal elements follow one another along the diagonal of B, and these diagonal elements of the B_i, $i = 1, 2, \ldots, r$, make up the diagonal elements of B. Each 0 in B is a block. The 0 in the ith row and jth column, $i \neq j$, of B stands for the zero matrix with the same number of rows as B_i and the same number of columns as B_j. If dim $V = n$, dim $W_i = n_i$, $i = 1, \ldots, r$ then B_i is an $n_i \times n_i$ matrix and B itself is an $n \times n$ matrix. Of course, $\sum_{i=1}^{r} n_i = n$.

Proof: Combine the r bases of the W_i to form a basis for V. Then T transforms each basis vector of each W_i into a vector of W_i, $i = 1, 2, \ldots, r$. With respect to the basis in W_i, T has the matrix B_i. Since V is a direct sum of the W_i, it follows that the matrix of T with respect to this basis for V is B above.

Definition. The block matrix formed in this way from the square matrices B_i, $i = 1, 2, \ldots, r$, is said to be the **direct sum** of these matrices,

$$B = B_1 \oplus B_2 \oplus \cdots \oplus B_r.$$

THEOREM 6

Let T be a linear operator on a finite-dimensional vector space V over F whose minimal polynomial $m(x) = p(x)^e$, where $p(x)$ is a monic ir-

Canonical Forms for Linear Transformations

reducible polynomial over F. Then

$$(5) \qquad V = Z_1 \oplus \ldots \oplus Z_r,$$

where the Z_i are T-cyclic subspaces of V.

Proof: We use induction on the dimension n of V. The theorem is clearly true for $n = 1$. In this case the minimal polynomial of T has the form $x - b$, where b is a scalar. Let us assume therefore that dim $V = n > 1$.

Choose a vector $\alpha_1 \in V$ such that $\alpha_1 p(T)^{e-1} \neq \bar{0}_V$. This is possible since, while $p(T)^e = 0$, $p(T)^{e-1} \neq 0$. Let Z_1 be the T-cyclic subspace generated by this vector α_1. If d is the degree of $p(x)$, then dim $Z_1 = de$ (see Theorem 1). If $de = n$ [this would be the case if and only if $p(x)^e$ is also the characteristic polynomial of T] then $V = Z_1$, and we are through, since V would itself be a T-cyclic space. Let us assume then that $de < n$.

Since Z_1 is a T-invariant subspace, T induces a linear operator T' on $V' = V/Z_1$, defined by

$$(\alpha + Z_1)T' = \alpha T + Z_1, \alpha \in V.$$

Hence $(\alpha + Z_1)T'^2 = (\alpha T + Z_1)T' = \alpha T^2 + Z_1$, and in general $(\alpha + Z_1)T'^k = \alpha T^k + Z_1$, and it follows that for any polynomial form $P(x) = a_0 + a_1 x + \cdots + a_m x^m$ we have

$$(\alpha + Z_1) P(T') = \alpha P(T) + Z_1.$$

This last equation makes evident two things: First, it shows $p(T')^e = 0'$ and hence that the minimal polynomial of T' on V' is a divisor of $p(x)^e$. Secondly, it shows that the minimal polynomial of the T-cyclic subspace generated by α is a multiple of the minimal polynomial of the T'-cyclic subspace generated by $\alpha + Z_1$.

Since dim $V' = $ dim $V - $ dim $Z_1 < $ dim V and since the minimal polynomial T' on V' is a power of the irreducible polynomial $p(x)$, the induction hypothesis applies to V' and T'. Therefore $V' = Z'_2 \oplus Z'_3 \oplus \cdots \oplus Z'_r$, where the Z'_i are T'-cyclic subspaces of V' and the minimal polynomials of T' on these subspaces have the forms $p(x)^{e'_2}, \ldots, p(x)^{e'_r}$, with $e \geq e'_2 \geq \cdots \geq e'_r$.

We are next going to prove that there are T-cyclic subspaces $Z_2, Z_3, \ldots, Z_r$ of V such that
  (i) for $i = 2, 3, \ldots, r$, Z_i and Z'_i are isomorphic,
  (ii) the minimal polynomials of T on Z_i and T' on Z'_i are the same, $i = 2, 3, \ldots, r$,
  (iii) $V = Z_1 \oplus Z_2 \oplus \cdots \oplus Z_r$.

Now the T'-cyclic subspaces Z'_i are quotient spaces of the form

V_i/Z_1, where V_i is a subspace of V and $Z_1 \subset V_i \subset V$. Hence the elements of Z_i' are cosets of the form $\alpha_i + Z_1$, $\alpha_i \in V_i$. Suppose Z_i' is generated by $\alpha_i + Z_1$, $i = 2, 3, \ldots, r$. We have designated the minimal polynomial of T' on Z_i' as $p(x)^{e_i}$. Hence

$$0' = (\alpha_i + Z_1)p(T')^{e_i} = \alpha_i p(T)^{e_i} + Z_1.$$

This means $\alpha_i p(T)^{e_i} \in Z$ and so $\alpha_i p(T)^{e_i} = \alpha_1 f(T)$ for some $f(T)$ where $f(x)$ is a polynomial form.

Since $\alpha_i + z_1 + Z_1 = \alpha_i + Z_1$ for any $z_1 \in Z_1$ it follows that $\alpha_i + z_1 + Z_1$ generates Z_i' for any vector z_1 of Z_1. Can we choose z_1 so that $(\alpha_i + z_1)p(T)^{e_i} = \overline{0}_V$? If so we shall show that $\alpha_i + z_1$ generates the required subspace Z_i.

Now $(\alpha_i + z_1)p(T)^{e_i} = \alpha_1 f(T) + z_1 p(T)^{e_i}$. Also since $\overline{0}_V = \alpha_i p(T)^e = \alpha_1 f(T) p(T)^{e-e_i}$ it follows that $p(x)^e$ must divide $f(x)p(x)^{e-e_i}$. This means that for some polynomial form $g(x)$, we have $f(x)p(x)^{e-e_i} = g(x)p(x)^e$, that is $f(x) = g(x)p(x)^{e_i}$. Hence if we choose $z_1 = -\alpha_1 g(T)$ then

$$[\alpha_i - \alpha_1 g(T)]p(T)^{e_i} = \overline{0}_V.$$

Now let Z_i be the T-cyclic subspaces of V that are generated by the vectors $\alpha_i - \alpha_1 g(T)$, $i = 2, 3, \ldots, r$. Since the minimal polynomial of T on Z_i is a multiple of the minimal polynomial $p(x)^{e_i}$ of T' on Z_i', this last equation above proves that $p(x)^{e_i}$ is the minimal polynomial of T on Z_i. Moreover, the mapping $\alpha_i - \alpha_1 g(T) \to \alpha_i - \alpha_1 g(T) + Z_1$ of $Z_i \to Z_i'$ is easily proved to be an isomorphism of these two cyclic subspaces. We have now proved (i) and (ii). Next we turn to (iii).

The Z_i' are generated by vectors that are not in Z_1 and $V' = V/Z_1 = Z_2' \oplus \cdots \oplus Z_r'$. It follows therefore from the way the Z_i are constructed out of the Z_i' that we can form the direct sum $W = Z_2 \oplus \cdots \oplus Z_r$, and that $W \cap Z_1 = 0$. Also the r isomorphisms of $Z_i \to Z_i'$ show that W and V/Z_1 are isomorphic spaces. Hence dim W = dim V/Z_1 = dim V − dim Z_1, that is dim $(Z_1 + W)$ = dim V (see Theorem 1, Chapter 4). Since $Z_1 + W$ is a subspace of V this implies $V = Z_1 \oplus W$. (See Exercise 6 at the end of Chapter 2). We therefore have

$$V = Z_1 \oplus Z_2 \oplus \cdots \oplus Z_r.$$

The Z_i are T-cyclic subspaces of V and this proves (iii). Hence the proof of our theorem is completed.

Corollary. Let T be a linear operator on a finite-dimensional vector space V. If Z is a T-cyclic subspace of V then there exists a T-invariant subspace W of V such that $V = Z \oplus W$.

Proof: This is, essentially, merely a restatement of the theorem.

If the degree of $p(x)$ is d and if $p(x)^{e_i}$ is the minimal polynomial of T on Z_i, for $i = 2, 3, \ldots, r$, where of course $e_1 = e$, then dim $Z_i = de_i$ and hence dim $V = d(e + e_2 + \cdots + e_r)$. Moreover, $e = e_1 \geq e_2 \geq \cdots \geq e_r$.

In particular, if the polynomial form $p(x) = x - \lambda$ then dim $V = e + e_2 + \cdots + e_r = n$ and the characteristic polynomial of T is $(x - \lambda)^n$.

There is therefore associated with the decomposition (5) a series of powers $p(x)^e, p(x)^{e_2}, \ldots, p(x)^{e_r}$ of the irreducible polynomial $p(x)$ in which $e \geq e_2 \geq \cdots \geq e_r$ and for each $i = 1, 2, \ldots, r$, $p(x)^{e_i}$ is the minimal polynomial of the restriction of T to the subspace Z_i.

In the next lemma we shall prove that these exponents $e, e_2, \ldots, e_r$ are unique for the linear operator T.

Lemma 2. The number of terms and the dimensions of the T-cyclic subspaces Z_i, $i = 1, 2, \ldots, r$, in the decomposition (5) of V are invariants for the linear operator T.

Proof: The proof is by induction on the dimension n of V. The lemma is trivially true for $n = 1$, for then V is itself a T-cyclic space. Assume the invariance for all vector spaces of dimension $< n$. Let

$$(5') \qquad V = Z'_1 \oplus \cdots \oplus Z'_s$$

be another decomposition of V into the direct sum of T-cyclic subspaces. Let Z'_i be generated by α'_i, $i = 1, 2, \ldots, s$.

We obtained Z_1 in (5) by choosing a vector α_1 such that $\alpha_1 p(T)^{e-1} \neq \overline{0}_V$. The dimension of Z_1 is de, where d is the degree of $p(x)$.

There must be some i for which $\alpha'_i p(T)^{e-1} \neq \overline{0}_V$, since if $\alpha'_i p(T)^{e-1} = \overline{0}_V$ for all i, then all

$$(\alpha'_i T^k) p(T)^{e-1} = (\alpha'_i p(T)^{e-1}) T^k = \overline{0}_V,$$

$i = 1, 2, \ldots, r$. This implies $\alpha p(T)^{e-1} = \overline{0}_V$ for all $\alpha \in V$, contradicting the fact that $p(x)^e$ is the minimal polynomial of T.

Hence without loss of generality, let us assume $a'_1 p(T)^{e-1} \neq \overline{0}_V$. Then dim $Z'_1 = de = $ dim Z_1, and the vector spaces $Z_2 \oplus \cdots \oplus Z_r$ and $Z'_2 \oplus \cdots \oplus Z'_s$ have the same dimension, and this dimension is less than n. Therefore, by the induction hypothesis, $r - 1 = s - 1$ and the dimensions of $Z_2, \ldots, Z_r$ are the same as those of $Z'_2, \ldots, Z'_r$ (though not necessarily respectively). We have proved the statement true for the vector space V of dimension n, and hence it is true for all finite dimensional vector spaces.

Combining Theorems 5 and 6 we see that a linear operator T on a finite dimensional vector space V over F whose minimal polynomial has the form $p(x)^e$, where $p(x)$ is a monic irreducible polynomial over F, determines a decomposition of V into T-cyclic subspaces, and that there exists a matrix representation of T in the form of the block matrix B of Theorem 5.

Exercise. A linear operator T on $V_4(R)$ is defined by $(x_1, x_2, x_3, x_4)T = (x_3 - x_2, x_1 + 2x_3 + x_4, 2x_3 + x_4, -5x_2 - 2x_4)$. Find (a) the matrix of T relative to the standard basis; (b) the minimal and characteristic polynomials of T. Express $V_4(R)$ as the direct sum of T-cyclic subspaces and prove that the matrix of T is similar to the (companion) matrix

$$\begin{bmatrix} 0 & 1 & 0 & 0 \\ 0 & 0 & 1 & 0 \\ 0 & 0 & 0 & 1 \\ -1 & 0 & -2 & 0 \end{bmatrix}.$$

Exercise. Find the same results as before for the linear operator T on $V_4(R)$ where R is the rational field, defined by

$$(x_1, x_2, x_3, x_4)T = (x_1 + x_3, x_2 + x_4, x_1 - x_3, x_2 - x_4).$$

10-3 THE RATIONAL CANONICAL FORM

In the general case the minimal polynomial of a linear operator T on a finite dimensional vector space V over F has the form

$$m(x) = p_1(x)^{e_1} p_2(x)^{e_2} \ldots p_k(x)^{e_k}, \; e_i > 0,$$

where the $p_i(x)$ are monic irreducible polynomials over F.

Let us thread our way through the theory as developed so far.

By Theorem 3, $V = W_1 \oplus \ldots \oplus W_k$, where the W_i are T-invariant subspaces and the minimal polynomial of T on W_i is $p_i(x)^{e_i}$.

By Theorem 6, for each W_i we have $W_i = Z_{i1} \oplus \ldots \oplus Z_{ir_i}$, where the X_{ij} are T-cyclic subspaces of V.

Thus V is the direct sum of a finite number, call it h, of T-cyclic subspaces Z_{ij}. With respect to each Z_{ij}, T has a minimal polynomial (see proof of Theorem 6) which is a divisor $p_i(x)^{e_{ij}}$ of $p_i(x)^{e_i}$.

If the vector α_{ij} generates Z_{ij}, then with respect to the basis α_{ij}, $\alpha_{ij}T, \ldots, \alpha_{ij}T^{e_{ij}-1}$ (the "cyclic" basis of Z_{ij}) the matrix of T is the companion matrix of its minimal polynomial $p_i(x)^{e_{ij}}$.

Let us choose the "cyclic" basis in each Z_{ij} and combine these bases to form a basis of V. Then, with respect to this basis of V it is evident, by Theorem 5, that the matrix of T has the form

Canonical Forms for Linear Transformations

(6) $$B = \begin{bmatrix} B_1 & 0 & 0 & \cdots & 0 \\ 0 & B_2 & 0 & \cdots & 0 \\ 0 & 0 & B_3 & \cdots & 0 \\ \vdots & & & & \vdots \\ 0 & 0 & 0 & \cdots & B_h \end{bmatrix},$$

where $V = Z_1 \oplus \cdots \oplus Z_h$ is the direct-sum decomposition of V into T-cyclic subspaces and B_i is the companion matrix of the minimal polynomial of T on the subspace Z_i.

Thus each B_i is a block and B is a block matrix. A matrix of the type B is said to be the **rational canonical form of the operator T**.

If A is an $n \times n$ matrix over a field F, then A is the matrix of a unique linear operator T on $V_n(F)$ with respect to some basis of $V_n(F)$. Hence we can state

THEOREM 7

A square matrix A over a field F is similar to a matrix B of rational canonical form. (B is called the **rational canonical form** of A.)

Since for a given linear operator T the decomposition of $V_n(F)$ into T-cyclic subspaces is unique (in the sense explained earlier) it follows that the rational canonical form B of the matrix A is essentially unique (that is except for the order of arrangement of the blocks B_i on the main diagonal of B).

Related to the decomposition $V = Z_1 \oplus \cdots \oplus Z_h$ of V into T-cyclic subspaces, is the set of minimal polynomials of A (the matrix of T) on the Z_i given by

$$p_1(x)^{e_{11}}, \ldots, p_1(x)^{e_{1r_1}},$$
$$p_2(x)^{e_{21}}, \ldots, p_2(x)^{e_{2r_2}},$$
$$\vdots$$
$$p_k(x)^{e_{k1}}, \ldots, p_k(x)^{e_{kr_k}},$$

where $e_{i1} \geq e_{i2} \geq \cdots \geq e_{ir_i}$, $i = 1, 2, \ldots, k$, and $r_1 + r_2 + \cdots + r_k = h$.

These constitute a complete set of invariants of the matrix A under similarity and are known as the **elementary divisors** of A.

We summarize our results in this final theorem of the section.

THEOREM 8 (Rational Decomposition Theorem)

A linear operator T on a finite-dimensional vector space V over F determines a decomposition of V into T-cyclic subspaces and there exists a matrix representation of T in the block form (6), in which the diagonal elements are the companion matrices of the minimal polynomials of the restrictions of T to the T-cyclic subspaces.

Example 4. In Example 3 we found that the minimal polynomial of the matrix $\begin{bmatrix} 1 & 1 & 0 & 0 \\ 0 & 0 & -2 & 4 \\ 0 & 0 & -1 & 1 \\ -1 & 0 & -4 & 0 \end{bmatrix}$ is $(x^2 + 2)(x^2 + 1)$. It follows then that the rational canonical form of this matrix is $\begin{bmatrix} 0 & 1 & 0 & 0 \\ -2 & 0 & 0 & 0 \\ 0 & 0 & 0 & 1 \\ 0 & 0 & -1 & 0 \end{bmatrix}$.

However, the alternative arrangement of the blocks on the main diagonal, given by $\begin{bmatrix} 0 & 1 & 0 & 0 \\ -1 & 0 & 0 & 0 \\ 0 & 0 & 0 & 1 \\ 0 & 0 & -2 & 0 \end{bmatrix}$ is also the rational canonical form of the given matrix.

It is easily verified that the T-invariant subspaces S_1 and S_2 are actually T-cyclic subspaces. In fact, S_1 can be seen to be generated by the vector $(1, 0, 2, 1)$ and S_2 to be generated by the vector $(1, 0, 1, 1)$. Hence $V_4(R) = S_1 \oplus S_2$ is a decomposition of $V_4(R)$ into T-cyclic subspaces.

EXERCISES

1. Find the rational canonical forms of the matrices

   (a) $\begin{bmatrix} 5 & 0 & 2 \\ -1 & 1 & 8 \\ 7 & 0 & 0 \end{bmatrix}$ (b) $\begin{bmatrix} 5 & 0 & 2 \\ -1 & 1 & 8 \\ -4 & 0 & 0 \end{bmatrix}$,

   (c) $\begin{bmatrix} \sin\theta & -\cos\theta \\ \cos\theta & \sin\theta \end{bmatrix}$ (d) $\begin{bmatrix} 1 & 3 & 0 \\ 0 & 0 & 2 \\ 2 & 0 & 0 \end{bmatrix}$

2. Fill in all the details for the proof of Theorem 3.

3. Find the rational canonical form of the linear operator T on $V_4(R)$ defined by
$$(x, y, z, u)T = (3x, 4x + 3y, 3z, x + 2u).$$
(Note that the minimal and characteristic polynomials of T are not equal.)

Canonical Forms for Linear Transformations

4. T is a linear operator on a finite-dimensional vector space V. Prove there exists a T-invariant subspace W of V such that $V = \text{im } T \oplus W$ if and only if $\text{im } T \cap \text{ker } T = 0$ and in this case prove that $\text{ker } T$ is the unique T-invariant subspace for which $V = \text{im } T \oplus \text{ker } T$.

5. Let A be a 7×7 real matrix with the minimal polynomial $(x^2 + 1)^2 (x + 2)$. Find the two possible rational canonical forms for this matrix with this minimal polynomial.

6. Given the linear operator T on $V_3(R)$ defined by $(x, y, z)T = (3x - y + 5z, -6x + 4y - 6z, -4x + 2y - 6z)$, find T-cyclic subspaces of $V_3(R)$ such that $V_3(R)$ is their direct sum.

7. Let G be a nonempty family of linear operators on a vector space V. A subspace U of V is called *G-invariant* if U is a T-invariant subspace for every $T \in G$. If there exists no proper subspace of V that is G-invariant, then we call G a *simple* or *irreducible* family of operators.

Let G and G' be nonempty irreducible families of operators on V and V' respectively. If f is a linear mapping of $V \to V'$ with the property that for any $T \in G$ there exists a $T' \in G'$ and for any $T' \in G'$ there exists a $T \in G$ such that $Tf = fT'$, prove that f is either the zero mapping or is an isomorphism.

10-4 THE JORDAN CANONICAL FORM

Suppose in particular that the minimal polynomial $m(x)$ of the linear operator T on a finite-dimensional vector space V over F has the form

$$(7) \quad m(x) = (x - \lambda_1)^{e_1} \cdots (x - \lambda_k)^{e_k}, \quad e_i > 0, \quad i = 1, 2, \ldots, k,$$

where the λ_i are distinct. This would always occur in the complex field or in any algebraically closed field.

We have seen that if Z is a T-cyclic subspace, generated by the vector α, then Z has a basis of the form $\alpha, \alpha T, \ldots, \alpha T^{e-1}$, $e > 0$, where e is the degree of the minimal polynomial of T as a linear operator on Z. Suppose now that this minimal polynomial is one of the form $(x - \lambda)^e$. Let us choose the basis $\beta_1, \beta_2, \ldots, \beta_e$ of Z defined by $\beta_1 = \alpha$, $\beta_2 = \alpha S, \ldots, \beta_e = \alpha S^{e-1}$, where $S = T - \lambda I$. (Since each $\beta_j = \alpha T^{j-1} +$ terms in αT^k, $k < j - 1$, it follows that the β's do form a basis.)

We have

$$\beta_1 T = \alpha T = \alpha S + \lambda \alpha = \beta_2 + \lambda \beta_1,$$

$$\beta_2 T = (\alpha S)T = \alpha S^2 + \lambda \alpha S = \beta_3 + \lambda \beta_2,$$

$$\cdots\cdots\cdots\cdots\cdots\cdots\cdots\cdots\cdots\cdots\cdots\cdots$$

$$\beta_{e-1} T = (\alpha S^{e-2})T = \alpha S^{e-1} + \lambda \alpha S^{e-2} = \beta_e + \lambda \beta_{e-1},$$

$$\beta_e T = (\alpha S^{e-1})T = \alpha S^e + \lambda \alpha S^{e-1} = \lambda \beta_e,$$

since $\alpha S^e = \alpha(T - \lambda)^e = \overline{0}_V$. Thus the matrix of T with respect to the β-basis is a matrix of the form

$$\begin{bmatrix} \lambda & 1 & 0 & 0 & \cdots & 0 \\ 0 & \lambda & 1 & 0 & \cdots & 0 \\ 0 & 0 & \lambda & 1 & \cdots & 0 \\ \multicolumn{6}{c}{\dotfill} \\ 0 & 0 & 0 & 0 & \cdots & \lambda \end{bmatrix}$$

where λ is an eigenvalue of T.

This type of matrix, in which all entries on the main diagonal are the same and all entries on the superdiagonal are 1 while all other entries are 0, is called a **Jordan matrix**.

Let us call the basis $\beta_1, \ldots, \beta_e$ above, the "Jordan" basis of the T-cyclic subspace Z. If we combine all the Jordan bases of the T-cyclic subspaces for $e = e_1, e_2, \ldots, e_k$ then, since their direct sum is V, we get a basis for V. With respect to this basis the matrix of T will therefore have the block form

(8) $$J = \begin{bmatrix} J_1 & 0 & 0 & \cdots & 0 \\ 0 & J_2 & 0 & \cdots & 0 \\ 0 & 0 & J_3 & \cdots & 0 \\ \multicolumn{5}{c}{\dotfill} \\ 0 & 0 & 0 & \cdots & J_k \end{bmatrix},$$

where each J_i is a Jordan matrix.

J is therefore the direct sum of the Jordan matrices J_i, $i = 1, 2, \ldots, k$,

$$J = J_1 \oplus J_2 \oplus \cdots \oplus J_k.$$

An eigenvalue λ_i appears as many times on the principal diagonal of J as its multiplicity as a root of the characteristic equation of T.

We have proved the following special case of Theorem 8:

THEOREM 9

If the minimal polynomial of a linear operator T on a finite-dimensional vector space V over F is a product of powers of linear factors then there exists a matrix representation of T in the Jordan canonical form (8).

Definition. The matrix J is called the **Jordan canonical form of the linear operator T**.

In terms of matrices we can state

Canonical Forms for Linear Transformations

THEOREM 10

If A is a square matrix over an algebraically closed field, then A is similar to a matrix of the form (8) and this matrix is called the **Jordan normal form** of A.

In particular if all the $e_i = 1$ in the minimal polynomial (7) of T then, and only then, is the Jordan canonical form a diagonal matrix with the eigenvectors λ_i as the diagonal elements. Here again each λ_i appears as many times on the principal diagonal as its multiplicity as a root of the characteristic polynomial of T. We have proved the important result:

Corollary. A linear operator T on a finite-dimensional vector space V has a diagonal matrix representation if and only if its minimal polynomial is a product of distinct linear factors. (T is said to be **diagonable**.)

In terms of matrices this last corollary reads as follows: a square matrix is similar to a diagonal matrix if and only if its minimal polynomial is a product of distinct linear factors.

Exercise. T is a projection on a finite-dimensional vector space V; that is, $T^2 = T$. Find the three possible forms for the minimal polynomial of T and prove that T is diagonable.

Exercise. If all the characteristic roots of a linear operator T on a finite-dimensional vector space are zero, prove that T is **nilpotent**; that is, $T^k = 0$ for some positive integer k.

If A is a square matrix over a field F, there may not exist a diagonal matrix that is similar over F to A. However, if F is algebraically closed then A is similar to a matrix in the Jordan normal form. The elements on the diagonal of this Jordan normal form are the eigenvalues of A, and each eigenvalue appears as many times on the diagonal as its multiplicity as a root of the characteristic polynomial.

The Jordan normal form of A is an example of what is called an *upper triangular matrix*; that is, a matrix whose entries below the principal diagonal are zero. The fact that for a linear operator T, on a finite-dimensional complex vector space, there exists a basis of V, with respect to which T is represented by such a triangular matrix, is important in the actual computational work with matrices. There is a straightforward algorithm for effecting the reduction of a matrix to this normal form but when carried out, it is a rather laborious task, and we offer below an alternative method. Thus we know the beginning, A, and the end, its Jordan normal form J, but the matrix P for which $PAP^{-1} = J$ is tedious to compute when the number n of rows and columns is large.

The Jordan canonical form J, if it exists, for the matrix A is essentially unique (that is, unique except for the order of the blocks J_i on the main diagonal). This follows, since this form is obtained from the essentially unique rational normal form, when the factors of the minimal polynomial are all linear.

Let A be an $n \times n$ matrix over a field F whose eigenvectors do *not* span $V_n(F)$. Assume all factors in the characteristic polynomial of A are linear, so that A has a Jordan canonical form

$$J = \begin{bmatrix} J_1 & 0 & 0 & \cdots & 0 \\ 0 & J_2 & 0 & \cdots & 0 \\ 0 & 0 & J_3 & \cdots & 0 \\ \vdots & & & & \vdots \\ 0 & 0 & 0 & \cdots & J_k \end{bmatrix},$$

and let $\lambda_1, \lambda_2, \ldots, \lambda_k$ be the distinct eigenvalues of A. Then a nonsingular matrix P exists such that $J = PAP^{-1}$. We present here an algorithm for computing the matrix P. Let $\epsilon_1, \epsilon_2, \ldots, \epsilon_n$ denote the standard basis for $V_n(F)$. Since P is nonsingular, the n vectors $\alpha_i = \epsilon_i P$ form a basis for $V(F)$ and the components of the vector α_i are the entries in the ith row of the matrix P. Thus

$$P = \begin{bmatrix} \alpha_1 \\ \alpha_2 \\ \vdots \\ \alpha_n \end{bmatrix}.$$

Our problem therefore is to determine the n vectors α_i. Suppose

$$J_1 = \begin{bmatrix} \lambda_1 & 1 & 0 & \cdots & 0 \\ 0 & \lambda_1 & 1 & \cdots & 0 \\ \vdots & & & & \vdots \\ 0 & 0 & 0 & \cdots & \lambda_1 \end{bmatrix}$$

is an $n_1 \times n_1$ matrix. Then

$$\epsilon_1 JP = (\lambda_1 \epsilon_1 + \epsilon_2)P$$
$$\epsilon_2 JP = (\lambda_1 \epsilon_2 + \epsilon_3)P$$
$$\cdots\cdots\cdots\cdots\cdots\cdots\cdots\cdots$$
$$\epsilon_{n_1} JP = \lambda_1 (\epsilon_{n_1} P)$$

Now $JP = PA$ and hence, in terms of the vectors α_i these equations become

(9)
$$\alpha_1 A = \lambda_1 \alpha_1 + \alpha_2$$
$$\alpha_2 A = \lambda_1 \alpha_2 + \alpha_3$$
$$\cdots\cdots\cdots\cdots\cdots$$
$$\alpha_{n_1} A = \lambda_1 \alpha_{n_1}$$

We use these equations, in which A and λ_1 are known, to solve for $\alpha_1, \alpha_2, \ldots, \alpha_{n_1}$. Start with the last equation and solve it for an eigenvector α_{n_1} associated with the eigenvalue λ_1. Substitute α_{n_1} in the next equation $\alpha_{n_1-1} A = \lambda_1 \alpha_{n_1-1} + \alpha_n$ and solve it for α_{n_1-1}. Working our way upwards through the system (9) we solve in succession for α_{n_1}, $\alpha_{n_1-1}, \ldots, \alpha_2, \alpha_1$ and can obtain in this way a set of n_1 linearly independent vectors.

Next we turn to

$$J_2 = \begin{bmatrix} \lambda_2 & 1 & 0 & \cdots & 0 \\ 0 & \lambda_2 & 1 & \cdots & 0 \\ \cdots & \cdots & \cdots & \cdots & \cdots \\ 0 & 0 & \cdots\cdots\cdots & & \lambda_2 \end{bmatrix}.$$

If J_2 is an $n_2 \times n_2$ matrix we obtain a set of n_2 equations similar to (9). They are

$$\alpha_{n_1+1} A = \lambda_2 \alpha_{n_1+1} + \alpha_{n_1+2}$$
$$\cdots\cdots\cdots\cdots\cdots\cdots\cdots$$
$$\alpha_{n_1+n_2} A = \lambda_2 \alpha_{n_1+n_2}.$$

Since $\lambda_1 \neq \lambda_2$ we solve this last equation for an eigenvector $\alpha_{n_1+n_2}$ distinct from α_{n_1}. Proceeding in the same way as for the system (9), we can obtain a set of n_2 linearly independent vectors such that all the vectors $\alpha_1, \alpha_2, \ldots, \alpha_{n_1+n_2}$ are linearly independent. Continuing in this way through $J_3, J_4, \ldots, J_k$ we obtain all the α_i, $i = 1, 2, \ldots, n$ and these form the matrix P.

Note that we can write the system (9) as

$$\alpha_i (A - \lambda_1 I) = \alpha_i, \quad i = 1, 2, \ldots, n_1 - 1$$
$$\alpha_{n_1} A = \lambda_1 \alpha_{n_1}.$$

No claim is made for brevity in solving for the α_i. As usual it works best when n is not large.

Exercise. If α is a solution of $\alpha_i(A - \lambda_1 I) = \alpha_{i+1}$, prove that $\alpha_i + \beta_i$ is also a solution, where β_i is an eigenvector belonging to the eigenvalue λ_1. Prove that α_i and β_i are linearly independent. Hence show that if a solution α_i is not independent of $\alpha_{i+1}, \ldots, \alpha_{n_1}$, then $\alpha_i + \beta_i$ is a solution independent of these vectors.

Example 5. Let us find a matrix P for which the matrix $A = \begin{bmatrix} 2 & 0 & 1 \\ 0 & 1 & 0 \\ 0 & -1 & 0 \end{bmatrix}$ and the matrix $J = \begin{bmatrix} 2 & 1 & 0 \\ 0 & 1 & 1 \\ 0 & 0 & 1 \end{bmatrix}$ are related by $J = PAP^{-1}$. We first solve $\alpha_3 A = \alpha_3$ for α_3 and we get $\alpha_3 = (0, 1, 0)$. Next solve $\alpha_2 A = \alpha_2 + (0, 1, 0)$ for α_2. We get $\alpha_2 = (0, 0, -1)$. Finally, we solve $\alpha_1 A = 2\alpha_1 + (0, 0, -1)$ and we obtain $\alpha_1 = (t - 1, -t, t)$ where t is arbitrary. Choose $t = 2$ and we get a vector $\alpha_1 = (1, -2, 2)$ that is independent of α_1 and α_2. Hence we can take

$$P = \begin{bmatrix} 1 & -2 & 2 \\ 0 & 0 & -1 \\ 0 & 1 & 0 \end{bmatrix}.$$

Example 6. The matrix $A = \begin{bmatrix} 3 & 4 & 1 \\ 0 & -1 & 1 \\ -1 & -3 & 1 \end{bmatrix}$ over the real field R has the characteristic equation

$$\begin{vmatrix} 3 - \lambda & 4 & 1 \\ 0 & -1 - \lambda & 1 \\ -1 & -3 & 1 - \lambda \end{vmatrix} = 0;$$

that is,

$$x^3 - 3x^2 + 3x - 1 = (x - 1)^3 = 0.$$

Its only eigenvalue is 1. The eigenvectors are the nonzero solutions of

$$(x_1, x_2, x_3) \begin{bmatrix} 3 & 4 & 1 \\ 0 & -1 & 1 \\ -1 & -3 & 1 \end{bmatrix} = (x_1, x_2, x_3),$$

$$(3x_1 - x_3, 4x_1 - x_2 - 3x_3, x_1 + x_2 + x_3) = (x_1, x_2, x_3).$$

Hence

$$3x_1 - x_3 = x_1$$
$$4x_1 - x_2 - 3x_3 = x_2$$
$$x_1 + x_2 + x_3 = x_3.$$

The solutions $x_3 = 2x_1$, $x_2 = -x_1$ of this sytem of equations yield vectors of the form $(x_1, -x_1, 2x_1) = x_1(1, -1, 2)$. Thus all eigenvectors are in the subspace spanned by the single vector $(1, -1, 2)$.

Hence A is not similar to a diagonal matrix. (Its eigenvectors do not span $V_3(R)$.) However, A is similar to a Jordan matrix. In fact the Jordan normal form of A is the matrix

$$\begin{bmatrix} 1 & 1 & 0 \\ 0 & 1 & 1 \\ 0 & 0 & 1 \end{bmatrix}.$$

Example 7. Consider the matrix

$$A = \begin{bmatrix} 0 & -1 & 0 & 0 \\ 1 & 0 & 0 & 0 \\ 0 & 0 & 0 & 1 \\ 0 & 0 & -1 & 0 \end{bmatrix}.$$

Its characteristic polynomial is $|A - \lambda I| = (\lambda^2 + 1)^2$. However,

$$A^2 = \begin{bmatrix} -1 & 0 & 0 & 0 \\ 0 & -1 & 0 & 0 \\ 0 & 0 & -1 & 0 \\ 0 & 0 & 0 & -1 \end{bmatrix},$$

and so

$$A^2 + I = \begin{bmatrix} 0 & 0 & 0 & 0 \\ 0 & 0 & 0 & 0 \\ 0 & 0 & 0 & 0 \\ 0 & 0 & 0 & 0 \end{bmatrix}.$$

Hence the minimal polynomial of A is $\lambda^2 + 1$.

Over the real field there is no Jordan normal form of the matrix A. However its rational canonical form is

$$B = \begin{bmatrix} 0 & 1 & 0 & 0 \\ -1 & 0 & 0 & 0 \\ 0 & 0 & 0 & 1 \\ 0 & 0 & -1 & 0 \end{bmatrix}.$$

A matrix P for which $PAP^{-1} = B$ is given by

$$P = \begin{bmatrix} 1 & 1 & 1 & 1 \\ 1 & -1 & -1 & 1 \\ 1 & 1 & 1 & 1 \\ 1 & -1 & -1 & 1 \end{bmatrix}.$$

Example 8. Consider the matrix

$$A = \begin{bmatrix} 0 & -1 & 0 & 0 \\ 1 & 0 & 0 & 0 \\ 1 & 2 & 0 & 1 \\ -1 & 3 & -1 & 0 \end{bmatrix}.$$

Its characteristic polynomial is

$$|A - \lambda I| = (\lambda^2 + 1)^2.$$

$$A^2 + I = \begin{bmatrix} -1 & 0 & 0 & 0 \\ 0 & -1 & 0 & 0 \\ 0 & 2 & -1 & 0 \\ 2 & -1 & 0 & -1 \end{bmatrix} + \begin{bmatrix} 1 & 0 & 0 & 0 \\ 0 & 1 & 0 & 0 \\ 0 & 0 & 1 & 0 \\ 0 & 0 & 0 & 1 \end{bmatrix} = \begin{bmatrix} 0 & 0 & 0 & 0 \\ 0 & 0 & 0 & 0 \\ 0 & 2 & 0 & 0 \\ 2 & -1 & 0 & 0 \end{bmatrix}.$$

Hence $A^2 + I \neq 0$. Of course $A^4 + 2A^2 + I = 0$ and the minimal polynomial of A is $(\lambda^2 + 1)^2$.

First let us take the base field to be the real field R, so that A is a matrix over R. Now $\lambda^2 + 1$ is a monic irreducible polynomial over R.

Since $A^2 + I \neq 0$, we can choose a vector $\alpha = (x_1, x_2, x_3, x_4)$ for which $\alpha(A^2 + I) = (2x_4, 2x_2 - x_4, 0, 0) \neq \bar{0}_V$. Suppose we make the choice $\alpha = (0, 0, 1, 1)$. Then $\alpha A = (0, 5, -1, 1)$, $\alpha A^2 = (2, 1, -1, -1)$, and $\alpha A^3 = (2, -8, 1, -1)$. It is easy to show that the vectors $\alpha, \alpha A, \alpha A^2, \alpha A^3$ are independent, and therefore they constitute a basis for the vector space $V_4(R)$. This proves $V_4(R)$ is a T-cyclic space, where T is the linear operator on $V_4(R)$, whose matrix is A with respect to the standard basis $(1, 0, 0, 0), (0, 1, 0, 0), (0, 0, 1, 0), (0, 0, 0, 1)$ of $V_4(R)$.

Hence, by Theorem 7, A is similar over R to the companion matrix

$$B = \begin{bmatrix} 0 & 1 & 0 & 0 \\ 0 & 0 & 1 & 0 \\ 0 & 0 & 0 & 1 \\ -1 & 0 & -2 & 0 \end{bmatrix}.$$

of the minimal polynomial $1 + 2\lambda^2 + \lambda^4$ of A. This means there exists a nonsingular matrix P over the real field R such that

$$PAP^{-1} = B.$$

If we write this last equation as

$$PA = BP,$$

then we can find such a matrix P by equating the 16 corresponding entries of the matrices PA and BP.

The reader will find that one such solution for P is the real matrix

$$P = \begin{bmatrix} 3 & 1 & 1 & 0 \\ 2 & -1 & 0 & 1 \\ -2 & 1 & -1 & 0 \\ 0 & 0 & 0 & -1 \end{bmatrix}.$$

Example 9. If the base field in Example 8 is taken to be the complex field C; that is, we regard the matrix A now to be a matrix over C, then the minimal polynomial of A factors into $(x^2 + 1)^2 = (x - i)^2(x + i)^2$, where $i = \sqrt{-1}$.

In this case A is therefore similar over C to the Jordan canonical form

$$J = \begin{bmatrix} i & 1 & 0 & 0 \\ 0 & i & 1 & 0 \\ 0 & 0 & -i & 1 \\ 0 & 0 & 0 & -i \end{bmatrix}.$$

This means there exists a nonsingular matrix P over C such that

$$PAP^{-1} = J.$$

Example 10. Consider the matrix $A = \begin{bmatrix} 1 & 2 & 1 \\ 1 & 1 & 3 \\ 2 & 1 & 1 \end{bmatrix}$. Its characteristic polynomial is $-\lambda^3 + 3\lambda^2 + 4\lambda + 7$. Clearly this polynomial is irreducible over the rational field (try $\lambda = \pm 1, \pm 7$). It has the one real root between $\lambda = 4$ and $\lambda = 5$, and hence it has two imaginary roots. Since the entries in A are rational, it follows that over the real (or rational) field R, the minimal polynomial of A is $\lambda^3 - 3\lambda^2 - 4\lambda - 7$.

If we select a nonzero vector α, say $\alpha = (1,0,0)$, then $\alpha A = (1,2,1)$, $\alpha A^2 = (5,5,8)$ and we see that these three vectors are independent and therefore form a basis for $V_3(R)$.

Thus A is similar over R to the companion matrix

$$B = \begin{bmatrix} 0 & 1 & 0 \\ 0 & 0 & 1 \\ 7 & 4 & 3 \end{bmatrix}$$

of the minimal polynomial of A.

A matrix P for which $PAP^{-1} = B$, that is $PA = BP$, is

$$\begin{bmatrix} 1 & -1 & 5 \\ 10 & 6 & 3 \\ 22 & 29 & 31 \end{bmatrix}.$$

Example 11. Let us consider a system of n first order differential equations in the complex field

(10) $\quad \dfrac{dx_i}{ds} = \sum\limits_{j=1}^{n} a_{ij} x_j, \quad i = 1, 2, \ldots, n.$

Let

(11) $\quad y_m = \sum\limits_{k=1}^{n} b_{mk} x_k, \quad m = 1, 2, \ldots, n$

be a nonsingular linear transformation of the old variables $x_1, \ldots, x_n$ to new variables $y_1, \ldots, y_n$. Let $A = (a_{ij})$, $B = (b_{mk})$, be the $n \times n$ matrices of (10) and (11) respectively. We are assuming these are matrices over the complex field. Write

$$X = (x_1, x_2, \ldots, x_n), \quad \dot{X} = \dfrac{dX}{ds} = \left(\dfrac{dx_1}{ds}, \ldots, \dfrac{dx_n}{ds} \right).$$

Then (10) can be written in the matrix form

(10') $\quad\quad\quad\quad\quad\quad\quad\quad\quad \dot{X} = XA$

and (11) in the same form becomes

(11') $\quad\quad\quad\quad\quad\quad\quad\quad\quad Y = XB.$

Hence $\dot{Y} = \dot{X}B$ and therefore $\dot{X} = \dot{Y}B^{-1}$. Multiply (10') on the right by B and we get $\dot{X}B = XAB$ and hence $\dot{Y} = XAB$. Now putting $X = YB^{-1}$, we have

(12) $\quad\quad\quad\quad\quad\quad\quad\quad\quad \dot{Y} = YB^{-1}AB.$

Thus the matrix of the new system is $B^{-1}AB$.

In this way we can often facilitate the solution of a system (10) of differential equations by applying to it a nonsingular transformation of variables. Since the new matrix is similar to the old one, it may be possible to produce a new system (12) of differential equations with a

Canonical Forms for Linear Transformations

diagonal matrix. In this case for each $i = 1, 2, \ldots, n$ we would obtain the system

$$\frac{dy_i}{ds} = c_i y_i, \qquad i = 1, \ldots, n$$

where the c_i are constants, and hence, solving, $y_i = k_i e^{c_i s}$ where k_i is an arbitrary constant.

Of course we may still need to return to the old variables $x_1, \ldots, x_n$, in which case we have to use $X = YB^{-1}$, which is the solution of (11') by Cramer's rule.

On the other hand, although it not always possible to reduce the complex matrix A to the diagonal form, we can always find the Jordan canonical form of A, and the new system of differential equations will probably be a great deal simpler to solve than the old one.

Example 12. Consider the real matrix $A = \begin{bmatrix} 3 & 4 & 1 \\ 0 & -1 & 1 \\ -1 & -3 & 1 \end{bmatrix}$. Its characteristic equation is $(x - 1)^3 = 0$. Its eigenvectors have the form $(s, -s, 2s)$, where s is an arbitrary real number. The subspace spanned by the eigenvectors is the one-dimensional subspace spanned by the vector $(1, -1, 2)$. Therefore, A is not similar to a diagonal matrix. However, since the characteristic polynomial has all linear factors, A is similar to a matrix of the Jordan canonical form. Hence A is similar to the matrix

$\begin{bmatrix} 1 & 1 & 0 \\ 0 & 1 & 1 \\ 0 & 0 & 1 \end{bmatrix}$. Let us now determine a nonsingular matrix P for which

$P^{-1}AP = \begin{bmatrix} 1 & 1 & 0 \\ 0 & 1 & 1 \\ 0 & 0 & 1 \end{bmatrix}$. To do this, we have nine equations, obtained by equating the entries of the two equal matrices

$$\begin{bmatrix} 3 & 4 & 1 \\ 0 & -1 & 1 \\ -1 & -3 & 1 \end{bmatrix} \begin{bmatrix} p_{11} & p_{12} & p_{13} \\ p_{21} & p_{22} & p_{23} \\ p_{31} & p_{32} & p_{33} \end{bmatrix} = \begin{bmatrix} p_{11} & p_{12} & p_{13} \\ p_{21} & p_{22} & p_{23} \\ p_{31} & p_{32} & p_{33} \end{bmatrix} \begin{bmatrix} 1 & 1 & 0 \\ 0 & 1 & 1 \\ 0 & 0 & 1 \end{bmatrix}.$$

Equating the entries in the first columns of these two products, we obtain

$$2p_{11} + 4p_{21} + p_{31} = 0,$$
$$-2p_{21} + p_{31} = 0,$$
$$-p_{11} - 3p_{21} = 0.$$

The matrix of this system has the rank 2 and we therefore get a single infinity of solutions. In other words, only two of these equations are independent, and solving, we have $p_{31} = 2p_{21}$, $p_{11} = -3p_{21}$. Let us choose $p_{21} = 1$. Then $p_{31} = 2$, $p_{11} = -3$. Proceeding now to the second columns, again equating entries, and substituting in these three values, we get

$$2p_{12} + 4p_{22} + p_{32} = -3$$
$$-2p_{22} + p_{32} = 1$$
$$-p_{12} - 3p_{22} = 2.$$

Again we find there are only two independent equations, and solving, we have $p_{12} = -2 - 3p_{22}$, $p_{32} = 1 + 2p_{22}$. Choosing $p_{22} = 1$, we get $p_{12} = -5$, $p_{32} = 3$. Continuing with the third columns, using these values, we find that if we select $p_{23} = 1$; then $p_{13} = -6$ and $p_{33} = 3$. Hence

$$P = \begin{bmatrix} -3 & -5 & -6 \\ 1 & 1 & 1 \\ 2 & 3 & 3 \end{bmatrix}.$$

Since det $P = -1$, we find

$$P^{-1} = \begin{bmatrix} 0 & 3 & -1 \\ 1 & -3 & 3 \\ -1 & 1 & -2 \end{bmatrix}.$$

It is now easy to verify that

$$P^{-1}AP = \begin{bmatrix} 1 & 1 & 0 \\ 0 & 1 & 1 \\ 0 & 0 & 1 \end{bmatrix}.$$

The matrix P defines the nonsingular linear transformation that would change variables and simplify the matrix A to the Jordan canonical form.

Example 13. Consider the system of first-order linear differential equations over the real field

$$\frac{dx}{ds} = 7x - 4y + 8z$$

$$\frac{dy}{ds} = x - 2y$$

$$\frac{dz}{ds} = -3x + 2y - 3z.$$

Canonical Forms for Linear Transformations

Form the matrix $A = \begin{bmatrix} 7 & -4 & 8 \\ 1 & -2 & 0 \\ -3 & 2 & -3 \end{bmatrix}$. Its eigenvalues are $-1, 1, 2$ and

A is similar to the diagonal matrix $B = \begin{bmatrix} 1 & 0 & 0 \\ 0 & -1 & 0 \\ 0 & 0 & 2 \end{bmatrix}$. With respect to the

matrix B, the system has the form

$$\frac{du}{ds} = u, \quad \frac{dv}{ds} = -v, \quad \frac{dw}{ds} = 2w.$$

The solutions are $u = ae^s$, $v = be^{-s}$, $w = ce^{2s}$ where a, b, c are arbitrary constants. We find $B = P^{-1}AP$, where $P = \begin{bmatrix} 12 & 2 & -4 \\ 4 & 2 & -1 \\ -7 & -1 & 2 \end{bmatrix}$. In matrix

form, we can write

$$\begin{bmatrix} dx/ds \\ dy/ds \\ dz/ds \end{bmatrix} = A \begin{bmatrix} x \\ y \\ z \end{bmatrix} \quad \text{and} \quad \begin{bmatrix} x \\ y \\ z \end{bmatrix} = P \begin{bmatrix} u \\ v \\ w \end{bmatrix}.$$

Hence

$$x = 12u + 2v - 4w = 12ae^s + 2be^{-s} - 4ce^{2s},$$
$$y = 4u + 2v - w = 4ae^s + 2be^{-s} - ce^{2s},$$
$$z = -7u - v + 2w = -7ae^s - be^{-s} + 2ce^{2s}.$$

EXERCISES

1. Find the rational canonical form of the real matrix

$$A = \begin{bmatrix} 1 & 2 & 1 \\ 1 & 1 & 3 \\ 2 & 1 & 1 \end{bmatrix}.$$

2. Find all the possible rational canonical forms for
   (a) a 6×6 real matrix whose minimal polynomial is $(x^2 + 1)^2(x - 3)$
   (b) a 4×4 real matrix whose minimal polynomial is $(x^2 + 2x - 1)(x + 2)$
   (c) a 5×5 real matrix whose minimal polynomial is x
   (d) a 5×5 real matrix whose minimal polynomial is $(x^2 + 2)^2$
   (e) a 15×15 real matrix whose minimal polynomial is $(x^3 + 2x^2 - 5)^3(x^2 + 1)^2$

3. Find the elementary divisors of the matrices in the illustrative examples 1–8.

4. (a) Find the eigenvalues of the real matrix

$$A = \begin{bmatrix} -5 & 2 & 6 & -3 \\ 18 & -11 & -29 & 14 \\ -8 & 7 & 16 & -7 \\ 2 & 0 & -1 & 0 \end{bmatrix}.$$

(b) Find a canonical form for A.
(c) Find the elementary divisors of A.

5. If $(x^2 + x + 1)^2$, $x^2 + x + 1$, and $x - 2$ are the elementary divisors of a real matrix A, find (a) the order of A; (b) the matrix A.

6. Find the Jordan canonical form of the real matrix

$$A = \begin{bmatrix} 1 & 0 & 0 & 1 & 0 \\ -2 & 3 & -1 & 1 & 1 \\ 0 & 0 & 2 & 0 & 1 \\ 0 & 0 & 1 & 2 & 0 \\ -1 & 1 & -1 & 0 & 0 \end{bmatrix}$$

7. Compute a matrix P for which PAP^{-1} is the Jordan canonical form of the matrix A in Exercise 3, regarded as a complex matrix.

8. A is a complex 6×6 matrix with the characteristic polynomial $(x + 2)^4 (x - 3)^2$.

(a) How many Jordan canonical forms of A are possible?
(b) Find the Jordan canonical form of A if the minimal polynomial of A is

(i) $(x + 2)^3 (x - 3)$
(ii) $(x + 2)^3 (x - 3)^2$
(iii) $(x + 2)^4 (x - 3)^2$

9. If the matrix of a linear operator is triangular, prove that the entries on the main diagonal are the eigenvalues of the operator.

10. Find the Jordan canonical form of each of the following linear operators on the complex vector space $V_3(C)$.

(a) $(x, y, z)T = (2ix - 2y, iz, 0)$
(b) $(x, y, z)T = (ix, x + iy, (3 - i)y + 2z)$

Find the bases of $V_3(C)$ for which the operators assume this form.

Chapter **11**

The Tensor Product of Vector Spaces

As an application (and extension) of our results in linear algebra, we introduce in this chapter a very important concept in multilinear algebra, the tensor product of two vector spaces. The tensor product has mushroomed into one of the most fertile of ideas in modern algebra, and it deserves an introductory and elementary treatment in these pages, even though we do not fully exploit its usefulness.

We have already seen how we can construct a new vector space out of two given vector spaces V and W over the same field, called the direct sum $V \oplus W$ of V and W. If V and W are finite dimensional vector spaces, say dim $V = n$, dim $W = m$, then dim $(V \oplus W) = n + m$. Here we are going to construct still another new type of vector space out of two given vector spaces V and W, called the **tensor product** $V \otimes W$ of V and W, and if dim $V = n$, dim $W = m$, then we shall see that dim $V \otimes W = nm$. However, the importance of the tensor product transcends this interesting fact. It is basic to the development of some very interesting and important algebras, one example of which is the exterior algebra of a vector space.

11-1 MULTILINEAR MAPPINGS

We have already introduced bilinear mappings in Chapter 9. They are the simplest forms of a multilinear mapping.

If V, W, Z are three vector spaces over the same field, a mapping f of $V \times W \to Z$ is **bilinear** if for each fixed $\alpha \in V$, f is a linear transformation of $W \to Z$ and if for each fixed $\beta \in W$, f is a linear transformation of $V \to Z$. Explicitly the bilinear mapping f satisfies the conditions:

For all α's $\in V$ and β's $\in W$ and x's $\in F$

(a) $(\alpha_1 + \alpha_2, \beta)f = (\alpha_1, \beta)f + (\alpha_2, \beta)f,$
(b) $(\alpha, \beta_1 + \beta_2)f = (\alpha, \beta_1)f + (\alpha, \beta_2)f,$
(c) $(x\alpha, \beta)f = x((\alpha, \beta)f),$
(d) $(\alpha, x\beta)f = x((\alpha, \beta)f).$

It is worth emphasizing that if f is bilinear then it does not follow that im f is a subspace of Z, as would be the case if f were a linear mapping. In fact, $V \times W$ is not even a vector space.

We mention here again the exponential notation $Z^{V \times W}$ which we shall sometimes use to denote the set of all mappings of $V \times W \to Z$.

In the same way, if V_1, V_2, V_3, Z are four vector spaces over the same field, then a mapping f of $V_1 \times V_2 \times V_3 \to Z$ is called **trilinear** if for each $\alpha_i \in V_i$, $i = 1, 2, 3$, the mapping of $V_i \to Z$, defined by $\alpha_i \to (\alpha_1, \alpha_2, \alpha_3)$, is linear. Clearly this concept generalizes to any finite number, $n > 0$, of vector spaces $V_1, V_2, \ldots, V_n, Z$ and the resulting mapping f is called **multilinear (an n-linear) mapping** of

$$V_1 \times V_2 \times \cdots \times V_n \to Z$$

if for each $\alpha_i \in V_i$, $i = 1, 2, \ldots, n$, the mapping of $V_i \to Z$, defined by

$$\alpha_i \to (\alpha_1, \alpha_2, \ldots, \alpha_n)f,$$

is linear.

The determinant function is an example of a multilinear mapping. If $\alpha_1, \alpha_2, \ldots, \alpha_n$ are n vectors in $V_n(R)$, where $\alpha_i = (a_{i1}, a_{i2}, \ldots, a_{in})$, $i = 1, 2, \ldots, n$, then

$$D(\alpha_1, \alpha_2, \ldots, \alpha_n) = \begin{vmatrix} a_{11} & a_{12} & \cdots & a_{1n} \\ a_{21} & a_{22} & \cdots & a_{2n} \\ \vdots & & & \vdots \\ a_{n1} & a_{n2} & \cdots & a_{nn} \end{vmatrix}$$

is linear in each α_i and represents an n-linear mapping of the n-fold cartesian product $V_n(R) \times \cdots \times V_n(R)$ into the scalar field R. It is an alternating multilinear function; that is, it changes sign if any two vectors α_i and α_j, $i \neq j$, are interchanged.

If V is a vector space over a field F, a k-linear mapping of the k-fold cartesian product $V \times \cdots \times V \to F$ is called a **k-form on V** or a **multilinear form on k vectors.** The determinant function is therefore an alternating multilinear form.

We shall use multilinear mappings to construct a new type of vector space called the tensor product.

EXERCISES

1. Let R be the real field and let f be a mapping of $V_2(R) \times V_2(R) \to V_2(R)$. For $\alpha = (a, b)$, $\beta = (c, d)$, let $\gamma = (\alpha, \beta)f$. For each of the following choices of vectors for γ, determine whether f is a bilinear mapping or not.

(1) $\gamma = (ad, bc)$
(2) $\gamma = (a + c, b + d)$
(3) $\gamma = (ac, bd)$
(4) $\gamma = (a + d, b + c)$
(5) $\gamma = (a, d)$
(6) $\gamma = (a + b, c + d)$

2. Prove that a mapping f of $V \times W \to Z$ is bilinear if and only if

$$(x_1\alpha_1 + x_2\alpha_2, y_1\beta_1 + y_2\beta_2)f = \sum_{i,j=1}^{2} x_i y_j (\alpha_i, \beta_j) f.$$

3. Prove the set of all k-linear forms on V, that is k-linear mappings of the k-fold cartesian product $V \times V \times \cdots \times V \to F$, where F is the scalar field, is a vector space over F.

4. Reexamine the discussion in Chapter 7 and then prove that every alternating multilinear form on n vectors of $V_n(R)$ must be the determinant form multiplied by some constant.

5. V is a vector space over F. Prove that the mapping ϕ of $V \times V^* \to F$ defined by $(\alpha, f)\phi = \alpha f$, $\alpha \in V$, $f \in V^*$, is bilinear.

6. U, V, W are vector spaces. Prove that the mapping f of Hom $(U, V) \times$ Hom $(V, W) \to$ Hom (U, W) defined by $(S, T)f = ST$ is bilinear.

11-2 THE TENSOR PRODUCT

Definition. Let V and W be vector spaces over the same field F. A vector space T and a bilinear mapping f of $V \times W \to$ T are called the **tensor product** (T, f) of V and W, if for each choice of a vector space Z over F, and for each choice of a bilinear mapping g of $V \times W \to Z$, there exists a unique linear transformation h of T $\to Z$ such that $f \circ h = g$.

This property of (T, f) is called the **Universal Factorization Property**; that is, (1) it is universal for all bilinear mappings g of $V \times W$ into an arbitrary vector space Z; (2) g factors into the form $g = fh$, where h is a unique linear transformation of T $\to Z$. It is a type of universal mapping property (a property extensively used in modern algebra) and is illustrated by diagram (1).

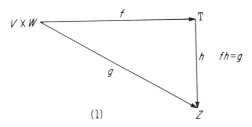

(1)

This means if $(\alpha, \beta) \in V \times W$, then $(\alpha, \beta)f \in$ T, $(\alpha, \beta)(fh) = ((\alpha, \beta)f)h \in Z$, and $((\alpha, \beta)f)h = (\alpha, \beta)g$. The mapping fh of $V \times W \to Z$ is a composite of the mappings f and h.

We leave temporarily in abeyance the question of whether such thing as the tensor product exists, and we first prove that it is unique up to an isomorphism, a result which justifies our calling it "the" tensor product.

We use the abbreviation U.F.P. for the phrase "Universal Factorization Property."

THEOREM 1

If (T,f) and (T',f') are tensor products of the same two vector spaces V and W over F, then there exists a unique vector space isomorphism of $T \to T'$ such that $fh = f'$.

Proof: Consider the diagrams.

Figure (2) illustrates the U.F.P. for (T,f). Referring to figure (1), it is seen that we are taking $Z = T'$ and $g = f'$. This yields a unique linear trans-

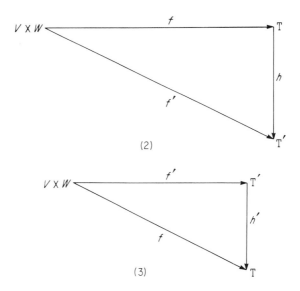

formation h of $T \to T'$ such that $fh = f'$. Figure (3) illustrates the U.F.P. for (T',f'). Here we are taking $Z = T$ and $g = f$. Again we obtain a unique linear transformation h' of $T' \to T$ such that $f'h' = f$. Hence $fhh' = f'h' = f$, where hh' is unique. Since $f1_T = f$, where 1_T is the identity map on T, it follows that $1_T = hh'$. (This is actually the uniqueness condition of the U.F.P. for (T,f) with $g = f$ and $Z = T$. Similarly, $f'h'h = fh = f'$, and similarly it follows that $h'h = 1_{T'}$ the identity map on T'. Now $hh' = 1_T$ proves h is surjective (Lemma 2 Chapter 1), and $h'h = 1_{T'}$ proves h is injective (Lemma 2, Chapter 1)

Hence the linear transformation h of $T \to T'$ is an isomorphism and $fh = f'$.

Because of this last theorem, it is conventional to speak of a tensor product of two vector spaces as being "essentially unique." From this it is but a modest step to speaking of it as "the" tensor product.

By a set G of generators of a vector space V we mean a subset of V, such that every vector of V can be expressed as a finite linear combination of vectors belonging to G.

Lemma 1. A linear transformation T on a vector space V is uniquely determined by its values on a set of generators of V.

Proof: Let G be a set of generators of V. If α is any vector of V, then $\alpha = x_1 g_1 + x_2 g_2 + \cdots + x_k g_k$, where the $g_i \in G$ and the x_i are scalars. Hence

$$\alpha T = (x_1 g_1 + \cdots + x_k g_k)T = x_1(g_1 T) + \cdots + x_k(g_k T).$$

This equation shows that αT, for every $\alpha \in V$, is determined by the values of T on G.

We shall use this lemma to prove that the uniqueness condition in the U.F.P. for the tensor product (T, f) of V and W is equivalent to the condition that im f generates the tensor product T.

THEOREM 2

If (T, f) is the tensor product of the vector spaces V and W, then im f generates T.

Proof: Denote by L the vector space generated by im f. Then $L \subset T$. The U.F.P. for (T, f) with $Z = L$ and $g = f$ (see figure 1) yields a unique linear transformation h of $T \to L$ such that $fh = f$. Also the U.F.P. for (T, f), with this time $Z = T$ and $g = f$, yields the unique homomorphism 1_T of $T \to T$ such that $f = f \circ 1_T$. Now h can be regarded as a homomorphism of $T \to T$ and we know $f = fh$. Hence by the uniqueness property, $h = 1_T$. Therefore $L = T$.

The converse of this theorem is true. For let V, W, T be vector spaces and let f be a bilinear map of $V \times W \to T$ such that im f generates T. Suppose that for any vector space Z and bilinear map g of $V \times W \to Z$, there is a linear transformation h of $T \to Z$ such that $fh = g$. Then h is unique. This follows from the fact that, since im f generates T, the linear transformation h is defined on the generators of T and hence is uniquely determined on all of T (Lemma 1).

We shall make use of the following lemma in the next section.

Lemma 2. Let V and V' be two vector spaces over the same field F and let T be an epimorphism (surjective linear transformation) of

$V \to V'$. If S is a set of generators of V then T maps S into a set S' of generators of V'.

Proof: Let α' be any vector of V'. Since T is surjective, there exists $\alpha \in V$ such that $\alpha T = \alpha'$. Since S generates V, $\alpha = x_1\beta_1 + \cdots + x_k\beta_k$ where the x_i are scalars and the β_i are vectors in S. Hence we have

$$\alpha' = \alpha T = x_1(\beta_1 T) + \cdots + x_k(\beta_k T).$$

Now the $\beta_i T$ belong to S' and so α' is a finite linear combination of vectors of S'. Since this is true of every vector of V', it follows that S generates V'.

Corollary. If the set S is a basis of V then its image S' is a basis of V' if and only if T is an isomorphism.

Exercise. Consider the following definition of the tensor product.

Let V and W be vector spaces over the same field F. Denote by S the vector space over F that is generated by the pairs (α, β), $\alpha \in V$, $\beta \in W$, considered as vectors. Thus, S will consist of the linear combinations of such pairs. Denote by H the subspace of S generated by all vectors of the form

$$(\alpha_1 + \alpha_2, \beta) - (\alpha_1, \beta) - (\alpha_2, \beta),$$
$$(\alpha, \beta_1 + \beta_2) - (\alpha, \beta_1) - (\alpha, \beta_2),$$
$$(x\alpha, \beta) - x(\alpha, \beta), \; x \in F,$$
$$(\alpha, x\beta) - x(\alpha, \beta).$$

Now form the quotient space $T = S/H$. Then T is called the **tensor product** of V and W. The vectors of T are therefore the cosets of H in T; that is, if $(\alpha,\beta) \in S$ then $(\alpha,\beta) + H$ is a vector of T. Of course $(\alpha,\beta) + H$ is the image of (α, β) under the canonical epimorphism of $S \to T = S/H$.

We use the symbol $\alpha \otimes \beta$ for $(\alpha, \beta) + H$; that is, write

$$\alpha \otimes \beta = (\alpha, \beta) + H,$$

and write

$$V \otimes W = T.$$

Now show that $V \otimes W$ together with the mapping f of $V \times W \to V \otimes W$ defined by $(\alpha, \beta)f = \alpha \otimes \beta$ has the U.F.P. on $V \times W$. This means that if Z is any vector space over F and if g is any bilinear mapping of $V \times W \to Z$, then a unique homomorphism h of $V \otimes W \to Z$ can be defined in terms of the mappings f and g, such that a consequence of this definition is that $fh = g$. This, combined with the discussion to follow in the next section, will reconcile these two definitions of the tensor product.

The Tensor Product of Vector Spaces

11-3 EXISTENCE OF THE TENSOR PRODUCT

We still are ignorant as to whether the tensor product actually exists or not. We shall prove that it does by constructing it.

Construction of the Tensor Product

Let V and W be two vector spaces over the same field F. Let S denote the set of all mappings of $V \times W \to F$ whose values are almost all (all except a finite number) zero. This means that if $s \in S$, then s is a function on $V \times W$ whose values are all zero, except at a finite number of points of $V \times W$. Thus S is a subset of $F^{V \times W}$. We have seen earlier that, with the usual definitions of addition of mappings and multiplication of a mapping by a scalar, S is a vector space over F.

For $(\alpha, \beta) \in V \times W$, let $\langle \alpha, \beta \rangle$ denote the function which is 1 at (α, β) and is 0 elsewhere.

Let H be the subspace of S spanned by all vectors (functions) of the form

(1) $\langle \alpha_1 + \alpha_2, \beta \rangle - \langle \alpha_1, \beta \rangle - \langle \alpha_2, \beta \rangle, \alpha_1, \alpha_2 \in V, \beta \in W,$
(2) $\langle \alpha, \beta_1 + \beta_2 \rangle - \langle \alpha, \beta_1 \rangle - \langle \alpha, \beta_2 \rangle, \alpha \in V, \beta_1, \beta_2 \in W,$
(3) $\langle x\alpha, \beta \rangle - x \langle \alpha, \beta \rangle,$
(4) $\langle \alpha, x\beta \rangle - x \langle \alpha, \beta \rangle, \alpha \in V, \beta \in W, x \in F.$

Now form the quotient space S/H and write $T = S/H$. Let j denote the natural epimorphism of $S \to T$; that is, for $s \in S$, $(s)j = s + H$.

Let f be the composite mapping of $V \times W \to T$ defined by $(\alpha, \beta) \to \langle \alpha, \beta \rangle \to \langle \alpha, \beta \rangle j$, that is

(5) $(\alpha, \beta)f = \langle \alpha, \beta \rangle j.$

We now undertake to prove that (T, f) is the tensor product of V and W. In order to do this we must prove that (T, f) has the U.F.P. on the set $V \times W$.

First of all we remark that f is bilinear. This follows when we apply j to the generators (1), (2), (3), (4) of H. For example we find by (1) that, since $\langle \alpha_1 + \alpha_2, \beta \rangle - \langle \alpha_1, \beta \rangle - \langle \alpha_2, \beta \rangle$ is a generator of H (and hence it belongs to H),

$$(\langle \alpha_1 + \alpha_2, \beta \rangle - \langle \alpha_1, \beta \rangle - \langle \alpha_2, \beta \rangle)j = 0$$

(where 0 stands for the zero vector of S/H). Since j is a homomorphism, this last equation can be written

$$\langle \alpha_1 + \alpha_2, \beta \rangle j = \langle \alpha_1, \beta \rangle j + \langle \alpha_2, \beta \rangle j.$$

Hence, by (5) we have

$$(\alpha_1 + \alpha_2, \beta)f = (\alpha_1, \beta)f + (\alpha_2, \beta)f.$$

The fact that f possesses the remaining properties of a bilinear mapping follows in the same way by use of (2), (3), and (4).

Exercise. Verify that the above mapping f is bilinear.

We now continue with our proof that (T,f) has the U.F.P. on $V \times W$.

First we derive a very useful form for an element (function) of the function space S. Let $s \in S$. Then the values of s are 0 except at a finite number $(\alpha_1, \beta_1), (\alpha_2, \beta_2), \ldots, (\alpha_n, \beta_n)$ of points of $V \times W$. We claim that we can express the function s as

(6) $\quad s = ((\alpha_1, \beta_1)s) \langle \alpha_1, \beta_1 \rangle + \cdots + ((\alpha_n, \beta_n)s) \langle \alpha_n, \beta_n \rangle$

where, by definition, $\langle \alpha_i, \beta_i \rangle$ is that element of S which has the value 1 at (α_i, β_i) and is 0 elsewhere. Here $(\alpha_i, \beta_i)s \in F$. It is the value of s at (α_i, β_i).

To prove our claim, first note that if (α, β) is not one of the n points (α_i, β_i), then $((\alpha, \beta)) \langle \alpha_i, \beta_i \rangle = 0$ for all $i = 1, 2, \ldots, n$, and hence every term of the right-hand side of (6) is 0 and therefore $(\alpha, \beta)s = 0$, as it should. On the other hand, for the point (α_i, β_i), we have

$$((\alpha_i, \beta_i)) \langle \alpha_j, \beta_j \rangle = 0, j \neq i, \text{ and } ((\alpha_i, \beta_i)) \langle \alpha_i, \beta_i \rangle = 1.$$

Hence at such a point the right-hand side of (6) becomes $(\alpha_i, \beta_i)s$· $((\alpha_i, \beta_i)) \langle \alpha_i, \beta_i \rangle = (\alpha_i, \beta_i)s$, as it should.

It is more convenient however to rewrite s in the form

(7) $\quad\quad\quad\quad\quad s = \Sigma(\alpha, \beta)s \langle \alpha, \beta \rangle$

where (α, β) ranges over $V \times W$. Since the values of s are 0 almost everywhere, the right-hand side is a finite sum.

Now to return to the proof that (T, f), as defined, has the U.F.P. on $V \times W$.

Let Z be any vector space over F and let g be any bilinear mapping of $V \times W \to Z$. Define a mapping h of $T \to Z$ by

(8) $\quad\quad\quad\quad (sj)h = \Sigma(\alpha, \beta)s (\alpha, \beta)g, s \in S.$

Clearly h is a homomorphism. For $s, s' \in S$,

$$\begin{aligned}((s + s')j)h &= \Sigma((\alpha, \beta)s + (\alpha, \beta)s')(\alpha, \beta)g \\ &= (sj)h + (s'j)h.\end{aligned}$$

Also,

$$(xs)jh = \Sigma x(\alpha, \beta)s(\alpha, \beta)g = x\Sigma(sj)h.$$

Thus h is a linear transformation. For $(\alpha, \beta) \in V \times W$, we have $(\alpha, \beta)f = \langle \alpha, \beta \rangle j$. Hence $(\alpha, \beta)fh = (\langle \alpha, \beta \rangle j)h = (\alpha, \beta)g$, by (8). Therefore $fh = g$.

The Tensor Product of Vector Spaces

All that is left for us to do is to prove the uniqueness of h. Assume h' is a second linear transformation of $T \to Z$ with the property that $fh' = g$. Then for any $s \in S$,

$$(sj)h' = [(\Sigma(\alpha,\beta)s\langle\alpha,\beta\rangle)j]h'$$
$$= \Sigma(\alpha,\beta)s(\langle\alpha,\beta\rangle jh')$$
$$= \Sigma(\alpha,\beta)s(\alpha,\beta)g = (sj)h.$$

Hence $h' = h$.

This completes the proof that (T, f), as constructed, has the U.F.P. on $V \times W$ and is therefore the tensor product of V and W.

We know, by Theorem 1, that any other form for the tensor product of V and W is a vector space isomorphic to the one we have constructed.

We now introduce the notation for tensor products. We write the tensor product T of V and W as

$$T = V \otimes W.$$

Also, we write

$$\alpha \otimes \beta = (\alpha, \beta) f, \alpha \in V, \beta \in W.$$

This is called the **tensor product of the vectors α and β**.

11-4 COMPUTATION IN $V \otimes W$

A very important observation is that the elements $\langle\alpha, \beta\rangle$ constitute a basis of the vector space S; that is, they form a set of independent generators of S. Since the mapping j of $S \to S/H$ is an epimorphism (but not an isomorphism) the elements $(\alpha, \beta) f = \langle\alpha, \beta\rangle j = \alpha \otimes \beta$ form a set of generators of $V \otimes W$ but not a basis of $V \otimes W$. The tensor product $V \otimes W$ is spanned by the elements of the form $\alpha \otimes \beta$. Hence the rules of computation in $V \otimes W$ are determined when we know the basic rules of computation for the elements $\alpha \otimes \beta$.

These basic rules for computation in $V \otimes W$ are derived from the forms (1), (2), (3), (4) for the generators of the subspace H of S. Take a generator of the form (1), for instance:

$$\langle\alpha_1 + \alpha_2, \beta\rangle - \langle\alpha_1, \beta\rangle - \langle\alpha_2, \beta\rangle.$$

Apply to it the natural epimorphism j of $S \to S/H = V \otimes W$. Since this generator belongs to H, we get

$$\langle\alpha_1 + \alpha_2, \beta\rangle j = \langle\alpha_1, \beta\rangle j + \langle\alpha_2, \beta\rangle j.$$

In terms of our new notation this becomes

(1') $\qquad (\alpha_1 + \alpha_2) \otimes \beta = \alpha_1 \otimes \beta + \alpha_2 \otimes \beta.$

In the same way we can see that the generators of the forms (2), (3), and (4) produce respectively the formulas

(2') $\quad \alpha \otimes (\beta_1 + \beta_2) = \alpha \otimes \beta_1 + \alpha \otimes \beta_2,$
(3') $\quad (x\alpha) \otimes \beta = x(\alpha \otimes \beta), \quad x \in F,$
(4') $\quad \alpha \otimes (x\beta) = x(\alpha \otimes \beta).$

Thus (1')–(4') are the basic rules of computation for the generators of the tensor product $V \otimes W$.

Exercise. Verify the formulas (2'), (3'), and (4').

A typical term of $V \otimes W$ can be expressed as the finite sum of generators. For such a term has the form

$$x_1(\alpha'_1 \otimes \beta_1) + \cdots + x_k(\alpha'_k \otimes \beta_k) = (x_1\alpha'_1) \otimes \beta_1 + \cdots + (x_k\alpha'_k) \otimes \beta_k,$$

where the x_i are scalars, the $\alpha'_i \in V$, and the $\beta_i \in W$. Since the $x_i\alpha'_i$ are vectors of V, we put $x_i\alpha'_i = \alpha_i \in V$ and hence a typical term in $V \otimes W$ has the form

$$\alpha_1 \otimes \beta_1 + \alpha_2 \otimes \beta_2 + \cdots + \alpha_k \otimes \beta_k.$$

It is important to realize that the elements of $V \otimes W$ are the cosets of H in S; that is, they are elements of the form $s + H$, $s \in S$. Hence equality of two elements in $V \otimes W$ merely implies that the difference of the two elements is an element of H. For instance, whereas in the vector space S, $\langle \alpha_1, \beta_1 \rangle = \langle \alpha_2, \beta_2 \rangle$ implies $\alpha_1 = \alpha_2$, $\beta_1 = \beta_2$; in the quotient space $V \otimes W = S/H$, $\alpha_1 \otimes \beta_1 = \alpha_2 \otimes \beta_2$ does not imply $\alpha_1 = \alpha_2$, $\beta_1 = \beta_2$. It does imply that $\langle \alpha_1, \beta_1 \rangle - \langle \alpha_2, \beta_2 \rangle \in H$, since $\alpha \otimes \beta = \langle \alpha, \beta \rangle + H$.

11-5 LINEAR TRANSFORMATIONS ON $V \otimes W$

Next let V, W, Z be three vector spaces over a field F. Let us denote by L the set of all linear transformations of the vector space $V \otimes W$ into a vector space Z. As we know from our earlier study, L is a vector space under the usual definitions of map additions and scalar multiplication of maps. In particular we recite the definition of the scalar multiple of a linear transformation of L. For $x \in F$, $T \in L$, and $\alpha \otimes \beta \in V \otimes W$, we have $(\alpha \otimes \beta)(xT) = x((\alpha \otimes \beta)T)$.

Denote by B the set of all bilinear mappings of $V \times W$ into Z. Using similar definitions of addition and scalar multiplication to those for L, it is very easy to show that B is likewise a vector space over the common field of V, W, and Z.

Exercise. Give the details and show how B is a vector space.

The Tensor Product of Vector Spaces

We now prove the following theorem:

THEOREM 3

The vector spaces B and L are isomorphic.

Proof: For $T \in L$ define a mapping ψ of $L \to B$ by $(T)\psi = fT$, where $(V \otimes W, f)$ is the tensor product of V and W. Thus f represents the bilinear mapping of $V \times W \to V \otimes W$ defined by $(\alpha, \beta)f = \alpha \otimes \beta$, $\alpha \in V$, $\beta \in W$. First we must verify that $fT \in B$. fT is the composite map $V \times W \xrightarrow{f} V \otimes W \xrightarrow{T} Z$. We find

$$((\alpha_1 + \alpha_2, \beta)f)T = (\alpha_1 \otimes \beta + \alpha_2 \otimes \beta)T = (\alpha_1 \otimes \beta)T + (\alpha_2 \otimes \beta)T$$
$$= ((\alpha_1, \beta)f)T + ((\alpha_2, \beta)f)T.$$

Hence $(\alpha_1 + \alpha_2, \beta)(fT) = (\alpha_1, \beta)(fT) + (\alpha_2, \beta)(fT)$. Similarly, we can prove the three remaining requirements (b), (c), (d) for bilinearity of the mapping fT. Thus $fT \in B$.

Next we show that ψ is a homomorphism. Now clearly $(T + T')\psi = f(T + T') = fT + fT' = (T)\psi + (T')\psi$, where $T, T' \in L$. Also, since $((\alpha, \beta)f)(xT) = (\alpha \otimes \beta)(xT) = x(\alpha \otimes \beta)T$, $\alpha \in V$, $\beta \in W$, it follows that $(xT)\psi = f(xT) = x(fT) = x((T)\psi)$. This shows ψ is a linear transformation.

ψ is surjective. For if $g \in B$ then by the U.F.P. for the tensor product $(V \otimes W, f)$, there exists a unique linear transformation $h \in L$ such that $g = fh$. Hence $g = (h)\psi$.

ψ is injective. For if $T \in \ker \psi$, then $(\alpha, \beta)fT = (\alpha \otimes \beta)T = \bar{0}_Z$. This defines the values of T on all the generators of $V \otimes W$ to be the zero vector of Z. Hence by Lemma 1, T is the zero linear transformation of L; that is, the zero vector of the vector space L. This proves ψ is injective. Hence we have shown that ψ is an isomorphism, which proves our theorem.

A very important immediate corollary of this last theorem is the following.

Corollary. There exists a linear transformation T of $V \otimes W \to Z$ having specified values $(\alpha \otimes \beta)T$, $\alpha \in V$, $\beta \in W$, on the generators $\alpha \otimes \beta$ of $V \otimes W$, if and only if the function Φ of $V \times W \to Z$, defined by $(\alpha, \beta)\Phi = (\alpha \otimes \beta)T$ is bilinear.

Exercise. $V_2(R)$ is the vector space of all pairs of real numbers (x, y). Let $\alpha = (x_1, y_1)$ and $\beta = (x_2, y_2)$ be arbitrary vectors of $V_2(R)$. Prove that the mapping T of $V_2(R) \otimes V_2(R) \to V_2(R)$ defined by

$$(\alpha, \beta)T = (x_1 y_2, x_2 y_1)$$

is linear.

This corollary provides a rather simple criterion for the linearity of a transformation T on the tensor product of two vector spaces. For it is not possible, in any simple way, to verify directly, by use of the two conditions for linearity, that a given mapping on the tensor product is actually linear. However, the real importance of the corollary is that it simplifies the theory of bilinear transformations into the familiar theory of linear transformations.

We now use this corollary to prove the next theorem.

Exercise. Prove that the tensor product of two vector spaces V and W over the same field F is isomorphic to the dual space of the vector space B of bilinear functions of $V \times W \to F$; that is, prove $V \otimes W \cong B^*$. (This is sometimes used as the definition of the tensor product.)

We next construct the tensor product of two linear transformations.

THEOREM 4

Let V, V', W, W' be vector spaces over a field F. Let T be a linear transformation of $V \to W$ and T' a linear transformation of $V' \to W'$. Then the mapping $T \otimes T'$ of $V \otimes V' \to W \otimes W'$ defined by

$$(\alpha \otimes \alpha')(T \otimes T') = (\alpha T) \otimes (\alpha' T'), \quad \alpha \in V, \alpha' \in V'$$

is a linear transformation.

Proof: This follows at once from the corollary, for the function Φ, defined by

$$(\alpha, \alpha')\Phi = (\alpha T) \otimes (\alpha' T'), \quad \alpha \in V, \alpha' \in V'$$

is easily proved to be bilinear on $V \times V'$, since T and T' are both linear.

Exercise. Supply the details in this proof.

Canonical Isomorphisms

Let V and W be two vector spaces over the same field F. A linear transformation T of $V \to W$ is said to be **canonical** (or **natural**) if, in its definition, only the properties of V and W as vector spaces are used and the definition is not dependent on some additional choice of bases or inner products in these spaces. In general the word canonical is applied to any mathematical object, not necessarily a linear transformation, if its definition or construction is unique in that it involves no arbitrary or free choices.

For example, if U is a subspace of a vector space V, the epimorphism j of $V \to V/U$, defined by $\alpha j = \alpha + U$, $\alpha \in V$, is canonical. Another example is the inclusion mapping i of a subspace U into a vector space V, defined by $\alpha i = \alpha$, $\alpha \in U$; it is canonical.

However, a linear transformation is also called canonical, if it is defined in terms of some choice but is independent of the particular choice.

For instance, an n-dimensional vector space over a field F is canonically isomorphic to the vector space $V_n(F)$. See Theorem 7, Chapter 3.

It is customary and very convenient to identify two vector spaces V and W that are canonically isomorphic and to write $V = W$, provided that no essential relationship gets lost in doing so. We shall frequently do this in the next theorems. These theorems deal with the properties of the tensor products of vector spaces.

EXERCISES

1. If α_1, α_2 form a basis of the vector space V, prove that $\alpha_i \otimes \alpha_j$, $i, j = 1, 2$ form a basis of $V \otimes V$ and hence that dim $V \otimes V = 4$.

2. The vectors α_1, α_2 form a basis for the vector space V. S and T are linear operators on V represented respectively by the matrices

$$A = \begin{bmatrix} a_{11} & a_{12} \\ a_{21} & a_{22} \end{bmatrix} \quad \text{and } B = \begin{bmatrix} b_{11} & b_{12} \\ b_{21} & b_{22} \end{bmatrix}$$

relative to this basis. Show that the linear operator $S \otimes T$ on $V \otimes V$ has the block matrix $\begin{bmatrix} a_{11}B & a_{12}B \\ a_{21}B & a_{22}B \end{bmatrix}$ relative to the basis $\alpha_i \otimes \alpha_j$, $i, j = 1, 2$, of $V \times V$. Here each block $a_{ij}B$ is a 2×2 matrix and the matrix of $S \otimes T$ is a 4×4 matrix.

11-6 ISOMORPHISMS OF TENSOR PRODUCTS

Let V be a vector space over a field F. The field F itself can always be regarded as a vector space over itself, of dimension one.

THEOREM 5

$$F \otimes V = V = V \otimes F.$$

Proof: The equality signs here denote canonical isomorphisms. Consider the linear transformation T of $F \otimes V \to V$ defined by $(x \otimes \alpha)T = x\alpha$, $x \in F$, $\alpha \in V$, and the linear transformation S of $V \to F \otimes V$ defined by $\alpha S = 1 \otimes \alpha$, $1 \in F$, $\alpha \in V$. (It is easily verified, by the corollary of Theorem 3, that both T and S are actually linear.)

Now TS is a linear transformation of $F \otimes V \to F \otimes V$. Also,

$$(x \otimes \alpha)(TS) = ((x \otimes \alpha)T)S = (x\alpha)S = x(\alpha S)$$
$$= x(1 \otimes \alpha) = x \otimes \alpha, \quad x \in F, \quad \alpha \in V.$$

This implies that TS is the identity map on $F \otimes V$. In the same way we can prove that ST is the identity map on V. Hence by Lemma 2 of Chapter 1, T (and also S, of course) is an isomorphism and it is canonical.

Similarly, it can be shown that $V = V \otimes F$.

Exercise. Prove that V and $V \otimes F$ are canonically isomorphic.

Let V and W be vector spaces over F. We next prove

THEOREM 6

$$V \otimes W = W \otimes V.$$

Proof: Again the equality sign denotes canonical isomorphism.

By the corollary to Theorem 3 we can easily see that the transformation T of $V \otimes W \to W \otimes V$ defined by $(\alpha \otimes \beta)T = \beta \otimes \alpha$, $\alpha \in V$, $\beta \in W$, is linear. (For the mapping $(\alpha,\beta)\Phi = \beta \otimes \alpha$ is seen to be bilinear.)

Similarly the mapping S of $W \otimes V \to V \otimes W$ defined by $(\beta \otimes \alpha)S = \alpha \otimes \beta$ is linear. Also, ST and TS are the identity maps on $W \otimes V$ and $V \otimes W$ respectively. Hence T is an isomorphism and moreover this isomorphism is canonical.

Let V, W, Z be three vector spaces over the field F. We shall next prove the associativity of the tensor product.

THEOREM 7

$$(V \otimes W) \otimes Z = V \otimes (W \otimes Z).$$

Proof: The mapping T_1 of $V \times W \times Z \to V \otimes (W \otimes Z)$ defined by $(\alpha,\beta,\gamma)T_1 = \alpha \otimes (\beta \otimes \gamma)$, $\alpha \in V$, $\beta \in W$, $\gamma \in Z$, is trilinear. This means that if any two of the vectors α,β,γ are held fixed, then T_1 is linear in the third one. The trilinearity follows at once from the basic computation formulas (1′), (2′), (3′), (4′).

Hence holding γ fixed, T_1 is a bilinear mapping of $(V \times W) \times Z \to V \otimes (W \otimes Z)$ and hence, by the corollary to Theorem 3, it determines a linear transformation T_2 of $(V \otimes W) \times Z \to V \otimes (W \otimes Z)$ where $(\alpha \otimes \beta, \gamma)T_2 = \alpha \otimes (\beta \otimes \gamma)$.

Now release γ and we see that T_2 is a bilinear mapping of $(V \otimes W) \times Z \to V \otimes (W \otimes Z)$, which again therefore determines a linear transformation T of $(V \otimes W) \otimes Z \to V \otimes (W \otimes Z)$ (Corollary to Theorem 3) such that

$$((\alpha \otimes \beta) \otimes \gamma)T = \alpha \otimes (\beta \otimes \gamma), \qquad \alpha \in V, \quad \beta \in W, \quad \gamma \in Z$$

A similar argument used on the function S_1 of $V \times W \times Z \to (V \otimes W) \otimes Z$, defined by $(\alpha,\beta,\gamma)S_1 = (\alpha \otimes \beta) \otimes \gamma$, $\alpha \in V$, $\beta \in W$, $\gamma \in Z$, proves that it leads to a linear transformation S of $V \otimes (W \otimes Z) \to (V \otimes W) \otimes Z$ such that $(\alpha \otimes (\beta \otimes \gamma))S = (\alpha \otimes \beta) \otimes \gamma$. Hence

$$\begin{aligned}((\alpha \otimes \beta) \otimes \gamma)(TS) &= [((\alpha \otimes \beta) \otimes \gamma)T]S \\ &= (\alpha \otimes (\beta \otimes \gamma))S \\ &= (\alpha \otimes \beta) \otimes \gamma, \quad \alpha \in V, \quad \beta \in W, \quad \gamma \in Z.\end{aligned}$$

Similarly,
$$(\alpha \otimes (\beta \otimes \gamma))(ST) = \alpha \otimes (\beta \otimes \gamma).$$
Since TS and ST are linear, these two results imply that TS and ST are the identity maps on $(V \otimes W) \otimes Z$ and $V \otimes (W \otimes Z)$ respectively. It therefore follows, as in the proofs of the two previous theorems, that T is an isomorphism. Moreover, it is canonical. This proves the theorem.

Definition. The **tensor product of a finite number of vector spaces** $V_1, V_2, \ldots, V_n$ is a vector space Υ and a multilinear mapping f of $V_1 \times \cdots \times V_n \to \Upsilon$, such that the pair (Υ, f) has the U.F.P. for $V_1 \times \cdots \times V_n$.

We have seen that the tensor product of 2 and 3 vector spaces exists. By induction on n, we can prove the following theorem.

THEOREM 8

The tensor product of any finite number n of vector spaces, $n \geq 2$, exists.

EXERCISES

1. Carry out the induction and prove Theorem 8.
2. If $S, T \in V^*$, prove $S \otimes T \in (V \otimes V)^*$.
3. If V is finite-dimensional, prove
$$(V \otimes V)^* = V^* \otimes V^*.$$

11-7 DIRECT SUMS

In this section we use the tensor product of direct sums to find the dimension of $V \otimes W$ where V and W are finite-dimensional vector spaces.

Lemma 3. Let V, V_1, V_2 be vector spaces over F. Then $V = V_1 \oplus V_2$ (direct sum) if and only if there exist homomorphisms π_i of $V \to V_i$ and σ_i of $V_i \to V$, $i = 1, 2$, such that
$$\sigma_i \pi_j = 0 \text{ (the zero homomorphism)}, i \neq j$$
$$= 1_{V_i} \text{ (the identity map on } V_i\text{)}, j = i, i = 1, 2.$$

Proof: If $V = V_1 \otimes V_2$ then for every $\alpha \in V$, $\alpha = \alpha_1 + \alpha_2$, $\alpha_1 \in V_1$, $\alpha_2 \in V_2$, and this decomposition is unique. Define the projections π_i of $V \to V_i$ by $\alpha \pi_i = \alpha_i$, $i = 1, 2$; and define the maps σ_i of $V_i \to V$, $i = 1, 2$, as being the canonical inclusions $\alpha_i \sigma_i = \alpha_i$, $\alpha_i \in V_i$. Then it is easy to see that the two conditions of the theorem are true.
Conversely, assume the two conditions hold. Define a mapping ϕ of $V \to V_1 \oplus V_2$ by $(\alpha)\phi = \alpha \pi_1 + \alpha \pi_2$, $\alpha \in V$, and a mapping ψ of $V_1 \oplus V_2 \to V$ by $(\alpha_1 + \alpha_2) = \alpha_1 \sigma_1 + \alpha_2 \sigma_2$, $\alpha_1 \in V_1$, $\alpha_2 \in V_2$. It is quite easy to verify that both ϕ and ψ are homomorphisms (linear transformations).

Moreover, $\psi\phi$ and $\phi\psi$ are the identity maps on $V_1 \oplus V_2$ and V respectively. This likewise is readily checked. Hence ϕ and ψ are canonical isomorphisms. Hence

$$V = V_1 \oplus V_2.$$

Exercise. In the last proof, show that ϕ and ψ are homomorphisms. Also prove that $\psi\phi$ and $\phi\psi$ are the identity maps indicated.

THEOREM 9

Let V_1, V_2, and Z be vector spaces over F. Then

$$(V_1 \oplus V_2) \otimes Z = V_1 \otimes Z \oplus V_2 \otimes Z.$$

Proof: This is again a canonical isomorphism. Let 1_Z denote the identity map on Z. As in the previous lemma, for $i = 1,2$, let π_i be the projections of $V_1 \oplus V_2$ on V_i, and σ_i the inclusions of V_i into $V_1 \oplus V_2$.

For $i = 1,2$, form the linear transformations $\pi_i \otimes 1_Z$ of $(V_1 \oplus V_2) \otimes Z \to V_i \otimes Z$ and $\sigma_i \otimes 1_Z$ of $V_i \otimes Z \to (V_1 \oplus V_2) \otimes Z$. Since $(\sigma_i \otimes 1_Z)(\pi_j \otimes 1_Z) = 0$ (zero homomorphism), $i \neq j$ and $(\sigma_i \otimes 1_Z)(\pi_i \otimes 1_Z) = 1_{V_i} \otimes Z$ (identity map on $V_i \otimes Z$), $i = 1, 2$, the two conditions in Lemma 3 are satisfied and hence this theorem is true.

THEOREM 10

Let V and W be finite-dimensional vector spaces over F. If $\alpha_1, \ldots, \alpha_n$ and $\beta_1, \ldots, \beta_m$ are bases for V and W respectively, then the nm vectors $\alpha_i \otimes \beta_j$, $i = 1,2,\ldots,n$, $j = 1,2,\ldots,m$ form a basis for $V \otimes W$, and

$$\dim V \otimes W = nm.$$

Proof: We first dispose of the special case where $n = m = 1$, since the general case will depend on it. If $\dim V = \dim W = 1$, let α_1 be a basis of V and β_1 a basis of W. Then every vector of $V \otimes W$ has the form $x(\alpha_1 \otimes \beta_1)$, where x is a scalar. This means that $\dim V \otimes W = 0$ or 1 (depending on whether $\alpha_1 \otimes \beta_1$ is the zero vector or not). The mapping T of $V \times W \to F$ defined by $(x(\alpha_1 \otimes \beta_1))T = x$ is seen to be linear and surjective. Since $\dim F = 1$, we have $\dim V \otimes W = 1$.

Now consider the general case. Let V_i be the vector space generated by the vector α_i, $i = 1, 2, \ldots, n$, of the basis V, and let W_i be the vector space generated by the vector β_i, $i = 1, 2, \ldots, m$, of the basis of W. Then

$$V = V_1 \oplus V_2 \oplus \cdots \oplus V_n$$
$$W = W_1 \oplus W_2 \oplus \cdots \oplus W_m.$$

Now by repeated use of the previous theorem we see that

$$V \otimes W = \sum_{i,j=1}^{i=n, j=m} V_i \otimes W_j.$$

The Tensor Product of Vector Spaces

By the first part of our proof, dim $V_i \otimes W_j = 1$ for $i = 1, 2, \ldots, n$, and $j = 1, 2, \ldots, m$. Hence from the property of a direct sum, we obtain

$$\dim V \otimes W = nm$$

and the $\alpha_i \otimes \beta_j$, $i = 1, 2, \ldots, n$, $j = 1, 2, \ldots, m$ constitute a basis for $V \otimes W$.

EXERCISES

1. V is the vector space of all $n \times n$ matrices over a field. If A, B are matrices in V, prove the mapping f of $V \otimes V \to V$ defined by

$$(A \otimes B)f = AB$$

is linear.

2. If V is a vector space over a field F prove that the mapping of Hom $(V, V) \otimes$ Hom $(V, V) \to$ Hom (V, V) defined by $S \otimes T \to ST$, where $S, T \in$ Hom (V, V), is linear.

3. If $\alpha_1, \alpha_2, \ldots, \alpha_n$ is a basis of the vector space V, find a basis of the vector space $V \otimes V \otimes V$.

4. V and W are vector spaces over a field F, and $\alpha_1, \alpha_2, \ldots, \alpha_k$ are linearly independent vectors in V. If $\beta_1, \beta_2, \ldots, \beta_k$ are vectors in W, prove that

$$\sum_{i=1}^{k} \alpha_i \otimes \beta_i = \overline{0}_{V \otimes W}$$

implies $\beta_i = \overline{0}_W$ for $i = 1, 2, \ldots, k$.

5. Prove that the mapping T of $V \otimes V^* \to R$ defined by

$$(\alpha \otimes f)T = \alpha f, \quad \alpha \in V, \quad f \in V^*,$$

is a linear transformation.

If V is finite-dimensional, find the kernel of T. Here V is a vector space over the real field R and V^* is its dual space.

6. U is a subspace of a vector space V. Prove that $U \otimes U$ is a subspace of $V \otimes V$. If j is the canonical epimorphism of $V \to V/U$, prove that the mapping J defined by

$$(\alpha \otimes \beta) J = (\alpha j) \otimes (\beta j), \quad \alpha, \beta \in V,$$

is the canonical epimorphism of $V \otimes V \to V \otimes V/U \otimes U$.

7. V_1, V_2, W_1, W_2 are vector spaces and $T_1: V_1 \to W_1$, $T_2: V_2 \to W_2$ are linear transformations. Prove that $T_1 \oplus T_2$ defined by

$$(\alpha_1 \otimes \alpha_2) T_1 \otimes T_2 = (\alpha_1 T_1) \otimes (\alpha_2 T_2), \quad \alpha_1 \in V_1, \quad \alpha_2 \in V_2,$$

is a linear transformation of $V_1 \otimes V_2 \to W_1 \otimes W_2$.

8. V is a finite-dimensional vector space and V^* is its dual space. T is an isomorphism of $V \to V^*$ mapping the basis $\alpha_1, \alpha_2 \ldots, \alpha_n$ of V into its dual basis $f_1, f_2, \ldots, f_n$ of V^*. Prove that the mapping $T \otimes T$ of $V \otimes V \to V^* \otimes V^*$ transforms the basis of $V \otimes V$ formed from $\alpha_1, \alpha_2, \ldots, \alpha_n$ into its dual basis in $V^* \otimes V^*$; that is, $T \otimes T$ preserves the duality of the bases.

9. V and W are finite-dimensional vector spaces over the same field. Prove that the vector spaces $V \otimes W$ and $\text{Hom}(V, W)$ are isomorphic. Find such an isomorphism.

10. A free module M over a commutative ring R with unit element is a module that has a basis (a linearly independent set of generators). If S is a set of basis elements of the free module M, prove that (M, f), where f is the inclusion mapping of $S \to M$, has the U.F.P. on S. This means that for any module X over R and any mapping g of $S \to X$, there exists a unique homomorphism h of $M \to X$ such that $f \circ h = g$.

11. Prove the converse of Exercise 10.

12. Prove that every finite-dimensional vector space has the U.F.P. on a set of basic elements.

13. If $\dim V = n$, find the dimension of $\text{Hom}(V \otimes V, V \otimes V)$.

14. If U is a subspace of a finite-dimensional vector space V, prove that $U \otimes U$ is always a subspace of $V \otimes V$.

On the other hand if, for instance α_1, α_2 is a basis of V, find a subspace of $V \otimes V$ that cannot be expressed in the form $U \otimes U$, where U is a subspace of V.

In general then a subspace of $V_1 \otimes V_2$ cannot necessarily be expressed in the form $U_1 \otimes U_2$ where U_1, U_2 are subspaces respectively of V_1, V_2.

15. V and W are finite-dimensional vector spaces. If $\alpha \neq \overline{0}_V$ and $\beta \neq \overline{0}_W$, prove $\alpha \otimes \beta \neq 0$ in $V \otimes W$. (Hint: We can choose α as a basic vector of V and β as a basic vector of W.)

16. V and W are finite-dimensional vector spaces. Prove that $\text{Hom}(V \otimes W, V \otimes W)$ and $\text{Hom}(V, V) \otimes \text{Hom}(W, W)$ are canonically isomorphic vector spaces.

Chapter **12**

The Exterior Algebra of a Vector Space

We have defined a group, a ring, a vector space, and a module, and in this chapter we introduce another type of algebraic system called an algebra. The concept of the exterior algebra of a finite-dimensional vector space is very useful and important in the applications of linear (actually multilinear) algebra to the calculus of functions of several variables. Although an algebra is a new type of algebraic system, it is in reality only a vector space with a multiplication of vectors defined. An algebra can also be described as a vector space that is also a ring.

12-1 ALGEBRAS

We begin with the formal definition of an algebra.

Definition. An **associative algebra** A over a field F is a vector space over F together with a linear mapping f of $A \otimes A \to A$ such that the multiplication defined by $\alpha\beta = (\alpha \otimes \beta)f$ is associative.

We shall assume that an algebra A has a **unit element** e; that is, there exists a vector $e \in A$ such that $e\alpha = \alpha = \alpha e$ for all $\alpha \in A$.

Observe that in an algebra the multiplication has therefore the following properties:
  (i) $\alpha(\beta + \gamma) = \alpha\beta + \alpha\gamma, \quad \alpha,\beta,\gamma \in A$
  (ii) $(\alpha + \beta)\gamma = \alpha\gamma + \beta\gamma$
  (iii) $x(\alpha\beta) = (x\alpha)\beta = \alpha(x\beta), \quad x \in F$.

Property (iii) is often called the **mixed associative law.** It relates scalar multiplication to ring multiplication.

If $\alpha\beta = \beta\alpha$ for all $\alpha,\beta \in A$, the algebra is said to be **commutative**.
The field F is trivially an algebra over itself.

Example 1. One of the most important examples of an algebra over a field F is the set of linear operators on a vector space V over F. This algebra was explained in Chapter 3. It is called the **operator algebra** of V.

Example 2. We have seen that the set of all polynomials $a_0 + a_1 x + \cdots + a_n x^n$ in a variable x over a field F is both a ring (the product of two polynomials is a polynomial) and a vector space, and hence the polynomials form an associative algebra over F. It is called the **polynomial algebra over F**.

Example 3. By specifying a suitable multiplication table for the basis vectors of a finite-dimensional vector space V over F we can construct a **finite-dimensional algebra** over F; that is, equip V with the structure of an algebra. For instance, if α and β form a basis of a two-dimensional vector space V, let the multiplication rules for this basis be given by the table

	α	β
α	α	β
β	β	α

Now define the general product by

$$(x\alpha + y\beta)(u\alpha + v\beta) = (xu + yv)\alpha + (xv + yu)\beta$$

where x, y, u, v are scalars.

Then it may be quickly verified that the vector space V has now become a two-dimensional commutative algebra with the unit element α.

Definition. An **ideal** in an algebra A is a vector subspace B of A such that for every $\alpha \in A$ and every $\beta \in B$, $\alpha\beta$ and $\beta\alpha$ belong to B.

Example 4. The set of all polynomials with zero constant terms—that is, polynomials of the form $a_1 x + a_2 x^2 + \cdots + a_n x^n$, is an ideal in the polynomial algebra.

If B is an ideal in an algebra A, then B is clearly itself an algebra and is therefore a subalgebra of A.

Let B be an ideal in the algebra A and designate by A/B the resulting quotient vector space. We can define a multiplication in A/B by

$$(\alpha_1 + B) \cdot (\alpha_2 + B) = \alpha_1 \alpha_2 + B, \qquad \alpha_1, \alpha_2 \in A.$$

It is easily verified that this multiplication converts A/B into an algebra. It is called the **quotient algebra** of A with respect to the ideal B.

Exercise. Supply the details that prove A/B is an algebra.

Let V_i, $i = 0, 1, 2, \ldots$ be a sequence of subspaces of a vector space V. V is said to be the **direct sum** of this sequence of subspaces, written

$$V = \sum_{i=0}^{\infty} \oplus V_i,$$

if given any $\alpha \in V$, there exists *only one* representation of α in the form $\alpha = \Sigma \alpha_i$, $\alpha_i \in V_i$, $i = 0, 1, 2, \ldots$ such that almost all (all but a finite number) $\alpha_i = \overline{0}_V$.

The Exterior Algebra of a Vector Space

Definition. A **graded vector space** is a vector space V and a direct-sum decomposition $V = \sum_{i=0}^{\infty} \oplus V_i$ of V into a sequence of subspaces of V.

This particular grading is often called a *regular grading*, to distinguish it from a grading where the index set of the family of subspaces is arbitrary.

Definition. A **graded algebra** is an algebra A that is a graded vector space $A = \sum_{i=0}^{\infty} \oplus A_i$ (the A_i are subspaces of A) such that $A_i A_j \subset A_{i+j}$; that is, such that the product of an element of A_i and of an element of A_j is an element of A_{i+j}.

An immediate consequence of this latter definition is that in general A_i, $i > 0$, is not a subalgebra of A, since the product of two elements of A_i is not in A_i but is an element of A_{2i}. However, A_0 is a subalgebra of A.

We shall assume that the unit element e of a graded algebra is in A_0 (see Exercise 2) and that $xe \to x$, $x \in F$, is an algebra isomorphism of $A_0 \to F$. This isomorphism is canonical, and for this reason we shall always take $A_0 = F$ in our grading of a graded algebra over F.

Example 5. The polynomial algebra P of all polynomials in x over a field F can be expressed as $P = \sum_{i=0}^{\infty} \oplus P_i$, where P_i is the subspace of the polynomials of the form ax^i, $a \in F$. The product of two such nonzero polynomials of degrees i and j is a polynomial of degree $i + j$ and belongs to P_{i+j}. Also $P_0 = F$. Clearly any polynomial of P has a unique representation as a finite sum of elements of the P_i. Hence P is the graded algebra $\sum_{i=0}^{\infty} \oplus P_i$.

Definition. If A and A' are two algebras over the same field F, then a vector space linear transformation T of $A \to A'$ is called an **algebra homomorphism** if T satisfies the additional requirements that

$$(\alpha\beta)T = (\alpha T)(\beta T), \quad \alpha, \beta \in A,$$

and eT is the unit element of A', where e is the unit element of A.

Thus, an algebra homomorphism preserves vector multiplication and the unit element.

We have used a universal mapping property to define the tensor product of two vector spaces. We shall now see how a universal mapping property (U.F.P.) can be used to define another algebraic system called the **tensor algebra** on a vector space. Its actual existence is demonstrated by constructing it.

Definition. A **tensor algebra** T on the vector space V over R is an associative algebra over R together with a linear mapping f of $V \to$ T such that for each choice of an associative algebra X over R and of a linear mapping g of $V \to X$, there exists a unique algebra homomorphism h of T $\to X$ such that $fh = g$.

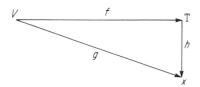

In the same way as for the tensor product of two vector spaces (see Theorems 1 and 2 of Chapter 11) we can show that the tensor algebra (T, f) on V is unique up to isomorphism and further that $f(V)$ and the unit element e of T form a set of generators of T.

It can be shown that the graded algebra

$$T = R \oplus V \oplus (V \otimes V) \oplus (V \otimes V \otimes V) \oplus \cdots$$

and the inclusion mapping f of $V \to$ T (defined by $\alpha f = \alpha$, $\alpha \in V$) have the required U.F.P. on V and hence that the pair (T, f) constitute (up to isomorphism) the tensor algebra on V. We shall later do this in detail for a special type of tensor algebra (the exterior algebra) that will concern us in this chapter.

EXERCISES

1. Prove that the continuous functions on the closed interval $[-1, 1]$ form an algebra over the real field.

2. Prove that the unit element of a graded algebra must be in A_0.

3. Let S be a subset of an algebra A that contains the unit element of A. Show that the elements of the subalgebra generated by S are the linear combinations of terms that are finite products of elements of S.

4. If S is a set of generators of an algebra A, prove that an algebra homomorphism on A is uniquely determined by its values on S.

5. Verify the three rules (i)–(iii) for multiplication in an algebra.

6. Define a subalgebra of an algebra.

7. Show that the multiplication in an associative algebra can be defined as a bilinear mapping g of $V \times V \to V$ such that the product $\alpha\beta = (\alpha, \beta)g$ is associative.

8. Is a subalgebra necessarily an ideal? Give an example.

9. Prove that the set of self-adjoint operators on a vector space V forms an algebra (the *self-adjoint operator algebra*).

The Exterior Algebra of a Vector Space

10. B is an ideal in an algebra A. Prove that the mapping j of $A \to A/B$ defined by $aj = a + B, a \in A$, is an algebra epimorphism.

11. Show that $V_3(R)$ with a product of vectors $\alpha = (x_1, x_2, x_3)$ and $\beta = (y_1, y_2, y_3)$ defined by

$$\alpha\beta = (x_2 y_3 - y_2 x_3, \quad y_1 x_3 - x_1 y_3, \quad x_1 y_2 - x_2 y_1)$$

is a real associative noncommutative algebra.

(Note that this product is the so-called "vector" or "cross-product" commonly denoted by $\alpha \times \beta$.)

12. A and A' are algebras and T is an algebra homomorphism of $A \to A'$. Prove that ker T is an ideal in A.

12-2 THE ALGEBRA OF ALTERNATING FORMS

In this section it is proved that, with suitable definitions of multiplication, the set of alternating real linear forms on q-fold tensor products $V \otimes V \otimes \cdots \otimes V, q = 1, 2, \ldots$ forms a graded algebra.

Let V be a real n-dimensional vector space with a basis $\beta_1, \ldots, \beta_n$. For $q \geq 2$, let $\otimes^q V = V \otimes \cdots \otimes V$ be the **q-fold tensor product**. The dual space $(\otimes^q V)^*$ is the real vector space of the real linear forms on $\otimes^q V$. Call $F \in (\otimes^q V)^*$ an **alternating real linear form** on V if its value $F(\alpha_1 \otimes \cdots \otimes \alpha_q)$, $\alpha_1, \ldots, \alpha_q \in V$, changes sign whenever any two of the vectors α_i are interchanged. The set of alternating linear forms clearly forms a subspace A_q of $(\otimes^q V)^*$. They are called **q-forms** or **q-covectors** on V.

Let

$$\alpha_i = \sum_{j=1}^n x_{ij}\beta_j, \quad i = 1, 2, \ldots, q$$

be any q vectors of V. Let $F \in A_q$. Then

(1) $\quad F(\alpha_1 \otimes \cdots \otimes \alpha_q) = F(\Sigma x_{1j}\beta_j \otimes \cdots \otimes \Sigma x_{qj}\beta_j)$

$$= \sum_{i_1, \ldots, i_q = 1}^n x_{1 i_1} \cdots x_{q i_q} F(\beta_{i_1} \otimes \cdots \otimes \beta_{i_q}).$$

Consider the terms in the summation (1). Since F is alternating, $F(\beta_{i_1} \otimes \cdots \otimes \beta_{i_q}) = 0$ whenever two of its β-vectors are equal; that is, whenever two of the subscripts $i_1, \ldots, i_q$ are equal. We can therefore drop such terms from the summation. Suppose then that the subscripts $i_1, \ldots, i_q$ are distinct. There are $q!$ terms in (1) with these particular subscripts in some order. Suppose now $i_1 < i_2 < \cdots < i_q$. Then the value of F for an even permutation of these subscripts is equal to $F(\beta_{i_1} \otimes \cdots \otimes \beta_{i_q})$, whereas the value of F for any odd permutation is $-F(\beta_{i_1} \otimes \cdots \otimes \beta_{i_q})$. This is true because F is alternating. The sum of all these $q!$ terms can be therefore written in the form

$$D_{(i_1,\ldots,i_q)}F(\beta_{i_1}\otimes\cdots\otimes\beta_{i_q}), \quad i_1<\cdots<i_q.$$

An examination of the $q!$ terms that make up $D_{(i_1,\ldots,i_q)}$ reveals, on comparison with (3) in Chapter 7, that $D_{(i_1,\ldots,i_q)}$ is the $q\times q$ determinant formed from the q rows, numbered $i_1, i_2, \ldots, i_q$ of the real $n\times q$ matrix

$$(2) \qquad (\alpha_1,\ldots,\alpha_q) = \begin{bmatrix} x_{11} & x_{21} & \cdots & x_{q1} \\ \vdots & & & \vdots \\ x_{1n} & x_{2n} & \cdots & x_{nn} \end{bmatrix}.$$

The columns of this matrix are the components $x_{i1},\ldots,x_{in}$ of the vectors $\alpha_i, i=1,2,\ldots,q$.

If we define a real-valued function $f_{(i_1,\ldots,i_q)}$ on $\otimes^q V$ by

$$(3) \qquad f_{(i_1,\ldots,i_q)}(\alpha_1\otimes\cdots\otimes\alpha_q) = D_{(i_1,\ldots,i_q)}$$

then $f_{(i_1,\ldots,i_q)}\in A_q$, since the determinant function is a real alternating linear function.

It follows then that every $F\in A_q$ can be expressed in the form

$$(4) \qquad F = \Sigma F(\beta_{i_1}\otimes\cdots\otimes\beta_{i_q})f_{(i_1,\ldots,i_q)}, \quad i_1<\cdots<i_q,$$

where the summation extends over the $\binom{n}{q}$ distinct sets of q subscripts $i_1,\ldots,i_q$ chosen from $1,2,\ldots,n$ and for which $i_1<\cdots<i_q$.

Observe that for the basis vectors $\beta_1,\ldots,\beta_n$ of V we have from (3)

$$(5) \qquad f_{(i_1,\ldots,i_q)}(\beta_{i_1},\ldots,\beta_{i_q}) = 1$$

while

$$(6) \qquad f_{(i_1,\ldots,i_q)}(\beta_{j_1},\ldots,\beta_{j_q}) = 0$$

if $j_1<j_2<\cdots<j_q$ is a set of subscripts distinct from $i_1<\cdots<i_q$. This last formula follows from the fact that if i_k, say, does not occur in $j_1, j_2,\ldots,j_q$ then the matrix corresponding to (2), formed for $(\beta_{j_1},\ldots,\beta_{j_q})$, contains all zeros in the i_k th row.

We also note that if $\alpha_1,\ldots,\alpha_q$ are linearly independent vectors and if $F\in A_q$, then $F(\alpha_1\otimes\cdots\otimes\alpha_q) = 0$. Thus in particular if $q>n$, $F=0$, the zero function (see Exercise 1).

The formula (4) shows that the $\binom{n}{q}$ vectors $f_{(i_1,\ldots,i_q)}, i_1<\cdots<i_q$, span A_q, but they are actually a basis of A_q, for let

$$(7) \qquad \Sigma x_{(i_1,\ldots,i_q)}f_{(i_1,\ldots,i_q)} = 0,$$

where $i_1 < \cdots < i_q$, $x_{(i_1,\ldots,i_q)} \in R$, and 0 stands for the zero function of A_q. Then
$$\Sigma x_{(i_1,\ldots,i_q)} f_{(i_1,\ldots,i_q)}(\beta_{j_1},\ldots,\beta_{j_q}) = 0$$
and by (5) and (6) this implies
$$x_{(i_1,\ldots,i_q)} = 0.$$
Hence all the coefficients $x_{(i_1,\ldots,i_q)}$ of (7) are 0 and therefore the $f_{(i_1,\ldots,i_q)}$ are linearly independent.

Moreover,
$$\dim A_q = \binom{n}{q}.$$

For $q = 1$, $A_1 = V^*$ and linear forms on V are alternating (by default); that is, they are 1-forms.

Now consider the real graded vector space A formed by the following direct sum of real vector spaces,

(8) $$A = R \oplus V^* \oplus \sum_{q=2}^{n} A_q.$$

R is the real field of scalars.

We define a product, denoted by a **wedge** $\wedge$, of a q-form and a p-form by first defining

(9) $$f_{(i_1,\ldots,i_q)} \wedge f_{(j_1,\ldots,j_p)} = f_{(i_1,\ldots,i_q,j_1,\ldots,j_p)}$$

where $i_1 < \cdots < i_q$ and $j_1 < \cdots < j_p$. This product is a $(q + p)$-form. It is zero if some j_s is equal to an i_m. If all the j's are distinct from the i's then the product is equal to either $f_{(k_1,\ldots,k_{q+p})}$ or $-f_{(k_1,\ldots,k_{q+p})}$, where $k_1 < k_2 < \cdots < k_{q+p}$, depending on whether an even or an odd permutation is required to put $i_1,\ldots,i_q, j_1,\ldots,j_p$ in the increasing order $k_1, k_2, \ldots, k_{q+p}$. Of course if $q + p > n$, then the product is zero.

The formula (9) defines the product of a basis vector of A_q and a basis vector of A_p. Hence for $F \in A_q$ and $G \in A_p$, we set

(10) $$F \wedge G = \Sigma F(\beta_{i_1} \otimes \cdots \otimes \beta_{i_q})$$
$$G(\beta_{j_1} \otimes \cdots \otimes \beta_{j_p}) f_{(i_1,\ldots,i_q)} \wedge f_{(j_1,\ldots,j_p)}.$$

This defines the product of any q-form and any p-form and the product is a $(q + p)$-form. With this multiplication we see that the vector space A defined by (8), becomes a real algebra. In fact the following multiplication rules of A can be verified.

(i) $(F \wedge G) \wedge H = F \wedge (G \wedge H)$, $F \in A_p$, $G \in A_q$, $H \in A_r$.

(ii) $F \wedge G = (-1)^{pq} G \wedge F$, where $F \in A_p$, $G \in A_q$.
(iii) $(F + G) \wedge H = F \wedge H + G \wedge H$.
$F \wedge (G + H) = F \wedge G + F \wedge H$.

If we define $x \wedge F = xF$, $x \in R$, then
(iv) $(xF) \wedge G = x(F \wedge G) = F \wedge (xG)$, $x \in R$.

It suffices to prove these rules for the basis vectors.

The property (ii) above of $\wedge$ makes A what is called an **anticommutative algebra**. We take the unit element of A to be $1 \in R$.

Definition. The algebra A is called the **algebra of alternating forms** on the vector space V.

EXERCISES

1. If $F \in A_q$ prove that $F(\alpha_1 \otimes \cdots \otimes \alpha_q) = 0$ when $\alpha_1, \alpha_2, \ldots, \alpha_q$ are linearly dependent vectors.

2. Prove that the rules (i)–(iv) for multiplication in A are true for the basis vectors of the vector spaces A_q.

3. Justify the statement that the multiplication rules (i)–(iv) are true for arbitrary elements of A if they hold for the basis vectors.

4. Prove that the set of all real linear forms is an algebra. Is the alternating algebra a subalgebra or an ideal in the algebra of all linear forms?

5. For the case $q = 1$ show that the equation (3) defines the basis of $A_1 = V^*$ to be the dual basis $f_1, f_2, \ldots, f_n$ to the basis $\beta_1, \beta_2, \ldots, \beta_n$ of V.

Now use (9) to define $f_i \wedge f_j$, $i, = 1, 2, \ldots, n$, and then prove that the product of two 1-forms is an alternating 2-form.

12-3 THE EXTERIOR ALGEBRA

The exterior algebra of a vector space is another example of a graded algebra. We shall define it as a special kind of tensor algebra. However, these abstractions will be followed by the actual construction of the exterior algebra of a finite-dimensional vector space. We shall also prove in this section that the algebra of alternating forms on a finite-dimensional vector space V and the exterior algebra of the dual space V^* are canonically isomorphic. In virtue of this fact we shall be able therefore to identify these two algebras. Finally, in the next section, we shall show that the differential forms in calculus form an exterior algebra.

Definition. The **exterior algebra** (E, f) of a vector space is the tensor algebra on V such that
  (a) $(\alpha f)^2 = 0$, for all $\alpha \in V$.
  (b) the arbitrary vector space homomorphism g of $V \to X$ must satisfy $(\alpha g)^2 = 0$ for all $\alpha \in V$.

The Exterior Algebra of a Vector Space

Then for any algebra X and any such homomorphism g of $V \to X$, there is a unique algebra homomorphism h of $E \to X$ such that $g = fh$ and for which $h(1)$ is the unit element of X.

It will be seen that this special property, $(\alpha f)^2 = 0$, $\alpha \in V$, of an exterior algebra (E, f) produces an algebra with special multiplication rules that are peculiar to it. We shall discover all these in the next section.

By proofs similar to those for the tensor product we can show that the exterior algebra (E, f) is unique up to isomorphism and also that the elements 1 and αf, $\alpha \in V$, form a set of generators of E.

EXERCISES

1. Prove that the exterior algebra (E, f) is unique up to isomorphism.
2. Prove that the elements 1 and αf, $\alpha \in V$, form a set of generators of E.

Let V be a real n-dimensional vector space with a basis $\beta_1, \beta_2, \ldots, \beta_n$. A direct generalization of Theorem 10, Chapter 11, and its method of proof shows that the n^q vectors

$$\beta_{i_1} \otimes \cdots \otimes \beta_{i_q}, \quad 1 \leq i_1 \leq n, \ldots, 1 \leq i_q \leq n,$$

constitute a basis for the vector space formed by the q-fold tensor product

$$\otimes^q V = V \otimes \cdots \otimes V.$$

Let U_q be the subspace of $\otimes^q V$ generated by all elements of $\otimes^q V$ of the form $\alpha_1 \otimes \cdots \otimes \alpha_q$, where $\alpha_i = \alpha_{i+1}$ for at least one subscript i. Write

$$\wedge^q V = \otimes^q V / U_q$$

for the quotient vector space $\otimes^q V / U_q$, and write

$$\alpha_1 \wedge \cdots \wedge \alpha_q = \alpha_1 \otimes \cdots \otimes \alpha_q + U_q.$$

We observe that the following rules are true for the "**exterior product**" $\wedge$:

(i) $\alpha_1 \wedge \cdots \wedge \alpha_q = 0$ (that is $\alpha_1 \otimes \cdots \otimes \alpha_q \in U_q$) if any two adjacent α_i are equal.

(ii) $\alpha_1 \wedge \cdots \wedge \alpha_q$ changes sign if two adjacent α_i are interchanged. For

$$\alpha_1 \wedge \cdots \wedge \alpha_i \wedge \alpha_{i+1} \wedge \cdots \wedge \alpha_q + \alpha_1 \wedge \cdots \wedge \alpha_{i+1} \wedge$$
$$\alpha_i \wedge \cdots \wedge \alpha_q = \alpha_1 \wedge \cdots \wedge (\alpha_i + \alpha_{i+1}) \wedge (\alpha_i + \alpha_{i+1}) \wedge$$
$$\cdots \wedge \alpha_q - \alpha_1 \wedge \cdots \wedge \alpha_i \wedge \alpha_i \wedge \cdots \wedge \alpha_q - \alpha_1 \wedge \cdots \wedge$$
$$\alpha_{i+1} \wedge \alpha_{i+1} \wedge \cdots \wedge \alpha_q = 0, \text{ by (i)}.$$

(iii) $\alpha_1 \wedge \cdots \wedge \alpha_q$ changes its sign if any two of the α_i are interchanged.

For any such interchange is effected by a succession of interchanges of adjacent α_i and the total number of these is always an odd number. (This can be easily checked by simple counting.) Thus the term changes sign an odd number of times.

(iv) $\alpha_1 \wedge \cdots \wedge \alpha_q = 0$ if any two of the α_i are equal. This follows at once from (iii).

These are the rules of computation in the vector space $\wedge^q V$.

We have seen that the n^q vectors $\beta_{i_1} \otimes \cdots \otimes \beta_{i_q}$ form a basis for $\otimes^q V$. If j is the canonical epimorphism of $\otimes^q V \to \wedge^q V$ given by $(\alpha_1 \otimes \cdots \otimes \alpha_q)j = \alpha_1 \wedge \cdots \wedge \alpha_q$, then (by Lemma 2, Chapter 11) the n^q vectors $\beta_{i_1} \wedge \cdots \wedge \beta_{i_q}$ constitute a set of generators of $\wedge^q V$. By (iii) and (iv) we see that actually the subset of the $\binom{n}{q}$ vectors $\beta_{i_1} \wedge \cdots \wedge \beta_{i_q}$, for which $i_1 < \cdots < i_q$, forms a set of generators of $\wedge^q V$. We shall soon prove that this latter set of generators is a basis for $\wedge^q V$.

Denote by $\wedge V$ the vollowing direct sum of vector spaces

$$\wedge V = R \oplus V \oplus \sum_{q=2}^{n} \wedge^q V,$$

where R is the real field. Thus $\wedge V$ is a real graded vector space.

We now define a multiplication in $\wedge V$ as follows: for $\alpha_1 \wedge \cdots \wedge \alpha_p \in \wedge^p V$ and $\gamma_1 \wedge \cdots \wedge \gamma_q \in \wedge^q V$, define $(\alpha_1 \wedge \cdots \wedge \alpha_p) \cdot (\gamma_1 \wedge \cdots \wedge \gamma_q) = \alpha_1 \wedge \cdots \wedge \alpha_p \wedge \gamma_1 \wedge \cdots \wedge \gamma_q$ and the product is a term in $\wedge^{p+q} V$. Of course if $p + q > n$ this product is 0. This multiplication is readily seen to be associative, and it is not hard to verify that this multiplication turns $\wedge V$ into a real associative graded algebra.

The graded algebra $\wedge V$ is called the **exterior algebra of** V. We shall prove later that it has the universal mapping property mentioned earlier in the definition. Before we do this, we examine the relationships among the three algebras: the algebra A of alternating forms on V, the exterior algebra $\wedge V$ of V and the exterior algebra $\wedge V^*$ of the dual V^* of V.

EXERCISES

1. If $\alpha_{i_1}, \ldots, \alpha_{i_q}$ is any permutation of $\alpha_1, \alpha_2, \ldots, \alpha_q$, prove $\alpha_{i_1} \wedge \ldots \wedge \alpha_{i_q} = \pm (\alpha_1 \wedge \ldots \wedge \alpha_q)$ depending on whether the permutation is even (plus sign) or odd (minus sign).

2. If $\alpha \in \wedge^i V$ and $\tau \in \wedge^j V$, prove
$$\gamma \tau = (-1)^{ij} \tau \gamma.$$

3. Verify that the rules (i)–(v) in Sec. 12-2 hold for the multiplication in $\wedge V$.
4. If dim $V = n$, prove dim $\wedge V = 2^n$.

Returning now to the algebra of forms in Sec. 12-2, let A_q be the real vector space of q-forms on $\otimes^q V$, where V is a real n-dimensional vector space with a basis $\beta_1, \ldots, \beta_n$. We have seen that the $\binom{n}{q}$ q-forms $f_{(i_1,\ldots,i_q)}$, $i_1 < \cdots < i_q$ form a basis for A_q.

Let $(\wedge^q V)^*$ be the dual space of $\wedge^q V$.
Define a mapping T of $(\wedge^q V)^* \to A_q$ by $T(\phi) = F$, $\phi \in (\wedge^q V)^*$, where F is given by

(11) $\qquad F(\alpha_1 \otimes \cdots \otimes \alpha_q) = \phi(\alpha_1 \wedge \cdots \wedge \alpha_q)$.

Note that $F = \phi j$ where j is the canonical epimorphism of $\otimes^q V \to \wedge^q V$. The four properties (i)–(iv) of the $\wedge$-product and the linear property of ϕ prove that $F \in A_q$. Moreover, T is a linear transformation. Clearly $F = 0$ implies $\phi = 0$. Also, every $F \in A_q$ defines, according to (11), a $\phi \in (\wedge^q V)^*$. Hence T is a canonical isomorphism and so we write

(12) $\qquad A_q = (\wedge^q V)^*$.

Now dim $A_q = \binom{n}{q}$, and so dim $(\wedge^q V)^* = \binom{n}{q}$. Since $\wedge^q V$ is a finite-dimensional vector space, it follows that dim $\wedge^q V = \binom{n}{q}$. We can therefore infer from this that the $\binom{n}{q}$ generators $\beta_{i_1} \wedge \cdots \wedge \beta_{i_q}$, $i_1 < \cdots < i_q$, of $\wedge^q V$ must be a basis of $\vee^q V$. We call this the **standard basis** of $\wedge^q V$.

Next consider the vector space $\wedge^q V^*$, where we have replaced V in $\wedge^q V$ by its dual V^*. Define a mapping S of $\wedge^q V^* \to (\wedge^q V)^*$ by

$$S(g_1 \wedge \ldots \wedge g_q)(\alpha_1 \wedge \cdots \wedge \alpha_q) = \begin{vmatrix} g_1(\alpha_1) & g_2(\alpha_1) & \cdots & g_q(\alpha_1) \\ g_1(\alpha_2) & g_2(\alpha_2) & \cdots & g_q(\alpha_2) \\ \cdots & \cdots & & \cdots \\ \cdots & \cdots & & \cdots \\ g_1(\alpha_q) & g_2(\alpha_q) & \cdots & g_q(\alpha_q) \end{vmatrix},$$

where $g_1, g_2, \ldots, g_q \in V^*$.

The reader can verify that S is well-defined. We shall show that S maps the standard basis of $\wedge^q V^*$ into the dual of the standard basis of $\wedge^q V$. This will prove that S is an isomorphism and we shall use this isomorphism as an identification, and write

(13) $\qquad \wedge^q V^* = (\wedge^q V)^*$.

Let $f_1, f_2, \ldots, f_n$ be the basis of V^* dual to the basis $\beta_1, \beta_2, \ldots, \beta_n$ of V. Then $f_i(\beta_j) = 0$ if $i \neq j$ and $f_i(\beta_i) = 1$. The standard basis of $\wedge^q V$ was defined to be the set of all $\beta_{i_1} \wedge \cdots \wedge \beta_{i_q}$, where $i_1 < i_2 < \cdots < i_q$. Hence the standard basis of $\wedge^q V^*$ is the set of all $f_{i_1} \wedge \ldots \wedge f_{i_q}$, $i_1 < \cdots < i_q$. Let us designate the basis of $(\wedge^q V)^*$ which is dual to the standard basis of $\wedge^q V$ by $f_{i_1 i_2 \ldots i_q}$, $i_1 < \cdots < i_q$. Then $f_{i_1 i_2 \ldots i_q}(\beta_{i_1} \wedge \cdots \wedge \beta_{i_q}) = 1$ and the value of $f_{i_1 i_2 \ldots i_q}$ is 0 at all the other basis vectors of $\wedge^q V$. Therefore

$$S(f_{i_1} \wedge \cdots \wedge f_{i_q})(\beta_{i_1} \wedge \cdots \wedge \beta_{i_q}) = \begin{vmatrix} 1 & 0 & 0 & \cdots & 0 \\ 0 & 1 & 0 & \cdots & 0 \\ \cdots & \cdots & \cdots & \cdots & \cdots \\ \cdots & \cdots & \cdots & \cdots & \cdots \\ 0 & 0 & 0 & \cdots & 1 \end{vmatrix} = 1,$$

while

$$S(f_{i_1} \wedge \cdots \wedge f_{i_q})(\beta_{j_1} \wedge \cdots \wedge \beta_{j_q}) = 0 \text{ if } j_1 < \cdots < j_q$$

is not the same as $i_1 < \cdots < i_q$. For example, if an i_k of $i_1 < \cdots < i_q$ does not occur in $j_1 < \cdots < j_q$ then the determinant will contain all zeros in the kth column.

We have proved that the $S(f_{i_1} \wedge \cdots \wedge f_{i_q})$, $i_1 < \cdots < i_q$ form the dual of the standard basis of $\wedge^q V$. Hence S is an isomorphism.

Combining (12) and (13), we have

$$\wedge^q V^* = A_q.$$

Thus we identify the two algebras and write

$$\wedge V^* = A.$$

That is, *the exterior algebra of the dual V^* of V is the algebra of alternating forms*, defined in Sec. 12-2.

Finally, we shall now prove the universal mapping property for $\wedge V$ and this will complete the proof that $\wedge V$ is the exterior algebra of V. We have

$$\wedge V = R \oplus V \oplus \sum_{q=2}^{n} \oplus \wedge^q V.$$

Define the mapping f of $V \to \wedge V$ as the inclusion mapping $f(\alpha) = \alpha$, for all $\alpha \in V$. We claim $(\wedge V, f)$ is the exterior algebra of V. Let X be a real algebra and let g be a linear transformation of $V \to X$ for which $(g(\alpha))^2 = 0$ for all $\alpha \in V$.

For $r + \alpha + \Sigma(\alpha_1 \wedge \ldots \wedge \alpha_q) \in \wedge V$, define a mapping h of $\wedge V \to X$ by

$$h(r + \alpha + \Sigma(\alpha_1 \wedge \ldots \wedge \alpha_q)) = re + g(\alpha) + \Sigma(g(\alpha_1) \wedge \ldots \wedge g(\alpha_q)),$$

where e is the unit element of X. The reader can without difficulty verify that h is an algebra homomorphism and that $h(1) = e$. Moreover, $h(\alpha) = hf(\alpha) = g(\alpha)$, for all $\alpha \in V$.

Hence $hf = g$.

Now 1 and the $f(\alpha)$, $\alpha \in V$, form a set of generators of $\wedge V$ and $h(1) = e$ and $hf(\alpha) = g(\alpha)$ define h on these generators. Since there exists only one algebra homomorphism with given values on a set of generators, it follows that h is unique. This completes the proof that $(\wedge V, f)$ is the exterior algebra on the n-dimensional vector space V.

We might add that if V is an arbitrary real vector space (that is, V is not necessarily finite-dimensional), then it can be proved, in much the same way, that $(\wedge V, f)$, where $\wedge V$ is the graded algebra

$$\wedge V = R \oplus V \oplus \sum_{q=2}^{\infty} \oplus \wedge^q V$$

and f is the inclusion mapping of $V \to \wedge V$, is the exterior algebra of V. The terms of $\wedge V$ are finite sums, that is sums whose terms are almost all zero.

It can be verified that $U = \sum_{2}^{\infty} \oplus U_q$ is the ideal in the tensor algebra T on the vector space V (used to define V in terms of the U.F.P.) that is generated by the elements of T of the form $\alpha \times \alpha$, $\alpha \in V$. The elements of U are finite sums of elements from the U_q. Hence we can describe the exterior algebra $\wedge V$ as the quotient algebra T/U.

EXERCISES

1. Show that if 2 elements of $\wedge^q V^*$ are equal then they map under S into equal elements of $(\wedge^q V)^*$; that is, S is well-defined.

2. Supply the details of the proof that S maps the standard basis of $\wedge^q V^*$ into the dual of the standard basis of $\wedge^q V$.

12-4 THE ALGEBRA OF DIFFERENTIAL FORMS

In the calculus of functions of several variables, such basic theorems, as the general divergence theorem and Stokes' theorem, involve the integration of differential forms, defined on certain sets in euclidean space called **manifolds**.

The reader is no doubt familiar with such a differential form as $f = f_1 dx_1 + \cdots + f_n dx_n$, where the f_i are real-valued functions defined on some domain D in $V_n(R)$. It is called a **differential form of de-**

gree 1. For each $\alpha = (x_1, x_2, \ldots, x_n)$ in the domain we can regard $f_1(\alpha), f_2(\alpha), \ldots, f_n(\alpha)$ as the components of a linear form ϕ (a covector) with respect to some basis in $V_n(R)^*$. Since f is determined by the f_i, we can define f as a function on a domain in $V_n(R)$ with values in $V_n(R)^*$; that is $f(\alpha) = \phi$, $\alpha \in D$.

For $q \geq 2$ a differential form of degree q is an alternating polynomial of degree q in the dx_i, $i = 1, \ldots, n$ with real-valued functions as coefficients. A precise definition of a differential form is the following, suggested by the above discussion.

Definition. A **differential form** of degree q is a function f with a domain in $V_n(R)$ and values in $V_n(R)^*$.

The values of f are therefore q-covectors.

An exterior product and an addition of forms of the same degree can be defined in the set of all differential forms, and this set will then be seen to take on the structure of an exterior algebra. In order to do this we shall need to make use of the notion of a module.

The module was defined in Sec. 2-4. The definition of the tensor product of two vector spaces, given in Chapter 11, applies equally well to the tensor product of two modules over a commutative ring with unit element. We need simply replace vector space by module, and field by ring. Furthermore, we have defined an algebra over a field and again this can be easily adapted to an algebra over a commutative ring.

Definition. An **associative algebra A over a commutative ring** with unit element is a module over this ring together with a linear mapping of $A \otimes A \to A$ such that the multiplication in A defined by this mapping is associative.

It is necessary to emphasize that, while the ring is assumed commutative, it does not follow that the multiplication in the algebra itself is commutative.

Moreover, the exterior algebra of a module can be defined just as was the exterior algebra of a vector space.

R, as usual, shall stand for the real field.

As before, D is any open set in the vector space V. We shall assume that all our mappings on D are of class C^k; that is, are k times continuously differentiable. This will lead to the differential forms of class C^k.

A real-valued function f on D is called a 0-form. Let $\mathfrak{F}$ denote the set of all such functions. For $f, g \in \mathfrak{F}$ and all $\alpha \in D$ let us define:

(i) $f + g$ by $(f + g)\alpha = f(\alpha) + g(\alpha)$, then $f + g \in \mathfrak{F}$,
(ii) fg by $(fg)\alpha = f(\alpha) \cdot g(\alpha)$, then $fg \in \mathfrak{F}$,
(iii) rf, for $r \in R$, by $(rf)\alpha = rf(\alpha)$, then $rf \in \mathfrak{F}$.

The Exterior Algebra of a Vector Space

Then $\mathcal{F}$ is seen to be a commutative algebra over R. It is therefore a commutative ring with unit element, the unit element being the function that maps every $\alpha \in D$ into 1. For each $r \in R$ the function $f \in \mathcal{F}$, for which $f(\alpha) = r$ for all $\alpha \in D$, is called a **constant function**. Clearly the subset of all constant functions of $\mathcal{F}$ forms a subring of $\mathcal{F}$ that is isomorphic to R. We express this by saying that the ring R is embedded in the ring $\mathcal{F}$.

We now return to the set of all differential forms on D which we propose to show is an algebra over the commutative ring $\mathcal{F}$. We first define a product in this set.

Let ω_p be a differential form of degree p, and ω_q a differential form of degree q. We define an **exterior product** on the set of all differential forms as follows:

Define a $(p + q)$-form $\omega_p \wedge \omega_q$ by

(14) $\qquad (\omega_p \wedge \omega_q)\alpha = \omega_p(\alpha) \wedge \omega_q(\alpha), \quad \alpha \in D.$

Since ω_p is a mapping of $D \to \wedge^p V^*$, it follows that $(\omega_p \wedge \omega_p)\alpha = \omega_p(\alpha) \wedge \omega_p(\alpha) = 0$ for every $\alpha \in D$. Hence $\omega_p \wedge \omega_p = 0$. This justifies our calling (14) an **exterior product**.

Next we define a "scalar" product in the set of differential forms, using the elements of the commutative ring $\mathcal{F}$ as the scalars.

For $f \in \mathcal{F}$, define a q-form $f\omega_q$ by

$$(f\omega_q)\alpha = f(\alpha) \cdot \omega_q(\alpha), \quad \alpha \in D.$$

Finally we observe that two differential forms of the same degree can be added in an obvious way and the sum is also a differential form of this same degree.

With these definitions it now becomes apparent that the set of **differential forms is an algebra over the commutative ring** $\mathcal{F}$. It has the structure of an exterior algebra.

Let us denote by W^* the $\mathcal{F}$-module of all functions of $D \to V^*$. (By an $\mathcal{F}$-module we mean a module over the commutative ring $\mathcal{F}$.) These functions are by definition the differential forms of degree one. We can see now that the algebra of differential forms is therefore the exterior algebra $\wedge W^*$ of the $\mathcal{F}$-module W^*.

We recapitulate our findings by emphasizing that, starting with a real finite-dimensional vector space V, we first form the exterior algebra of the dual space V^* (dim V^* = dim V). The differential forms on an open subset D of V are then defined as mappings on D with values in the $\wedge^q V^*$, $q = 0, 1, 2, \ldots$. Finally, we find that the differential forms themselves form an algebra over the ring $\mathcal{F}$ of real-valued functions on D and this algebra is the exterior algebra of the $\mathcal{F}$-module W^* of all functions on D to V^*.

Interestingly enough, the definition and properties of determinants can be based very simply and efficiently on the properties of an exterior algebra of a finite dimensional vector space. We outline the procedure.

Let T be a linear operator on the real n-dimensional vector space V. The U.F.P. of the exterior algebra $(\wedge V, f)$ of V determines a unique algebra homomorphism of $\wedge V \to \wedge V$, which we denote symbolically by $\wedge T$, and which is an extension of T. Consider the accompanying diagram

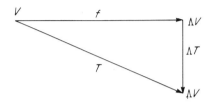

where f denotes, as usual, the inclusion mapping and, since $V \subset \wedge V$, where T is regarded as a vector space homomorphism of $V \to \wedge V$. By the U.F.P. there exists a unique algebra homomorphism $\wedge T$ of $\wedge V \to \wedge V$ such that $f \circ (\wedge T) = T$. Thus if $\alpha \in V$, $\alpha f = \alpha$, and $\alpha(\wedge T) = \alpha T$.

Now $\wedge^n V$ is a vector space of dimension $\binom{n}{n} = 1$. If $\alpha_1, \alpha_2, \ldots, \alpha_n$ is a basis for V, then we know $\alpha_1 \wedge \alpha_2 \wedge \cdots \wedge \alpha_n$ is a basis for $\wedge^n V$. Hence

$$(\alpha_1 \wedge \alpha_2 \wedge \cdots \wedge \alpha_n) \wedge T = x(\alpha_1 \wedge \alpha_2 \wedge \cdots \wedge \alpha_n),$$

where x is a scalar (here a real number), since this is the only possible form for a linear operator on a one-dimensional vector space.

Clearly, if some other basis $\beta_1, \ldots, \beta_n$ for V is used, we still get

$$(\beta_1 \wedge \cdots \wedge \beta_n) \wedge T = x(\beta_1 \wedge \cdots \wedge \beta_n);$$

that is, the scalar x is the same. Hence the value of x is dependent only on T. (Prove this statement!)

Definition. The real number x is called the *determinant* of the linear operator T, and we write $\det T = x$.

We have therefore defined a mapping of $\text{Hom}(V, V) \to R$ given by $T \to \det T$. In other words det is a function on $\text{Hom}(V, V)$ to the real field R.

We leave it as a "research" exercise for the reader to prove that this is our familiar determinant function defined in Chapter 7. In fact it can be easily proved directly from its definition above that

(i) $\det 1_V = 1$

(ii) det $T = 0$ if and only if T is singular (i.e., ker T contains a non-zero vector)
(iii) det $(S \circ T) = (\det S)(\det T)$, $S, T \in \text{Hom}(V, V)$
(iv) if T is nonsingular, then
$$\det(T^{-1}) = (\det T)^{-1}.$$

Example 15. Let dim $V = 2$ and let α_1, α_2 be a basis for V. Let T be a linear operator on V. Then $\alpha_i T = \sum_{j=1}^{2} a_{ij}\alpha_j$, $a_{ij} \in R$, $i = 1, 2$. For $\alpha, \beta \in V$, let $\alpha \wedge \beta$ be any element of $\wedge^2 V$. Then

$$\alpha = \sum_{i=1}^{2} x_i \alpha_i, \quad \beta = \sum_{i=1}^{2} y_i \alpha_i, \quad x_i, y_i \in R.$$

Hence $\alpha \wedge \beta = (x_1 y_2 - x_2 y_1)\alpha_1 \wedge \alpha_2$. Moreover,
$$\alpha T \wedge \beta T = (\alpha \wedge \beta) \wedge T \in \wedge^2 V.$$

A simple calculation yields
$$\alpha T \wedge \beta T = (a_{11} a_{22} - a_{12} a_{21})(x_1 y_2 - x_2 y_1)\alpha_1 \wedge \alpha_2.$$

Hence
$$(\alpha \wedge \beta) \wedge T = \begin{vmatrix} a_{11} & a_{12} \\ a_{21} & a_{22} \end{vmatrix} (\alpha \wedge \beta).$$

Another interesting fact that can be proved is that the exterior algebra provides an easy test of whether some given finite set of vectors of V is independent or not, and whether two given sets, containing the same number of vectors, span the same subspace of V.

Linear algebra is principally useful in calculus in the theory of functions of several variables, their integration over manifolds, and their various kinds of derivatives. If V and W are two inner-product vector spaces over the real field R, the continuity and differentiation of two main types of functions are studied:

(i) real-valued functions defined on certain subsets U of V; these are called **scalar-valued functions** of vectors and are mappings of $U \rightarrow R$,
(ii) functions defined on certain subsets of V with values in W; these are called **vector-valued functions** of vectors.

Of course R itself can be regarded as a one-dimensional inner product vector space over itself, by taking, say, ordinary multiplication as inner product. Thus real-valued functions of a single real variable (defined usually on some interval) can be regarded as of type (i). The same is true of real-valued functions of several variables, for we can put

$$f(x_1, x_2, \ldots, x_n) = f(\alpha),$$

where

$$\alpha = (x_1, x_2, \ldots, x_n)$$

is a vector in $V_n(R)$, expressed in terms, say, of the standard basis.

The "classical" definitions of limit and continuity for such functions depend on the properties of n-dimensional ($n = 1, 2, 3, \ldots$) euclidean space, as a metric space (absolute value and distance), but not regarded as a vector space. Since a real inner-product vector space has these same properties (the "absolute value" of a vector is its length, and the distance between two vectors is defined) it is a metric space. It is to be expected then that the concepts of limit and continuity can be readily defined in the more general context of inner-product vector spaces. In this way we can give definitions of derivatives for functions of the two types above, and their properties (the derivatives turn out to be linear transformations on the vector spaces) enable us to take advantage of the theory developed for vector spaces and linear transformations on vector spaces.

For the advantages to be derived by fully exploiting linear algebra as a tool in the calculus of functions of several variables, the reader is referred to H. K. Nickerson, D. C. Spencer, and N. E. Steenrod, *Advanced Calculus* (Princeton, N.J.: Van Nostrand, 1959).

EXERCISES

1. Define the sum of two differential forms of the same degree.

2. Prove that the subring of $\mathfrak{F}$ consisting of the constant functions is isomorphic to R.

3. Define the tensor product of two modules over the same commutative ring.

4. Verify that $\mathfrak{F}$ is a real commutative algebra.

5. Verify that the set of all differential forms is an algebra over the commutative ring $\mathfrak{F}$.

6. Give the details and verify that the algebra of differential forms is the exterior algebra of W^*.

7. Verify that $\wedge V = T/U$.

8. For each q, prove that the set of all differential forms of degree q is a module over $\mathfrak{F}$.

9. Deduce the four properties (i)–(iv) of the determinant function as it is defined in this chapter.

10. An element of an exterior algebra $\wedge V$ of a vector space V over F is called **decomposable** if it can be expressed as the exterior product $\beta_1 \wedge \beta_2 \wedge \cdots \wedge \beta_n$, $n \geq 1$, of vectors β_i of V, otherwise it is said to be **indecomposable**.

The Exterior Algebra of a Vector Space

If $\dim V = 4$ and $\alpha_1, \alpha_2, \alpha_3, \alpha_4$ is a basis for V, find a basis and the dimension of $\wedge^2 V$. Prove that an element $\sum_{i,j=1}^{4} x_{ij} \alpha_i \wedge \alpha_j$ of $\wedge^2 V$ is decomposable if and only if

$$x_{12} x_{34} - x_{13} x_{24} + x_{14} x_{23} = 0.$$

11. Find a subalgebra S of the real field R that is not an ideal in R.

12. If A is an algebra over F, and B is a subalgebra of A that is not an ideal, prove that the quotient vector space A/B is not an algebra.

13. Describe the elements of the quotient algebra P/K, where K is the ideal of all polynomials with constant term 0, in the polynomial algebra P over R. Prove P/K is isomorphic to the algebra R.

14. Define the exterior algebra of a finite-dimensional vector space, without using the U.F.P.

15. If $\alpha, \beta \in V$, prove $\alpha \wedge \beta + \beta \wedge \alpha = 0$. [*Hint:* Compute $(\alpha + \beta) \wedge (\alpha + \beta)$.]

Bibliography

There have been many books written on linear algebra. We offer the following list of books recommended for complementary and supplementary reading.

Gelfand, I. M. *Lectures on Linear Algebra.* New York: Interscience Publishers, Inc., 1962.

Greub, W. H. *Linear Algebra.* 3d ed. New York: Springer-Verlag, Inc., 1967.

Halmos, P. R. *Finite-Dimensional Vector Spaces.* 2d ed. Princeton, N.J.: Van Nostrand Co., Inc., 1958.

Hoffman K., and R. Kunze. *Linear Algebra.* Englewood Cliffs, N.J.: Prentice-Hall, Inc., 1961.

Hohn, F. E. *Elementary Matrix Algebra.* 2d ed. New York: The Macmillan Co., 1964.

Schreier, O., and E. Sperner. *Modern Algebra and Matrix Theory.* 2d ed. New York: Chelsea Publishing Co., 1959.

Smiley, M. F. *Algebra of Matrices.* Boston: Allyn and Bacon, Inc., 1965.

At a somewhat more advanced level...

Artin, E. *Geometric Algebra.* New York: Interscience Publishers, Inc., 1957.

Bourbaki, N. *Éléments de Mathématique.* Paris, France: Hermann. Book 2, Chapter 2, *Linear Algebra.*

Cater, F. S. *Lectures on Real and Complex Vector Spaces.* Philadelphia, Pa.: W. B. Saunders Co., 1966.

Jacobson, N. *Lectures in Abstract Algebra.* Vol. 2, *Linear Algebra.* Princeton, N.J.: Van Nostrand, 1953.

For the use of linear algebra in calculus and differential geometry, the reader is referred to the following:

Fleming, W. H. *Functions of Several Variables.* Reading, Mass.: Addison-Wesley Co., Inc., 1965.

Nickerson, H. K., D. C. Spencer, and N. E. Steenrod. *Advanced Calculus*. Princeton, N.J.: Van Nostrand, 1959.

And for additional material in algebra...

Ames, D. B. *An Introduction to Abstract Algebra*. Scranton, Pa.: International Textbook Co., 1969.

Godement, R. *Algebra*. Boston: Houghton Mifflin Co., 1968.

Jacobson, N. *Lectures in Abstract Algebra*, Vol. 1, *Basic Concepts*. Princeton, N.J.: Van Nostrand, 1951.

Rédei, L. *Algebra*. New York: Pergamon Press, 1967.

Index

Adjoint matrix, 143
Adjoint operator, 170, 176
Affine group, 57
Affine line, 61
Affine mapping, 59
Affine space, 58
Affine subspace, 60
Affine transformation, 57, 59
Algebra, 56, 255, 268
Algebra homomorphism, 257
Algebraically closed field, 163
Algebra of alternating forms, 262
Algebra of differential forms, 267, 269
Alternating form, 259
Angle, 104
Associative binary operation, 8
Automorphism, 44

Basis, 33
Bidual, 120
Bijective mapping, 4
Bilinear form, 186
Bilinear function, 185
Bilinear mapping, 124, 237
Binary operation, 7
Block matrix, 215

Canonical homomorphism, 67
Canonical injections, 75
Canonical isomorphism, 120, 248
Canonical projections, 75
Cartesian product, 2
Cayley-Hamilton theorem, 167
Characteristic of a field, 14
Characteristic polynomial, 156, 157, 158
Characteristic roots, 163
Codomain, 2
Cofactor, 143
Column rank, 130
Column space, 129
Commutative binary operation, 8
Companion matrix, 210

Complementary subspace, 73
Components, 35
Composite mapping, 3
Congruent matrices, 188
Conjugate of a matrix, 174
Conjugate transpose, 174
Constant transformation, 54
Coordinates, 21
Coset, 64
Covector, 259
Cramer's rule, 152
Cycle, 126

Dependent vectors, 30, 31
Determinant, 140
Determinant of a transformation, 148
Determinantal rank, 191
Diagonable operator, 160, 225
Differential form, 268
Dimension, 34, 58
Direct sum, 71, 72, 256
Direct sum of matrices, 216
Disjoint sets, 1
Distance, 102
Division algorithm, 14
Divisor of zero, 14
Domain, 2
Dual basis, 118
Dual space, 117

Eigenspace, 162
Eigenvalue, 155, 157
Eigenvector, 155, 157
Element, 1
Elementary divisors, 221
Elementary matrix, 133
Elementary row operation, 132
Empty set, 1
Endomorphism, 44
Epimorphism, 44
Equivalence class, 69
Equivalence relation, 68

Index

Equivalent quadratic forms, 189
Equivalent matrices, 135
Euclidean group, 113
Even permutation, 128
Exact sequence, 75, 76
Extension of a mapping, 6
Exterior algebra, 262, 264
Exterior product, 263, 269

Factor space, 65
Field, 13
Field of rational numbers, 13
Field of real numbers, 13
Finite dimensional, 33, 256
Finite induction axiom, 10
Finite induction axiom (second form), 11
Finite induction principle, 10
Full linear group, 56, 190
Function, 2

Graded algebra, 257
Graded vector space, 257
Group, 12

Hermitian form, 205
Hermitian matrix, 174
Hermitian operator, 176
Homogeneous system of equations, 150
Homomorphism, 42
Hom (V, W), 55, 96

Identity matrix, 86
Identity mapping, 4
If and only if, 1
Image, 2, 4, 42
Inclusion mapping, 6
Independent vectors, 30
Infinite dimensional, 33
Injective mapping, 3
Inner product, 100, 174
Inner product vector space, 100
Intersection, 1
Inverse element, 9
Inverse mapping, 4
Inverse matrix, 132, 147
Invertible mapping, 4
Invertible matrix, 89, 131
Invertible transformation, 47
Involution, 70
Irreducible polynomial, 14
Isometric spaces, 111
Isometry, 110
Isomorphism, 42

Jordan canonical form, 224, 225
Jordan matrix, 224

Kernel, 44

Length of a vector, 102, 175
Linear combination, 27
Linear form (functional), 117
Linear operator, 44
Linear sum, 28
Linear transformation, 42

Mapping, 2
Matrix, 80
Matrix of a quadratic form, 187
Matrix of a system of equations, 149
Metric space, 102
Minimal polynomial, 166, 210
Minor, 143
Module, 38
Monic polynomial, 14
Monomorphism, 44
Multilinear form, 238
Multilinear mapping, 238

Neutral element, 9
Nonsingular matrix, 89
Nonsingular transformation, 51
Norm, 102, 175
Normalized vector, 104
Normal operator, 181
Normal subgroup, 57
Null space, 209

Odd permutation, 128
Operator (linear), 44
Orthogonal basis, 105
Orthogonal complement, 105
Orthogonal group, 112
Orthogonal matrix, 115
Orthogonal operator, 112
Orthogonal projection, 108
Orthogonal vectors, 104, 175
Orthonormal basis, 105

Permutation, 125
Polynomial, 14
Polynomial algebra, 256
Polynomial function, 209
Positive definite form, 201, 205
Positive definite matrix, 201
Positive definite operator, 201
Primary decomposition theorem, 214
Prime integer, 11

Principal axes, 200, 206
Product of matrices, 84
Projection, 5, 70, 73

Quadratic form, 188
Quadratic function, 187
Quotient algebra, 256
Quotient space, 65

Range, 2
Rank of a hermitian form, 205
Rank of a matrix, 130
Rank of a quadratic form, 191, 195
Rank of a transformation, 131
Rational canonical form, 221
Rational decomposition theorem, 221
Relation, 68
Restriction of a mapping, 5
Ring, 12
Row-equivalent matrices, 132
Row rank, 130
Row space, 129

Scalar, 16
Scalar multiplication, 16
Schwarz inequality, 102
Self-adjoint operator, 170
Signature, 196
Similar matrices, 91
Similar operators, 91
Singular transformation, 51
Spectral theorem, 180, 182
Split exact sequence, 76
Square matrix, 86
Standard basis, 35

Standard inner product, 101, 175
Subgroup, 57
Subset, 1
Subspace, 25
Subspace generated by a set, 27
Sum of matrices, 83
Surjective mapping, 4
Sylvester's law of inertia, 195
Symmetric bilinear function, 185
Symmetric group, 126
Symmetric matrix, 171, 185
Symmetric operator, 170

T-cyclic subspace, 209
Tensor algebra, 258
Tensor product, 239, 242, 245, 251
T-invariant subspace, 181, 208
Translation, 56
Transpose of a matrix, 82
Transpose of a transformation, 121
Transposition, 127

Union, 2
Unit (identity) element, 9
Unitary matrix, 176
Unitary operator, 176
Unitary space, 175
Universal factorization property (U.F.P.), 239

Variable, 14
Vector, 16
Vector space, 16

Wedge, 261
Well-defined, 123
Well-ordering principle, 10